# Towards Human Work: Solutions to Problems in Occupational Health and Safety

# Towards Human Work: Solutions to Problems in Occupational Health and Safety

Edited by

**M. Kumashiro**

Department of Ergonomics,
University of Occupational and Environmental Health,
1-1 Iseigaoka, Yahatanishi-ku,
Kitakyushu, 807 Japan

**E.D. Megaw**

Ergonomics Information Analysis Centre,
School of Manufacturing and Mechanical Engineering,
University of Birmingham,
Birmingham B15 2TT, England

**Taylor & Francis**

| UK  | Taylor & Francis Ltd, 4 John St., London WC1N 2ET |
| USA | Taylor & Francis Inc., 1900 Frost Road, Suite 101, Bristol, PA 19007 |

Copyright © Taylor & Francis Ltd 1991

*All rights reserved. No part of this publication may be reproduced, stored in a retrieval system, or transmitted, in any form or by any means, electronic, electrostatic, magnetic tape, mechanical, photocopying, recording or otherwise, without the prior permission of the copyright owner.*

**British Library Cataloguing in Publication Data**

A catalogue record for this book is available from the British Library.

**Library of Congress Cataloging in Publication Data is available**

Cover design by Barking Dog Art, Tunbridge Wells, Kent

Printed in Great Britain by Burgess Science Press, Rankine Road, Basingstoke, Hampshire

# CONTENTS

*Preface* ............................................................. ix
*Acknowledgements* .................................................. xi

## PART I    Ergonomics in Perspective                                1

1  Ergonomics from Past to Future: An Overview ..................... 3
   *Brian Shackel*
2  A View of Ergonomics in Japan ................................. 18
   *Masamitsu Oshima*
3  The Development and Current Role of the Japanese Ergonomics
   Research Society .............................................. 25
   *Takao Ohkubo*
4  The Application and Development of Ergonomics in China ........ 31
   *Demao Lu and Guoming Feng*
5  Trends in Occupational Ergonomics in Korea .................... 35
   *Sang-Do Lee*
6  Ergonomics Reaches Those Occupations That Some Beers Should
   Not Reach? .................................................... 42
   *Ted Megaw*
7  The Correct Use of Ergonomics and the Trend of Social Change
   and Development ............................................... 49
   *Sadao Sugiyama*

## PART II    Health and Safety                                      53

8  Ergonomics and Design Factors in the Promotion of Health and
   Safety ........................................................ 55
   *Ilkka Kuorinka*
9  The Organization of Comprehensive Health Care from the Point of
   View of Stress Management and Mental Health ................... 60
   *Hiroshi Sakamoto*
10 What We Should Do for Occupational Health and Safety: Some
   Personal Views from an Occupational Physician ................. 65
   *A. Ward Gardner*
11 Human-Robot Interaction: An Overview of Perceptual Aspects of
   Working with Industrial Robots ................................ 68
   *Waldemar Karwowski*
12 The Application of Ergonomics to the Protection of
   Firebrick-Making Workers ...................................... 75
   *Xiaoxiong Liu*
13 Scenario Analysis of Wood-Bamboo Furniture Manufacturing
   Accidents ..................................................... 80
   *A. Wen-Shung Ma, Mao-Jiun J. Wang and Frank S. Chou*
14 Human Errors in Maintenance ................................... 85
   *Shang H. Hsu*

| PART III | Posture and Musculoskeletal Disorders | 91 |
|---|---|---|
| 15 | Musculoskeletal Disorders in the Workplace<br>Margaret I. Bullock | 93 |
| 16 | Occupational Stress at the Workplace in Industrially Developing Countries: Case Studies in China and Thailand<br>Houshang Shahnavaz, Shihan Bao and Pranee Chavalitsakulchai | 99 |
| 17 | The Benefits of Ergonomics Intervention with Particular Reference to Physical Strain<br>David Stubbs and Peter Buckle | 108 |
| 18 | What Is an Acceptable Load on the Neck and Shoulder Regions during Prolonged Working Periods?<br>Arne Aaras | 115 |
| 19 | Perceived Workload and Musculoskeletal Strain among Health Care Workers: A Four-Year Follow-Up Study<br>Tuulikki Luopajarvi and Clas-Hakan Nygard | 126 |
| 20 | Working Conditions and Occupational Influences on Low Back Pain among Japanese Truck Transportation Workers<br>Akinori Hisashige and Shigeki Koda | 133 |
| 21 | A Field Survey on Practical Load Carrying Limits<br>Albert Soued | 142 |
| 22 | Skill, Excess Effort, and Strain<br>Michael Patkin and John Gormley | 145 |
| 23 | Consequences of Variability in Postures Adopted for Handling Tasks<br>Christine M. Haslegrave | 151 |
| 24 | Seated Posture and Workstation Configuration<br>Marvin J. Dainoff and James Balliett | 156 |
| 25 | The Effect of Six Different Kinds of Gloves on Grip Strength<br>Mao-Jiun J. Wang | 164 |
| 26 | A Method to Measure the Forces Exerted by the Fingers When Writing with a Ball-Point Pen<br>Yoshio Ishida | 170 |
| PART IV | Ergonomics Methodology | 177 |
| 27 | Ergonomics Fieldwork: An Action Programme and Some Methods<br>E. Nigel Corlett | 179 |
| 28 | Field Trials on Network Maintenance Systems: An Ergonomic Approach<br>Rebecca Orring | 186 |
| 29 | The Use of Scaling Techniques for Subjective Evaluations<br>Martin G. Helander and S. Mukund | 193 |
| PART V | Stress and Mental Workload | 201 |
| 30 | Psychosocial Components of Ergonomic Measures: Methodological Issues<br>Barbara G. F. Cohen, Kevin E. Coray and Chaya S. Piotrkowski | 203 |

| | | |
|---|---|---|
| 31 | Multivariate Analysis of Mental Stress and Sinus Arrhythmia .......... <br> *Kyung S. Park and Dhong H. Lee* | 210 |
| 32 | Stress Moods and the Stressors of Workers in Korean Heavy Industries ................................................. <br> *Yoo-Jin Seo and Masaharu Kumashiro* | 216 |
| 33 | A Study on the Hygiene and Ergonomics of Medical Staff in a Hospital in China ........................................... <br> *Tianlin Li, Xiufen Zhang, Ailan Feng, Lijuan Zheng, Zhengxiang Wu, Chunfa Zhang and Zhenying Fu* | 222 |
| 34 | Relationship of Subjective Symptoms of Fatigue to Attitude to Life and Stress .............................................. <br> *Yukio Hiraoka, Junko Tanaka, Masahide Oda, Hisanori Okuda and Hiroshi Yoshizawa* | 226 |
| 35 | The Link between Stress and Attitude towards Life ................ <br> *Masahide Oda, Junko Tanaka, Yukio Hiraoka, Hisanori Okuda and Hiroshi Yoshizawa* | 234 |
| 36 | An Examination of the Mental Workload of Design Work in Offices ................................................... <br> *Takeshi Aoyama and Mamoru Umemura* | 243 |
| 37 | A Study on the Evaluation of Simple Workload by a Thermal Video System ............................................. <br> *Yoshinori Horie* | 250 |

## PART VI    Workplace Evaluation    253

| | | |
|---|---|---|
| 38 | Differences in Workload and the Influence of Age on Press and Welding Workers............................................. <br> *Koki Mikami, Soichi Izumi and Masaharu Kumashiro* | 255 |
| 39 | Ergonomics Problems among Kitchen Workers in Nurseries ......... <br> *Eiji Shibata, Yuichiro Ono, Jian Huang, Naomi Hisanaga, Yasuhiro Takeuchi, Midori Shimaoka and Shuichi Hiruta* | 263 |
| 40 | Workload of Workers in Supermarkets ....................... <br> *Koya Kishida* | 269 |
| 41 | A Comparison between New and Current Methods of Debranching Trees .......................................... <br> *T. Klen, E. Ahonen, Unto Kononen, Juhani Piirainen and Juha Venalainen* | 280 |

## PART VII    Environmental Ergonomics    287

| | | |
|---|---|---|
| 42 | An Ergonomics Evaluation of Cleanroom Work ................. <br> *Yeong-Guk Kwon* | 289 |
| 43 | Heat Transfer Characteristics of Industrial Safety Helmets .......... <br> *John D. A. Abeysekera, Ingvar Holmer and Christer Dupuis* | 297 |
| 44 | The Lowest Limit of Environmental Temperature for Office Work ....... <br> *Tianlin Li, Zunyong Liu, Haichao Huang and Yongzhong Yu* | 304 |
| 45 | Cabin Attendants' Working Environment ..................... <br> *Rebecca Orring and Lena Erneling* | 308 |
| 46 | A Study on the Visual Problems Experienced by Paint Inspection Workers Using Measures of Visual Accommodation ................ <br> *Naofumi Hirose, Shinobu Akiya, Susumu Saito, Kimiko Koshi and Sasitorn Taptagaporn* | 316 |

## PART VIII   VDT Ergonomics — 325

**47**  The Effect of VDT Data Entry Work on Operators ............ 327
*Chuansi Gao, Rongtai Cai, Lei Yang, Guogao Zhang, Demao Lu and Qiyuan She*

**48**  Technological Change and Work Related Musculoskeletal Disorders: A Study of VDU Operators ............ 333
*Choon-Nam Ong, J. Jeyaratnam and W. C. Kee*

**49**  The Effects of the Visual Conditions of VDT Viewing on Pupil Size ............ 340
*Sasitorn Taptagaporn and Susumu Saito*

**50**  The Effects of VDT Polarity and Target Size on Pupil Area ............ 346
*Masaru Miyao and Sin'ya Ishihara*

**51**  The Significance of Changes in CFF Values During Performance on a VDT-Based Visual Task ............ 352
*Tsuneto Iwasaki and Shinobu Akiya*

**52**  The Complexity of VDT Work Content and Its Relation to Visual and Postural Load: What an Organisational Analysis Can Tell ............ 358
*Gunnela Westlander and E. Aberg*

**53**  Ergonomics at S W I F T T ............ 364
*F. W. Darby*

## PART IX   Process Industries — 371

**54**  Ergonomics Considerations for the Design of a CRT-Based Process Control System ............ 373
*Min Keun Chung, Jae H. Choi and Eui S. Jung*

**55**  Human Engineering Analysis of the Chernobyl Accident ............ 380
*Vladimir M. Munipov*

**56**  An Evaluation of Operator Workload in Nuclear Power Plants ............ 387
*Yoshiaki Hattori, Ju-Ichiro Itoh, Teruaki Tomizawa and Katsuhiko Iwaki*

**57**  Development of the Team Activity Description Method (TADEM) ............ 393
*Kunihide Sasou, Akihiko Nagasaka and Takeo Yukimachi*

## PART X   Job Design — 401

**58**  Macroergonomics: A Sociotechnical Systems Approach for Improving Work Performance and Job Satisfaction ............ 403
*Hal W. Hendrick*

**59**  Operator Control in Modern Manufacturing ............ 409
*John R. Wilson*

## Closing Remarks — 415

**60**  Some Future Directions for Ergonomics ............ 417
*E. N. Corlett*

Author Index ............ 423
Keyword Index ............ 425

# Preface

This book is an edited collection of 60 papers that were presented at the Xth UOEH International Symposium and the 1st Pan-Pacific Conference on Occupational Ergonomics. The main theme of this conference was *searching for solutions to occupational problems*. The main purpose of the papers on this general topic was to examine the health and safety problems of workers in the industrial workplace from the interdisciplinary perspective of human factors engineering and to find concrete solutions to those problems.

A look at the progress of ergonomics in the three decades since the IEA was founded in 1960 shows that there have been steady advances in the theoretical and empirical activities of ergonomists around the world as they have sought to establish a field of systematic ergonomics research. With the large-scale development of social and industrial structures during that time, the areas covered by ergonomics have had to reach beyond human-machine systems to include human-information systems, with their emphasis on software systems, and to human-society systems, thus covering the total system that encompasses the entire human environment. As society's needs have grown, people have come to expect more from ergonomics. In particular, ergonomics must now play a vital role in the field of occupational health and safety. Unfortunately, however, for some people the hope that ergonomics would be the key to solving problems in that field has faded.

A major problem is that ergonomics is not yet a fully mature discipline. Despite its interdisciplinary nature, it often tends to examine current problems only from the point of view of individual specializations.

For ergonomics to make a useful contribution to occupational health and safety, it must attempt to solve problems directly. In this context, this book is a collection of studies from many different fields related to ergonomics problems in the workplace; its goal is to demonstrate the benefits of adopting a solution-oriented ergonomics approach in which answers to problems are evaluated from an interdisciplinary perspective. The authors seek a shift from passive to active ergonomics in order to make the field more practical and useful in industry and work.

Another important feature of this book is the large number of papers from Asian countries such as China, Taiwan, South Korea and Japan. These should provide valuable insights into the recent activities in occupational ergonomics in Asia. Thus the reader may learn about ergonomics concepts originally based on Western science and technology but with the added flavour of the culture of the East.

This book can be summarized as follows. It presents a search for ergonomics problems related to occupational health and safety in Asia, and it discusses ergonomics problems in several occupational fields. In particular, physical work continues to be performed by many people in a variety of settings and will continue to do so in the future. Thus problems related to musculoskeletal injuries will frequently occur. In this regard, actual workloads and possible methods of reducing them are discussed. Studies are included which survey industrial fatigue and stress caused by working under excessively high and low loads (in general, jobs involving repetitive movements or the lifting of heavy objects), and methods for evaluating these problems are presented.

One problem of recent interest concerns computer operators. Research conducted in Japan, the United States, China, Singapore, and South Korea on improving the working environment for such workers confirms the hidden potential of applying ergonomics widely to productive work and safety design in areas of new technology.

Another important area of new technology that has raised concern is the nuclear power industry. Methods are discussed here for improving the reliability of work in nuclear power plants, and references are made to the causes of the Chernobyl nuclear power plant disaster, which is still fresh in the memory of ergonomists throughout the world. The Chernobyl accident has taught us, for example, that appropriate operations management and accident control must be implemented in order to operate such systems safely, and one method of achieving this goal is to apply both information feedback and correct ergonomics.

As in the case of traditional production-oriented engineering research, papers in this volume show how design and research that consider human factors in work can boost productivity and safety, reduce worker stress, and improve the human aspects of work.

This book also discusses the topics that are the focus of many other works in ergonomics: work equipment and the work environment (both physical and organizational). Among the topics covered in this area are protective clothing and its effect on performance, technical standards for equipment and worker safety, efficient maintenance operations, work in low-temperature locations, and work in cleanrooms and air-conditioned environments. Another important topic is the effect of automated production systems on industrial workers; many papers consider the impact that automation has on both the minds and bodies of workers.

Other papers provide important clues for the development of a concept of occupational ergonomics. Since ergonomics is a human-centred discipline that deals with the interrelations between individuals or groups and their surroundings, it covers every aspect of the human environment. From this comprehensive point of view, ergonomics studies both the interactions of individuals and the influence of groups and organizations. This concept of ergonomics as an interdisciplinary field has been attracting increasing interest around the world.

Ergonomics research must now include both ecological and ethical considerations. These elements have often been ignored in the past because ergonomists focused only on the factors immediately facing them and forgot the general perspective. The purpose of this book is to present a comprehensive view of contemporary problems and solutions for ergonomics in the workplace in the expectation that some of the solutions will in fact be used. We hope our readers will include university personnel, researchers, and company employees (e.g. industrial physicians, production managers, occupational safety and hygiene managers, planners and designers, and so on). The task for the future is to develop ways of understanding and solving such problems so that a balance is struck among the long-term, systematic, and holistic aspects. It is our hope that this book will provide the reader with some hints on how to proceed.

# Acknowledgements

The Xth UOEH International Symposium and the 1st Pan-Pacific Conference on Occupational Ergonomics (Kitakyushu, Japan, 10-13 July 1990) was supported by the following sponsors:

The Japan Ergonomics Research Society
International Ergonomics Association
Human Engineering Society of Korea
The Ergonomics Society, United Kingdom
Safety & Environmental Protection Research Institute of China
Japan Association of Industrial Health
The Japanese Rehabilitation Medicine Academy
Japan Industrial Management Association
Japan Society of Health Science
Foundation for Total Health Promotion
The Japanese Association of Stress Science
Mitsubishi Research Institute Inc.
City of Kitakyushu
The University of Occupational and Environmental Health, Japan
Occupational Health Promotion Foundation.


Especially, we wish to express our thanks to the following members of the advisory committee:

Dr Kenzaburo Tsuchiya (President, UOEH, Japan)
Dr Nigel Corlett (Chief Editor, Applied Ergonomics, UK)
Dr Ward Gardner (Occupational Health Consultant, UK)
Dr Masamitsu Oshima (President, The Japan Ergonomics Research Society, Japan)
Dr Sadao Sugiyama (Professor, Kwansei University, Japan)

For the publication of the present proceedings, our first debt with this book is to our contributing authors, all of whom have responded to our various requests with great patience, in some cases, within a very limited time.

We are indebted to Dr Shinji Miyake and Ms Chie Nakatani at the Department of Ergonomics in the University of Occupational and Environmental Health, Japan, also, to Ms Christine Stapleton from the Ergonomics Information Analysis Centre at the University of Birmingham for their many hours of editorial assistance in the preparation of this book.

# Part I
# Ergonomics in Perspective

# 1. ERGONOMICS FROM PAST TO FUTURE: AN OVERVIEW

## Brian Shackel

*Department of Human Sciences and HUSAT Research Institute
University of Technology, Loughborough
Leicestershire LE11 3TU, England*

**Abstract**

This paper aims to present a synoptic overview of the past, present and some possible futures in ergonomics. This aim is very ambitious; therefore, the review has to be very selective, and there are many omissions to be regretted. After a general introduction, the need for ergonomics is discussed, and three reasons are presented to support the view that ergonomics must now be considered an essential discipline for many parts of industry. The development of ergonomics is briefly traced, and then the scope of the subject is outlined in terms of various categories and dimensions. Having thus presented a user-centred framework, this is then used to structure a series of examples to illustrate ergonomics in action. Finally, some tentative speculations are offered to indicate some of the possibilities in the future.

**Keywords:** *history of ergonomics; human-machine interface; information technology; leisure; system design; vehicle design; workplace; accident prevention; future of ergonomics*

## Introduction

In view of the pioneering nature of this meeting, both in Japan and in the Pacific world, I thought it would be appropriate to try to provide an overview of the past, present and future of ergonomics. Of course, that is far too ambitious for the time available; therefore, I ask you to forgive me for the many omissions which are inevitable and for the very selective way in which I must present my review and my examples to you.

### The craftsman and his tools

In prehistoric times men made their weapons and tools for themselves. Efficiency, and the fitting of the tool to the operator, were vital; if a man made a bad weapon and could not use it well enough, then after the first fight with a hungry animal there was one less designer and maker of bad tools in the world. The tradition of the craftsman making his own tools, or selecting them carefully before buying until he finds those which 'feel right' (that is, which fit him well), has lasted through the ages and still applies to hand tools. But few operators can try out, and throw aside if not suitable, the bigger machines of today - such as lathes or bulldozers or airliners - until they find one which fits them!

## The operator and his machine

Therefore in our factories and homes, in the air and on the roads, we are using tools and machines designed for our use by others with little advice from us about how they should be made to fit our own peculiarities. This does not of course mean that the machines will necessarily be badly designed or that we shall not be able to use them. But it may well mean that we cannot use them as well as the designer could himself, because what fits him may not fit us.

Moreover, design engineers are mostly, and quite rightly, concerned to improve the mechanical and electrical performance of their machines. But they sometimes forget that what matters most is how well the machine works in conjunction with the operator who has to use it day after day in his routine work. Now, there is nothing especially novel about emphasising this need for a good fit between man and machine; but the difference with ergonomics is that it provides a systematic scientific approach to the problem.

## Why the specialist - is not common sense enough?

This question would not be asked if the subject of ergonomics were other than human. But because everybody is human, each person tends to think that he automatically knows most of what needs to be known about humans when creating a working situation, i.e. that common sense is enough. Whereas in fact the amount of specialist knowledge now available about the human operator is such that there is probably no one human, even the specialist himself, who can learn it and know it all. If a problem arises which is primarily in, say, chemistry or quality control, a person quickly acknowledges if it is outside the limits of his own general knowledge and calls upon the chemist or quality control specialist for assistance. But if he has a problem involving humans, he still tends to use his own subjective opinion or to consult the subjective opinions of others instead of summoning a specialist, perhaps because the existence of specialists in ergonomics is not yet widely known.

Accidents, errors, poor quality and low output are the usual symptoms of a problem which, amongst other things, may need ergonomics attention. When a human-machine mismatch in industry today results in such symptoms, the cost of not finding a cure is often considerable but may not be so high as to compel action. But as humans in the future are progressively used less for their muscle power and more and more for their ability to process information and make decisions, so the human is in control of more assets and more output, and the cost of the mismatch grows in proportion. As with the military problems in the past, so now with modern industrial equipment, the demand upon the operator grows as the complexity grows. The more the operator is stressed, the greater is the need to ensure a good match between human and machine to minimize the risk of error and maximise accuracy, output and human satisfaction. These are the primary reasons why the specialist knowledge from ergonomics is needed.

## *Why is ergonomics essential?*

There are three basic reasons why ergonomics is essential for modern industry, especially in relation to product and system design.

First, the complexity and sophistication of modern industrial technology sets higher demands upon the human operators and controllers; but complexity also causes designers to be too busy with technical problems either to deal with the human factors properly or to learn enough about how to deal with them.

Second, there is a time and space barrier. The complexity of modern technology also separates designer and user, and thus usually prevents effective feedback from the user to improve the design. Therefore the ergonomist is an essential link who operates as a sort of preventive and predictive feedback channel.

Third, there is the separation of responsibilities and the cost consequences. Another problem which seems to follow from the complexity is that often the designer, manufacturer/marketer, buyer and user are separate. They may well be in separate organizations and certainly will have separate aims and criteria. The designer (and engineer) will aim for a good machine solution, and will expect to spend all his budget costs on technical machine factors; the manufacturer/marketer will aim to cut the capital cost (but not necessarily the running costs); the buyer will aim to pay a low price and will expect savings to come from the purchase (perhaps by staff reductions); the user will aim to minimise his personal loss (of skill, earnings, etc.) due to the new machine or method of working. The separation between them may often cause each of these four people not to use ergonomics, because they cannot see the cost justification within their own cost limits. Only the manager in charge of the user sees the final result, where the true cost of training and of inefficiency, if the design is not ergonomic, can exceed any savings in purchase cost.

Because each sector of responsibility is separate, as noted above, the cost-benefit evaluation of ergonomics can often be difficult to prove (but for references and examples see Beevis & Slade, 1970; Corlett & Coates, 1976; Shackel, 1987; Corlett, 1988). Therefore, a recent new concept may be helpful. Organisations using new systems (i.e. the 'buyers') are beginning to realise that the running and repair costs (including selection, training, maintenance, and labour turnover) may far exceed the capital costs. Some are beginning to ask the manufacturer/marketer not only to sell machines or systems at the capital purchase price, but also to guarantee the total running costs not to exceed some annual value over an agreed 'life' usage; this is called total system-life costing. This development should be strongly supported by ergonomists, because it will show more clearly the cost-benefit value of ergonomics.

## *How did ergonomics develop?*

In Great Britain, what is now called ergonomics had its beginning in the scientific study of human problems in ordnance factories during World War I. This kind of work continued under the Industrial Health Research Board between the wars. World War II led to greater emphasis not merely on matching men to machines by selection and training, but also, much more than previously, to the designing of equipment so that its operation was within the capacities of most normal people. This fitting the job to the man increased considerably the collaboration of engineers in certain fields with biological scientists. This collaboration, beginning primarily with military problems, because it was there particularly that operators were pushed to their limits, continued after the war and led to the formation if 1949 of the Ergonomics Research Society.

Similar developments occurred in other countries, leading in the USA to the formation of the Human Factors Society in 1954. On the international scene, the formation meeting which accepted the first constitution and rules of the International Ergonomics Association (IEA) was held during the Annual Conference of the Ergonomics Research Society in Oxford in 1959; the first international conference of the IEA was held in Stockholm in 1961. The IEA now has 16 member societies in nations around the world.

## The scope of ergonomics

### What is ergonomics?

The principle purpose or philosophy of ergonomics 'is **not** primarily to improve productivity or output or human methods of doing work; these are quite properly the main aims of other disciplines. The prime purpose of ergonomics is to study and understand the situation of people at work and play, and thus to be able to improve the whole situation for the people. Of course this knowledge may also be used to assist with productivity, and at times managers may need to be persuaded to use ergonomics by the expectation of some such benefits, but the main thrust of ergonomics remains always user-centred.

Ergonomics is defined as the study of the relation between people and their occupation, equipment and environment, and particularly the application of anatomical, physiological and psychological knowledge to the problems arising therefrom. This definition is in two parts and clearly describes both a science and a technology. Thus, the scope of work in ergonomics must clearly embrace both research and practical application.

### Some dimensions in ergonomics

In the past the main focus of ergonomics has been human work and the work situation in general. However, today ergonomics is focused very widely along many dimensions; I can only mention some of these dimensions to emphasise the breadth of our field, which in essence is relevant to most human activity.

This breadth is indicated in Table 1. In the left hand column are very broad areas of human activity, that is work, leisure and sport. Even from the early days of ergonomics, there have been studies in the second and third categories, for example on the ease of use of leisure items (such as camping and caravanning equipment and hi-fi music reproduction systems) and upon some sports aspects.

Table 1. *Some aspects within the scope of ergonomics*

| Activity types | Framework | Interaction levels |
|---|---|---|
| | Human | |
| Work | Machine | Data input |
| Leisure | Workspace | Operating control |
| Sport | Environment | Supervisory control |
| | System | |

The middle column shows a common framework in ergonomics for studying the situation with a user-centred orientation, working outwards from the human and considering the interaction of the person with the machine and immediate tasks, with the workspace and the environment surrounding him, and finally with the total organisation and system within which the human is working.

The third column points out some of the varying levels of interaction the user may have with a system, from the specialised but prescribed work of inputing data, to the next level up of controlling the detailed operations, and up to a further level of broad supervisory control of the whole system.

Within the area of work ergonomics (see Table 2) we often distinguish various types of work, each of which will have its own characteristics and lead to a different assembly of ergonomics issues.

Table 2. *Some aspects within work ergonomics*

| Type of work | Example | Some ergonomics issues |
|---|---|---|
| Heavy work | Mining | Energy expenditure |
|  | Metal work | Work-rest schedules |
| Light work | Product assembly | Boredom |
|  | Office typing | Manual dexterity |
| Mental work | Chemical plant supervision | Vigilance Motivation |

For example, heavy work, such as that involved in the mining and metal industries, typically involves issues of human energy expenditure and work-rest schedules to avoid excessive physical fatigue. Light work, such as product assembly or office typing, involves manual dexterity and may well, if the cycle time is short and the challenges in the work are few, bring the problem of boredom, carelessness and perhaps in consequence accidents. Mental work such as may be involved in the general supervision of a chemical plant, will involve issues of vigilance during the monitoring of the automatic operations, and the importance of maintaining motivation. There has, of course, already been much work reported in the literature on all the above issues.

Again, if we consider one level of the interaction paradigm, namely human-machine interaction, we may note (see Table 3) that the type of technology and the level of interaction of the human with it will lead to different aspects and areas of ergonomics work.

Table 3. *Some areas of work within human-machine interaction*

| | |
|---|---|
| Workstation ergonomics | Direct physical interaction |
| Systems ergonomics | Remote control and supervisory interaction |
| Cognitive ergonomics | Direct intellectual interaction |

So, for example, the direct physical interaction of the user with the machine will involve the application of workstation ergonomics to the design, as shown by Shackel (1969) and Kvalseth (1983). The level of remote control and supervisory interaction will involve both workstation ergonomics and more especially systems ergonomics, issues such as dealt with extensively by Sheridan & Johannsen (1976) and Damodaran et al. (1980). Further, the recent growth and widespread usage of advanced micro-computers with sophisticated applications programs has led to a new type of interaction of the human with the machine, in the form of a direct intellectual interaction with the sophisticated software programs; this is leading to the growth of an extensive new area of work usually known as cognitive ergonomics.

The above are only a few of the dimensions covered by ergonomics and of the categories which have been used to structure our field of work for easier explanation. However, the field is now so broad that any category scheme is more illustrative than comprehensive, and even an encyclopaedic textbook, edited by Salvendy (1987), can no longer cover the whole field.

However, these are all classifications for the scientist and researcher. How can the manager in industry decide whether ergonomics is relevant to his work and what policy should an industrial manager follow with regard to ergonomics?

### Where is ergonomics relevant in modern industry?

On the shop floor there are many situations where scientifically proven methods of selecting and training operators would yield worthwhile economic improvement. Organisationally, it would seem that these situations can best be dealt with by personnel and work study departments making use of technical data and consultant advice from ergonomics where relevant. In many companies it may not be good economics to employ a full-time ergonomics specialist for this work alone.

When new machines and task situations are being designed or considered for installation, ergonomics becomes more particularly relevant, and the cost of obtaining such knowledge, especially if it is sought early enough in the design stage, becomes relatively insignificant in proportion to the typical capital cost involved and the potential savings to be expected.

The same arguments apply with much greater force as complexity increases and the design of larger production units and systems is considered. A continuous flow production line is a typical example of a modern industrial situation requiring some thorough system design. The bigger such a line is, the greater is the importance of adequate feedback and feedforward of information from one stage to another to maintain everything within limits; but very frequently one finds that only the most primitive means of communication are available between operators, who are now usually at much greater distances from each other. Only thorough and extensive study of what information each operator needs from the machine and each other, and of the way that the system should be designed to present this information most simply and easily, can safeguard against lengthy commissioning time and the risk of costly shut-downs. The emphasis of the ergonomic approach on dynamic rather than static investigation and planning could be very fruitful here.

### What would be correct management policy?

Only broad generalizations are possible, because so much depends upon the nature of a company's product and organization. The manufacturers of complex capital goods are more likely to find ergonomics help of practical and economic value than are manufacturers of simple products. On the other hand, any manufacturer contemplating major capital expenditure on new plant or new production facilities, particularly envisaging a major increase in the complexity of his manufacturing processes, might well save much trouble later by getting ergonomics assistance early enough.

In general, the best criterion is probably the financial size of the company compounded with the amount of man-machine conjoint working time involved, either (1) in the consumer's use of the product, or (2) on the production floor, according to which aspect of ergonomics application is being considered. For instance, on the question of ergonomics applied to the design of the company's products, manufacturers of machine-tools, motor cars, furniture and household appliances have much greater need for ergonomics knowledge than manufacturers of electrical generators, tyres, submarine cables and water tanks. Again, on the question of ergonomics applied production methods, manufacturers with a high ratio of labour to other costs (such as in the boot and shoe industry), with special environmental problems (such as in steel or tyre making), or with important safety considerations (such as passenger transport services), can expect significant improvements in efficiency and safety by applying ergonomics knowledge.

Given a need for ergonomics suggested by the above criteria, the scale of ergonomics activity required and the manner of obtaining it will then depend on the size and turnover of the company. In general, it would seem that only large companies, say of more than 1,000 employees, or those with a large turnover, say of more than £50,000,000 are likely to feel justified in employing a full-time specialist on their staff. Most managements would be better served by seeking consultant advice when appropriate. This, however, points to the need for one important action. When a decision has been taken to seek ergonomics advice only when required, it is essential to establish, within the normal organisation of the company, a clear definition of who is responsible for watching over the general running of the company to detect when and where ergonomics assistance would be economically advisable. Some one or more persons should be given the necessary training and charged with the responsibility of watching for important ergonomics problems and, when they arise, with the duty of seeking the specialist advice required. In various industries and companies this function is carried out sometimes by the medical officer, sometimes by the work study department, sometimes by the production manager, and sometimes by the chief engineer or his staff.

Therefore, in relation to ergonomics, a management decision is required, in a company of any size, upon the following three points:
1. What areas of our factory and/or products might benefit from the application of ergonomics knowledge?
2. Does our requirement justify a full-time specialist or a consultant called in as necessary?
3. If a consultant is to be used at any time, what department or person shall be charged with the duty of watching for significant ergonomics problems and calling in a consultant in good time?

## *Some examples of ergonomics in action*

I shall now try to bring to life the above description of the scope of ergonomics by presenting to you some illustrative examples of ergonomics in action. The theme I shall follow starts with the human at the centre, in terms of safety aspects, and then moves outwards in turn to some illustrations of ergonomics work on human-machine interaction, human-workplace interaction, human-environment interaction and finally systems ergonomics.

### Safety

The emphasis of ergonomics with regard to safety is upon the design of tools and equipment, so as to try to produce equipment which is so well adapted to the operator that he never makes a mistake (e.g. see Shackel, 1959).

Much of the accident prevention work by safety officers and national safety organisations is concentrated upon the operator and upon safe usage. I do not in any way suggest that this emphasis is wrong, but there is a tendency to place responsibility more on the user and operator, and I do suggest that this needs to be balanced by an equally strong emphasis towards the designer.

For example, the official British report upon an accident with a compressed air pneumatic road-drilling tool states "the operating trigger of this particular compressed air tool was fitted on top of the operating handle, so that it would be depressed by the palm of the hand when the tool had to be operated. When, however, the whole tool was turned upside down pressure from the ground depressed the trigger and the tool commenced to operate. This type of accident could be avoided by the provision of a guard over the trigger but the basic cause of the accident was incorrect use of the tool". This analysis surely excuses the designer far too easily. Mis-use of such robust and heavy tools is inevitable and should be allowed for by the designers; unguarded triggers are certain to result in an accident eventually.

By contrast, a noteworthy ergonomic achievement is the Airstream Helmet, a piece of protective headwear developed by ergonomists in the British Steel Corporation. After a very thorough ergonomic design and development programme, eventually over 2,000 workers in British Steel, primarily those working on coke ovens, were issued with this device which protected their eyes, face, head and lungs. Apart from the direct safety benefit, i.e. reduced risk of injury to eyes, faces, etc., there were many indirect benefits to health such as improved sleep for some people and improved appetite for others. The device received a Design Council Award for innovative design, and the manufacturer received the Queen's Award to Industry for the export performance of the device, which has now been sold in every continent in quantities far exceeding any other protective respirator.

## Human-machine interface

Turning to the first sector of interaction of the human with his working situation, I have three examples of major contributions by ergonomics.

The first concerns the display of information in the context of a high stress and high-responsibility skilled performance situation, namely the piloting of aircraft and airliners. The first instruments for displaying altitude to pilots were rather crude, but in those days even airline flying was a fair weather occupation. The standard altimeter into even the 1960s was the three-pointer device which had been shown as early as 1945 to cause quite a high error rate in reading under stressful or test conditions. Pioneering work in Britain and the U.S.A. by ergonomists proved the advantages of the digital altimeter to be about 60% in speed of reading and, above all, a reduction from 13% errors to zero errors (Rolfe, 1969). At least two airliner accidents in the 1960s were officially attributed to the misreading of the altimeter by the pilot, and the complete loss of the aircraft involved cost many millions of pounds. It is likely that by now this new design of ergonomic altimeter must have prevented several crashes and saved both money and lives, but of course it is happy we shall never be able to quantify this success.

Ergonomics was first brought to the machine-tool industry by Gibbs (1952). At that time the displays attached to the basic controls on lathes and other such machines were typically a scale round the edge of a cylindrical drum on the axle of the crank manipulated by the operator to control the lead screw and the tool bar, etc. Gibbs realised that these scales were not very readable and, above all, that they did not present the information in a way compatible with the digital display on the typical workshop drawing. He therefore invented a digital indicator unit which effectively placed a digital micrometer directly upon the control shaft operated by the crank handle. An industrial comparison of lathes fitted with scales and with digital position indicators showed a 41.9% improvement in favour of the digital indicator for machine usage time, and a 35.4% saving in total time for whole jobs (Beevis & Slade, 1970). Since that time, digital indicators, nowadays in an electronic form, have become the standard display device on almost all machine tools.

The third example relates to what is probably the most sophisticated piece of machinery used by the most number of people in the world today, the motor car. The design of cars for ease of use, comfort and convenience has improved markedly in the last 25 years or so. It seems justified to claim that this improvement has been helped and made more rapid by the unbiased and outspoken reports produced by the British Consumers' Association and equivalent organisations in Europe and North America. For 20 years I had the interesting experience of being a consultant to the motoring test unit of the Consumers' Association, and in that capacity I supervised and did a fair proportion of the work involved in providing an ergonomics evaluation of each group of cars under test four times per year (Shackel, 1980a, b; Brigham, 1980). The procedure involved a detailed questionnaire about all aspects of the car filled in by between 15 and 20 test drivers on the staff of the motoring test unit, together with the two ergonomics advisers, during the course of each car being driven for a minimum of 10,000 miles. An appraisal of the ergonomics contribution reads as follows "the evaluation by C. A. (Consumers' Association) of the human factors contribution and of the HFQ programme is positive and unequivocal. In cost-benefit terms it is one of the cheapest components of the car testing activity, the external paid-out to C. A. being about £800 annually from a total car testing budget of £250,000 ".

**Human-workspace interaction**

My first example of workspace follows from the previous and is concerned with the seat design in cars. We all know that car seats are often rather uncomfortable. To design a satisfactory seat for most people is not easy, because of the very wide range of body size and shape contained in even the middle 80% of the normal population not including the extremes. But success can be achieved; in the July 1963 issue of 'Motoring Which?' on page 83, the Consumers' Association rated the Renault very highly, and in the April 1965 issue (page 43) it stated "In July 1963 we said that the Renault R8 seat was almost certainly the most comfortable we had ever tested. Since then it has been improved by being given an adjustable squab. Most drivers were quite lyrical about the comfort and support the seat gave them even on very long runs".

It so happens, and I have checked and found that the CA testers did not know this, that Renault was at that time the only European car manufacturer to have its own ergonomics department.

Along with seats and chairs, the desk and visual display terminal are particularly important parts of the modern workspace, especially in offices. Surveys have revealed the many ergonomics problems at the average office computer workspace. Many books are already available dealing with the ergonomics issues of this subject, for example, Cakir et al. (1980) and Grandjean (1987), so I shall not dwell further upon this topic despite its importance.

Finally, the redesign of the control room for Esso at London's Heathrow Airport provides a good example of the need first to analyse the whole operation and the task of the controller, before specifying the design of his desk and related equipment (Shackel & Klein, 1976). This was a case where the control room had far outgrown its original design and had become so cluttered that at times the controller was actually losing important information. After a full analysis of the whole task situation, a new concept was developed to display all the relevant decision information on and just above a 'state-board'; and as a result a totally new workspace could be designed. These new solutions served and satisfied the whole working situation for at least 14 years.

## Human-environment interaction

Under this heading space limitation only permits two examples of the results of ergonomic improvements to the environmental working conditions.

It is not easy to isolate the variables precisely and thus prove the advantages of achieving a better ergonomic match of the environment to the human user. However, this illustration of improvements by reducing the noise levels comes from a study carried out at Dresden (Hartig, 1962). One case compares two otherwise identical offices, one of which was quietened by being fitted with a "sound swallowing ceiling and wall-clothing material", and in both of which everything they could measure was measured for a year. The results were compared and in the quiet office there were:

9% higher production
29% less typing errors
52% less errors on calculating machines
37% less sick leave
47% less staff changes.

This paper also states that, in 1960, 1 million deutschmarks were paid in compensation for occupational deafness.

Turning to psychological and performance effects of environmental factors, one of the best examples proving the importance of routine ergonomics was reported by DeMarco & Lister (1985). In this study - really a competition - 166 programmers (in 83 pairs) from 35 organisations took part; each pair worked at their own workplace, using the same support environment and the same language, and performed the same one-day programming and implementation task working to the same specification. Careful time records were kept of their work and working environment. To detect any correlation between environment and performance, the programmers were divided into four groups based upon performance; the average performance of those in the top 25% group was 2.6 times better than that of those in the bottom 25% group. The environmental ergonomics for the top group were significantly better than for the bottom group, as shown in Table 4.

The authors conclude that "changing the workplace from that typical of a lower-25% performer to that of an upper-25% performer offers the potential of a 2.6 to 1 improvement in the time to perform a complex programming activity".

Table 4. *Results from DeMarco & Lister (1985)*

| Environmental factor | Top 25% | Bottom 25% |
|---|---|---|
| Dedicated floor space | 78 sq ft | 46 sq ft |
| Acceptably quiet workspace | 57% yes | 29% yes |
| Acceptably private workspace | 72% yes | 19% yes |
| Can you silence your phone? | 52% yes | 10% yes |
| Can you divert your calls? | 76% yes | 19% yes |
| Do people often interrupt you needlessly? | 38% yes | 76% yes |
| Does your workspace make you feel appreciated? | 57% yes | 29% yes |
| Average performance:- | 2.6 ratio to | 1 |

## Systems ergonomics

Under this heading the ergonomist must be concerned with most human aspects of a major system and must be able to start from a global orientation toward the whole system. While many ergonomists have been involved with the design of large systems there are relatively few case studies of such work in the literature. The redesign of the Esso control room at London airport, already mentioned, is an example of a full system study first providing the necessary background for the subsequent detailed work on the console and the workspace and environmental features.

One of the few documented examples of ergonomics applied throughout the design of a large non-military system is the automation of meat handling in the London Docks (Shackel, 1971). Although not involved from the very beginning, the ergonomics team was brought in before the full system contract had been placed, and gradually gained acceptance and became an integrated part in the total design process. In addition to conducting an ergonomics system analysis, they were involved with routine ergonomics design of equipment, workspace and environment, and then they had to conduct an emergency research programme to deal with a problem which had not been foreseen in the original specification and which was first detected by the ergonomics system analysis. Finally, the ergonomics team was commissioned to produce a complete training programme for the whole plant.

System design work is not fundamentally different, for ergonomists, from other types of design. Many of the routine machine, workspace and environment aspects are similar; but the important difference lies in the opportunity, and indeed the necessity to work more globally and to be able to influence all human aspects throughout the system. At times, of course, this is most challenging but also it can be most satisfying.

## *Ergonomics into the future*

### Ergonomics decades in Europe and North America

I believe it is not an undue distortion or over-simplification to suggest that the four decades of ergonomics from the foundation of the Ergonomics Research Society in 1949 can be identified as in Table 5.

*Table 5. Focus over the years*

| | |
|---|---|
| 1950s | Military ergonomics |
| 1960s | Industrial ergonomics |
| 1970s | Consumer ergonomics |
| 1980s | Computer ergonomics |
| 1990s | Information ergonomics |
| | Leisure ergonomics |
| | Space ergonomics |

The preponderance of work in each decade was in the areas named; the reason is that the major sources of funding for both applied research and application work came from the listed areas. I should, of course, emphasise that the given type of work does not end at the end of a decade; the work still carries on, perhaps at a lower level, while work in the next decade develops and perhaps exceeds the level of work in the former decade. I believe this is a fair summary of the past, but of course that does not necessarily mean that the same will happen in the next decade.

## Changing structure of work and the workforce

However, there is certainly one major change which has already happened in North America and Europe, and which suggests a major growth in information ergonomics in the 1990s. The industrialised world is fast moving from a manufacturing society towards being an information society.

This change occurred during the 1970s and early 1980s, and is now being recognized more widely. For example, in the USA between 1940 and 1980 the percentage of blue-collar workers decreased from 40% to 32%, whereas the white-collar percentage increased from 31% to 52% (OTA, 1985). The process has been similar in other countries, and today more than 50% of Britain's working population work in offices (McConnell, 1986).

This change in the structure and location of the largest proportion of the workforce has been accelerated by the change in the nature and type of work caused by the very rapid growth in computing power.

Chris Evans (1979) illustrated this growth by comparing it with the more familiar motor-car:

> "But suppose," he said, "that the automobile industry had developed at the same rate as computers and over the same period: how much cheaper and more efficient would the current models be? If you have not already heard the analogy, the answer is shattering. Today you would be able to buy a Rolls Royce for £1.35, it would do 3 million miles to the gallon, and it would deliver enough power to drive the Queen Elizabeth II. And if you were interested in miniaturisation, you could place half a dozen of them on a pinhead."

## The growth of information technology

It should be emphasised that most workers in ergonomics are aware of the importance of looking ahead to foresee future problems likely to arise for the human users as a result of advances in technology. As long ago as 1955, Mackworth (1955) discussed the relevance of research then to the problems expected when automation would come. Again, with automation still hardly arrived, Welford (1960) presented a survey report on the aspects of ergonomics likely to be relevant to the design of automatic equipment, and on the human problems likely to arise from automation.

Some of the early concepts about automation have been fundamentally changed by the remarkable growth in power and reduction in size of the basic computing elements, as noted above. This has fundamentally changed the predominant types of users and their expectations (for reviews of these developments see Gaines, 1984, and Shackel, 1985). The advent of the microcomputer, in widespread use from 1980, has already caused great growth in the use of computing for many different purposes by non-socialists of all types from bank clerk to business executive, from librarian to life insurance salesman and from secretary to stockbroker and traveller.

The result of this rapid growth is that the designers of computers are no longer typical of or equivalent to the users, and the users have become much more critical and selective because of problems of usability they have experienced. With this growth in the importance of usability, to the extent that IBM now appears to make usability as important as functionality (Shackel, 1986), the ergonomics aspects have become paramount. Illustrations and case studies of the importance of ergonomics have been presented in many publications; see, for example, National Electronics Council (1983) and Shackel (1987).

As the emphasis moves from computing to more comprehensive information technology, so also many issues arise with regard to the adaptation of the new technology to the whole organisation; some of the current and future issues here have been discussed by Blackler (1988) and by Dray (1988).

Another aspect entirely of information technology in relation to ergonomics, in contrast applying ergonomics knowledge to the usage of information technology, is the adaptation and innovative usage of information technology to advance further upon existing ergonomics problems. A particularly good example of this usage in ergonomics is the Loughborough Anthropometric Shadow Scanner (LASS). This device enables very rapid and precise scanning and measurement of the external body contours in three-dimensional form (Bell, 1987). In time, it may be possible to do this so rapidly that people in future may be able to enter a special tailors' shop, be measured in about one minute, and then wait about 20 minutes while a total automation system produces a well fitting suit or dress from cloth which they have chosen at the same time.

**Information, space and leisure ergonomics?**

In relating our ergonomics research to the growth of technology, one especially important factor is timescale. There is little value in applied research if it is overtaken before completion by basic changes in the related equipment or environment. Therefore, we need to look ahead and consider what may be the general trends. The possibilities just noted are merely a few examples of many under the heading information ergonomics.

One of the other major fields of technical development in the 1990s and beyond is space exploration. Already the human factors discipline has been considerably involved in current space programmes, and we may expect to see a growing specialism in space ergonomics. However, I would expect space ergonomics to remain as a somewhat specialised sub-section of our field for quite some time yet.

On the other hand, I would expect that leisure ergonomics may become a large sub-section. With the advent of advanced technology, many are predicting a large growth in leisure time available. For example, I understand it is generally estimated that, if we could change attitudes and solve the distribution of real wealth acceptably in Britain, then we could achieve our present production levels with everyone working only about 75% of the time. This must surely lead to a major field of study in the ergonomics of leisure activities and of domestic, entertainment and educational products. Of course, many such products may well involve information technology (IT) components, but the ergonomics aspects will not be those of the IT system. Therefore, all those who do not wish to work in information ergonomics will have many traditional ergonomics problems to study, and indeed the amount of such work will probably grow considerably with the growth of the leisure population.

## Conclusion

The fundamental reason why I, as an ergonomist, welcome the information age is that at least we can foresee machines doing the tedious work which no-one wants, with people able to concentrate on what they prefer and only they can do well for each other. The focus will be upon all the activities where person-to-person interaction is the principal aspect.

Finally, we must develop a true synergy and symbiosis between ergonomists, computer professionals, architects, industrial designers, managers and ministry officials. I believe that the potential will exist to allocate many boring, dangerous, undignified or meaningless functions and tasks to the machine, and thus to enable people to be released and to grow. But I am convinced that this will not be achieved successfully by the computer scientists, engineers and technologists alone - architects, industrial designers, ministry officials and managers generally have a fundamental part to play in cooperation with scientists and practitioners in ergonomics. Are we all ready to meet this challenge?

## References

Beevis D & Slade I M 1970 Ergonomics - cost and benefits. Applied Ergonomics, 1.2, 79-84.

Bell J 1987 Automation beckons the garment industry. New Scientist, 115 no.1573, 42-43.

Blackler F 1988 Information technologies and organisation: Lessons from the 1980s and issues for the 1990s. Journal of Occupational Psychology, 61, 113-127.

Brigham F R 1980 Computerisation and subsequent further development of the human factors questionnaire. In: DJ Oborne and JA Levis (eds) Human Factors in Transport Research, pp 420-427. London, Academic Press.

Cakir A, Hart D J & Stewart T F M 1980 Visual Display Terminals. Chichester, Wiley. ISBN 0-471-27793-2.

Corlett E N 1988 Cost benefit analyses of ergonomics and work design change. In: DJ Oborne (ed) International Review of Ergonomics. London, Taylor & Francis.

Corlett E N & Coates J B 1976 Costs and benefits from human resources studies. International Journal of Production Research, 14.1, 135-144.

Damodaran L, Simpson A & Wilson P 1980 Designing Systems for People. NCC Publications, National Computing Centre, Oxford Road, Manchester. ISBN 0-85012-242-2.

DeMarco T & Lister T 1985 Programmer performance and the effects of the workplace. Proceedings of the 8th International Conference on Software Engineering, 28-30 August, pp 268-272. Institute of Electrical and Electronics Engineers, 445 Hoes Lane, Piscataway, NJ 08854, USA. ISBN 0-8186-8620-0.

Dray S M 1988 From tier to peer: Organisational adaptation to new computing architectures. Ergonomics, 31.5, 721-725.

Evans C 1979 The Might Micro. London, Gollancz. ISBN 0-575-02708-8.

Gaines B R 1984 From Ergonomics to the Fifth Generation: 30 years of Human-Computer Interaction Studies. Proceedings of the INTERACT '84 IFIP International Conference on Human-Computer Interaction, pp 1-5. Amsterdam, North-Holland.

Gibbs C B 1952 A new indicator of machine tool travel. Occupational Psychology, 26.4, 234-243.

Grandjean E 1987 Ergonomics in Computerised Offices. London, Taylor & Francis. ISBN 0-85066-350-4.

Hartig H 1962 Larmbekambfung in Der Industrie. Dresden Institut fur Arbeitsokonomie und Arbeitsschutzforschung, 15, 139. Quoted in Beevis and Slade 1970.

Kvalseth T O (ed) 1983 Ergonomics of Workstation Design. London, Butterworths. ISBN 0-408-12653-6.

Mackworth N H 1955 Work Design and Training for Future Industrial Skills (Sir Alfred Herbert Lecture). London, Institution of Production Engineers.

McConnell M 1986 The workplace - investment or overhead. Mind Your Own Business, July/August, 24-26.

National Electronics Council 1983 Human Factors in Information Technology. National Electronics Council, UK. ISBN 0-9508590-0-1, distributed by Wiley, Chichester, UK.

OTA 1985 Automation of America's Offices. US Congress, Office of Technology Assessment, Report OTA-CIT-287. Library of Congress No. 85-600623; Washington DC, US Government Printing Office.

Rolfe J M 1969 Human factors and the display of height information. Applied Ergonomics, 1.1, 16-24.

Salvendy G 1987 Handbook of Human Factors. New York, Wiley.

Shackel B 1959 Machine Design for Safety. National Industrial Safety Conference, Scarborough; Royal Society for the Prevention of Accidents, 52 Grosvenor Gardens, London, SW1.

Shackel B 1969 Workstation analysis - turning cartons by hand. Applied Ergonomics, 1.1, 45-51.

Shackel B 1971 Human factors in the PLA meat handling automation scheme. International Journal of Production Research, 9.1, 95-121.

Shackel B 1980a 'Motoring Which?' - 18 years of human factors in comparative car testing - an historical review. SAE Technical Paper Series No. 800332; Society of Automotive Engineers, 400 Commonwealth Drive, Warrendale, PA 15096, USA.

Shackel B 1980b Ergonomics in comparative car testing for 'Motoring Which?' In: DJ Oborne and JA Levis (eds) Human Factors in Transport Research, pp 411-419. London, Academic Press.

Shackel B 1985 Ergonomics in information technology in Europe - a review. Behaviour & Information Technology, 4.4, 263-287.

Shackel B 1986 IBM makes usability as important as functionality. The Computer Journal, 29.5, 475-476.

Shackel B 1987 Human factors for usability engineering. In: ESPRIT'87 - Achievements & Impact; Part No.2 pp 1019-1040. Edit. by C.E.C. Directorate General of T.I.I.I. Amsterdam, Elsevier. ISBN 0-444-70332-2.

Shackel B & Klein L 1976 Esso London Airport refuelling control centre redesign - an ergonomics case study. Applied Ergonomics, 7.1, 37-45.

Sheridan T B & Johannsen G 1976 Monitoring Behaviour and Supervisory Control. New York, Plenum Press. ISBN 0-306-32881-X.

Welford A T 1960 Ergonomics of Automation. D.S.I.R. Series No. 8. London, Her Majesty's Stationery Office.

# 2. A VIEW OF ERGONOMICS IN JAPAN

**Masamitsu Oshima**

*Medical Information System Development Center
10F, Landic Akasaka Building
2-3-4 Akasaka, Minato-ku
Tokyo 107, Japan*

**Keywords:** *history of ergonomics; methodology; information technology; ageing; system characteristics; future of ergonomics*

In an attempt to ascertain when ergonomics first appeared in Japanese society, the date of the foundation of the Japan Ergonomics Research Society in 1964 provides us with a clue. In Europe, the Ergonomics Research Society, in England, was founded in 1949, followed by the Human Factors Society of America in 1957. Subsequently, in 1961, the International Ergonomics Association (IEA) was founded. It was thereafter that the Japan Ergonomics Research Society, in Japanese 'Ningenkougakkai', was first founded.

Like a communicable disease, if such an expression can be used, the epidemic started in Europe, passed over to America, and then to Japan. A long time has elapsed since the Japanese Society was founded in 1964. In fact, we celebrated our 25th anniversary last year.

Ergonomics covers areas such as forms (shapes), weight, muscle force, speed, controllability, man-machine systems, tolerance, safety, reliability, stress, sensation, physical conditions, functions and so forth. It is true that these conditions must be studied both independently as individual factors and also collectively as a whole. On the other hand, when we consider that the human being is an integrated being and not the accumulated mass of all these conditions, we come to the conclusion that these conditions must also be integrated in one way or another. I am afraid we have not yet succeeded in finding solutions to how they should be integrated.

I would first like to outline the activities of the Japan Ergonomics Research Society. The Society has 11 subcommittees, actively engaged in specific research areas to meet the Society's primary objective. These subcommittees include:

| | |
|---|---|
| Clothing | Anthropometric Principles |
| Aeronautical Ergonomics | Underwater Activities |
| Handicapped Ergonomics | Safety and Human Reliability |
| Industrial Design | Organization |
| Manual Operation | Interface Amenity |
| Dentistry. | |

Most subcommittees operate for 5 years, with the exception of those intended for educational services which tend to continue for a longer period of time. One example is the Clothing Subcommittee which has been in operation for more than 20 years.

In order for ergonomics to be a science, it should have an established methodology. Whether or not it has a methodology is often questioned. Here I would like to present a list of possible methodologies.

They include:
* To determine the validity of the man-machine-information-environment system from the physical and mental reactions of man against the system. Man-machine system is more commonly referred to, however, I would suggest the incorporation of information and environment as well. The validity of the total system should be determined by assessing the physical and mental reactions.
* To determine the system validity by processing information obtained through man-machine integration. Since humans integrate information within the brain, each piece of information can be taken out to be processed for system validity assessment.
* To approach problems from the negative aspects. In other words, to try to eliminate such negative elements as fatigue, operational difficulties, problems or accidents.
* To assess operational difficulty by subjective judgement as well as by measuring the time required for operation and error frequency in user trials. If machines are awkward to handle, the time required for handling is lengthened and error frequency rises.
* To determine the optimum condition at the furthest extreme from the worst condition. Using a cup or a bowl as an example, once the borderlines beyond which handling difficulty arises are decided, the optimum conditions exist at the farthest point from any boundaries - approximately somewhere around the centre.
* To analyze the currently applied system from the viewpoint of its interaction with man.
* To make interdisciplinary approaches.
* To incorporate scientific principles. The existence of light and heavy is an irrefutable fact. Such physical principles as the gravitational influence of an object on the ground should be incorporated into the study of ergonomics as a scientifically proved principle.
* To apply system matching. Since the human being is a system, it is important to pursue proper matching of the human system with the machine system.
* To employ ecological analysis. Take for example a chair. Chairs are used in many different countries. Global analysis of the different usage of chairs would be one method for the study.
* To compute transfer-index, transfer-function. The transfer-index computed between the human sitting height and the seat-ceiling distance of a car, and the transfer-function, which is an equation, would provide one method for analysis. The American car-sitting height index, by the way, is 128 or 1.28, if you choose, 28 being a clearance.

As you see, there are several possible methodologies in the study of ergonomics, which provide some grounds for ranking ergonomics among other scientific disciplines.

Technology assessment can be attempted by looking at the effects of technological conditions on various human factors. Considering that technology assessment should include ergonomics and an analysis of human factors, the following conditions can be listed for assessment:
1. Conditions causing physiological load including mental and physical fatigue.

2. Conditions making work tasks incompatible with physiological characteristics.
3. Conditions that demand excessive physiological and psychological stress.
4. Conditions causing adverse environments.
5. Hazardous conditions or conditions threatening to be hazardous.
6. Conditions causing adverse health effects.
7. Conditions causing discomfort, arousing feelings of discomfort.
8. Conditions causing pain.
9. Conditions causing toxic effect.
10. Conditions with undue load on life (midday recess, breaks, vacations, holidays).
11. Conditions demanding constrained inactivity.
12. Conditions restricting physiological requirements.
13. Conditions restricting ingestion of nutrients, food.
14. Conditions limiting space.
15. Conditions afflicting family life.
16. Conditions prolonging and inconveniencing commuting.
17. Conditions restricting information and correspondence.
18. Conditions restricting free recreation.
19. Conditions deteriorating the quality of life and work.
20. Conditions that interfere with the operational facility.

As I have frequently pointed out, the human being is a system just like the machine system, the information system, or the environment system. Conditions for an agreeable matching of these systems include: that the systems involved should be mutually compensating, that they should be free from unsuitability, that they should be free from either too much stress or too little stress, that they should be free from discomfort and problems, that they should not cause system imbalance, that they should follow simultaneous cycles, that they should not interfere with the system balance, that they should ensure stable combination, that their sequential composition and their parallel composition should both fulfil the objective and adapt to the environment, that they should not limit themselves to partial usage only.

It is only proper that ergonomics should meet the changes brought forth with human progress. In primitive society, the human struggle with nature was observed. This advanced into an industrial society, a post-industrial society and then into the informationalized society of today. Some insist that we are already in an information society, while others choose the term informationalized - heading towards informationalization. Setting aside the terminology, it is time for ergonomics to bring information into focus.

Our modern society has been given many different names: such as Post-industrial Society, Informationalized Society, Information Society, M.E. Society or My Electronics Society (meaning to make the electronics society one's own), Urbanized Society, High-tech and High-touch Society, Technostress Society and so forth. Of all these terms, Informationalized Society is the one to which ergonomics should attach itself.

In the Informationalized Society many different kinds of new media have come into existence: CATV (Cable Television), Videotex, VAN (Value Added Network), Teletext, Facsimile, Satellite Broadcasting, ISDEN (Integrated Services Design Network), High-definition Picture Television, Package System, INS (Information Network System), to name but a few. These new media had their origins in communication technology and are the products of the informationalized society. It is important for ergonomics to view these media from the human perspective: how they should be viewed or explained in their relationship with the human being.

The introduction of new media has brought forth a variety of changes in their relationship with human beings. The provision of information has changed from unilateral to bilateral. The small range of choice has been replaced by a wide range of choice. We used to have only one TV channel, now we can choose from several channels or several tens of channels if satellite broadcasting is counted. Thus, instead of a single choice, multiple choice is now available. Multi-layer provision of information has replaced single-layer information. The quality of information has been improved. A batch system in information transmission has been replaced by on-line immediate information transmission. Instead of replacing the video tapes, on-line call for centre video is available. The information service has become diverse. Remote-controllability is ensured. Information communication has increased. Information processing has changed from being one-sided to many-sided. Robotization is available for multi-purposes. Translation or interpretation of foreign languages is possible. You can talk Japanese on the phone and the message will be conveyed in English. The likelihood of data renewals has increased. The amount of information transmitted has increased greatly. Information processing has been replaced by knowledge information processing. Thanks to the satellite broadcasting, horizontal communication is supplemented by vertical communication. In this way, new media have introduced great changes in their relationship with the human being.

Now let me discuss the influence of the Informationalized Society on the individual's work task. First there is a shift from analogue to digital. Some elderly people complain that an analogue watch is harder to see, while children show a preference for digital watches. Next, there is the increase in the usage of keys for input and displays for output. Further, direct input of data by touching the display is possible now. Other influences include: dependence on the computer for information processing, dependence of memory on computers, increased demand for efficient referencing, increased machine operations, increased opportunities for automatic decision making assisted by decision support systems, increase in the variety of work tasks performed with visual displays, and the rise in the inadaptability phenomenon. The inadaptability phenomenon produced by the computer society can be illustrated, for example, by: temptation to flee from the workplace, fatigue, worsening human relationships, decrease in efficiency, social pathology phenomenon, accidents, minor complaints and loss of enthusiasm.

In the Informationalized Society, information can be both advantageous and disadvantageous to the human being. The undesirable aspects of information include:
- Information can be used to entrap a specific individual.
- Information can be used to disclose someone's privacy.
- Information can be used to plot conspiracy.
- False information is sometimes diffused purposefully.
- Source of information can be tracked back in search of human connections.
- Information is sometimes kept back intentionally.
- Delicate expressions can be applied to imply the presence of some truth in what is informed without conveying the true information.
- Information can be used to establish an alibi.
- Mismanagement of the information channels can leak information to the conflicting persons concerned.
- Coded information can be decoded.
- Information can be misused for slanderous purposes.

- Information can be conveyed to induce panic.
- Only one-sided information can be transmitted.

Computer systems influence human beings in many ways. These include: mechanical way of thinking, loss of flexibility, uniformity, data judgement without considering their background, lack of mental training for problem solving, loss of creativity, lack of consideration to human communication as the important part of communication with machines, rise in computer allergy, the onset of technostress, the influence of television, changes in values, in ideology and in ethics, living in an artificial environment removed from Nature, the increase in computer crimes including computer viruses, increased preference for digital displays, the development of mental depression or neurosis when unloaded from the burden of completing the required task.

Thus, we are faced with a lot of negative effects in our Informationalized Society. What, then, can we do to restore humanity? Here are some of the suggestions of philosophers:

1. Pursue a proper matching of the human-system with the computer-system (Oshima).
2. Determine the location as a human support system (Oshima). The term 'diagnosis system' is often used in place of the term 'diagnosis-supporting system', missing the concept that it is the doctor who is really making the diagnosis.
3. Make human-computer interaction to suit human characteristics (Oshima).
4. Introduce occasional changes in work tasks.
5. Introduce occasional breaks between work tasks.
6. Activate the work place.
7. Encourage a return-to-Nature phenomenon (Oshima). As man used to be an aquatic animal, I would like to suggest a close association with Nature through woods bathing or seabathing, for instance.
8. Encourage wise usage of leisure hours.
9. Increase the chances of associating with Nature.
10. Release workers from hazardous work, a highly polluted work environment, or from work damaging to health (Masamura). These hazardous tasks should be left to robots.
11. Shorten working hours and increase work-sharing (Masamura).
12. Provide users and engineers with more chances of conversing with one another (Masamura).
13. Establish information democracy (Masuda).
14. View from both the lighter and darker perspectives (Oshima).
15. Broaden the chances of closer human contact.
16. Create life worth living (Masuda).
17. Search for harmonious interactions between man, equipment, and environment from an ergonomics viewpoint (Oshima).
18. Find solutions to VDT problems.

At the beginning of my paper, I pointed out that the human being is a system. I would like to illustrate the characteristics of a human system by listing the different terms given to the human system: Open System (open meaning both in time and space), Facilitation System, Facilitation-inhibition System, Adaptive Control System, Output-input System, Cybernetic System, Self-fuelling System (metabolism and fuelling coming to the same end), Self-repairing System (the injuries being healed naturally), Self-procreation System (as in the birth of human beings), Closed Loop

System, Organized System, Information Processing System, Self-optimization System. It is important for ergonomics to find ways of ensuring a harmonious relationship between the human system and the machine system, the environment system, or the information system.

Comparison of the human system with the machine system would give you an even better idea of the human system. The following is a list of distinguishing characteristics:

Human System (Soft System)
- Having characteristics of homeostasis, constancy
- Having a double feedback system
- Endowed with the quality of adaptation (long-term performance fluctuation)
- Performance fluctuates by levels of tension, attention, consciousness
- The ranges for fluctuations are wide
- Performance fluctuates by motivation
- The total system is maintained at a normal level through proper balancing
- No replacing of low performance parts
- Can be a victim of errors, illusions
- Requiring great redundancy
0. Causing disorganization, dysfunction
1. Interchanges of series and their parallel units during operation
2. Having great inertia
3. Having a 24-hour rhythm
4. Being a language system
5. Having a double control system of nerves and endocrines
6. Superior pattern recognition.

Machine System
1. Stabilization of performance
2. Designed to incorporate a feedback system
3. Short-term performance fluctuations induced by control
4. Little fluctuation in performance owing to automatic control
5. Not subjected to a motivation component
6. System balancing is subjected to automatic control
7. Low performance parts are replaceable
8. Free from errors, illusions
9. Requires little redundancy
10. Does not cause disorganization, dysfunction
11. Fixed composition of units in parallel with the series
12. Having little inertia
13. Not having a 24-hour physiological rhythm
14. Being a touch system, a code system
15. Mainly a single control system
16. Inferior pattern recognition.

In order to seek ways of applying terms to represent human beings as an integration of these systems, terminologies starting with the prefix 'homo' are used in the trial: Homo sapiens, Homo faber, Homo technics, Homo symbolics, Homo politics, Homo sexualis, Homo economics, Homo ludens, Homo imaginale, Homo socius. Each term is not sufficient to represent every aspect of the human being. It is for ergonomics to find a term to embrace all the above.

Human perspectives and the human way of thinking are constantly changing. Ergonomics should never lag behind these changes. It should focus on the human

being, renew perspectives and keep pace with human progress. In order to do this, the following human shortcomings need to be taken into account:
1. Too much attention to the individual function, while too little perspective to man as a system
2. Failure to view the human-machine system as a system matching the human and the machine
3. Lack of awareness for including human emotions in ergonomics studies
4. Little advancement in the ergonomics of human thoughts dealing with the different stages of human thinking
5. Failure to establish a methodology for system balancing
6. Insufficient awareness of the idea of 'human back-up technology'
7. Lack of means to grasp the human in the time passage process.

Is modern technology advanced enough to compensate for human defects? What support does it offer to make up for the deterioration in the human awareness level, or the confusion, or fluctuations? Are there means to compensate human inadvertency, the limit in the human ability to predict, human response including time required in responding, human illusion, human memory, too much self-confidence, limits of human assumptions and thinking, human injuries? Some say that I am crazy to make such demands, because these are problems inherent in man and therefore should be solved by man. However, in my opinion, we are in an age where the human can plan, and once plans are made, technologies to back up the human should be developed to facilitate the execution of the plan.

Ageing produces negative effects on man. Health promotion of the elderly is becoming an urgent problem in our country. Some resort to such tangible means as taking health foods, dieting or taking vitamin drinks. Seeking ways more essential to humanity, human amenity ranks superior to such tangible means, and this leads to the issue of mental health. Some of the defects, or negative effects, that ageing brings to man and their possible solutions are listed:

| | |
|---|---|
| Isolation in human relationship | -- Establishing new relationships |
| Increase in individual differences | -- Taking philosophical views |
| Functional deterioration | -- Enriching experience, becoming more enthusiastic, adaptable, enduring |
| Increasing tendency to ill health | -- Thorough physical check in a hospital health control (early diagnosis, early treatment) |
| Increased inactivity | -- Trying to be active, finding motivation for activities |
| High accident potential | -- Ensuring a safety control system |
| Lowering of work potential | -- Recuperation, adaptation, increasing abilities, improving human factors |
| Easily fatigued, low stress endurance | -- Fatigue prevention, increasing stress endurance. |

There are many examples of the human search for amenity. Human amenity should be pursued with due consideration to human physiology and human characteristics. In search of this positive contribution to humanity, ergonomics should strive to survive.

# 3. THE DEVELOPMENT AND CURRENT ROLE OF THE JAPANESE ERGONOMICS RESEARCH SOCIETY

**Takao Ohkubo**

*Ergonomics and Physiology
College of Industrial Technology
Nihon University, Japan*

**Keywords:** *history of ergonomics; future of ergonomics; research in ergonomics*

## Introduction

As is well known among ergonomists worldwide, ergonomics is the science of optimizing the relationship between man and machine or man and his environment. It can also be defined as an applied science because it can respond to and directly solve the various demands of society. Because of its characteristics, ergonomics is an interdisciplinary science covering a broad area comprising medicine, psychology, technology, physical science etc.

It is obvious that the characteristics of ergonomics vary depending on the country, its climatic conditions, its industries and so on.

Table 1 shows the research-related activities of the Japanese Ergonomics Research Society (JERS) since its foundation in 1963. The actual introduction of ergonomics into Japan, in a real sense, however, came immediately after World War II with the publication of books written by W. E. Woodson, E. J. McCormick and A. Chapanis. "NINGEN-KOGAKU", the Japanese name for ergonomics is a direct translation of the English words "Human Engineering" which was the term used by the Americans in those days. The Japanese term "NINGEN- KOGAKU" is now the established term among Japanese people and covers the two areas of ergonomics and human factors. However, the era of the first ergonomics-related research activities dates back to the end of the 19th century and the early 20th century when scientific management methods were introduced by F.W. Taylor into the U.S.A. in 1911, when the Kaiser-Wilhelm Institut fur Arbeitsphysiologie was established in Germany in 1913, and when the Industrial Fatigue Research Board was founded in the U.K. in 1916.

All those research activities, in the fields of applied experimental psychology, work physiology and industrial hygiene, were more or less closely related to present day ergonomics and human factors.

If we look at the period of establishment of ergonomics or human factors societies in the world, we can see national societies were founded in the U.K. in 1950, in Germany in 1953, in the U.S.A. in 1957, the IEA was established in 1960, and societies were founded in Holland in 1962, and in both France and Japan in 1963.

The research activities relating to present day ergonomics in Japan started long before the establishment of the JERS society.  For example, the Kurashiki Institute of Labour Science was established in 1921, supported financially by Magosaburo Ohhara, with very fruitful research by the staff at the Institute, whose first head was Gito Teruoka.

*Table 1. Ergonomics or human factors related research trends for the past thirty years in Japan*

---
(1) Research age based on labour science (-1950)
   In this period, most research activities concentrated on how to increase or improve safety, adaptability or aptitude of workers, fatigue and efficiency at work based on medicine and psychology.
(2) The dawn of a new age of ergonomics research (1951-1960)
   In the years before the establishment of the Japanese Ergonomics Research Society (JERS) and the periods connecting human science with engineering, ergonomics research aimed to improve the relationship between man and machines, equipment, tools, facilities and physical environmental conditions based mainly on psychology and engineering.
(3) The first development age of hardware ergonomics research (1961-1970)
   With the rapid progress in improving or developing various kinds of machines and equipment because of the microelectronics revolution, work concentrated on physical workload, working areas or working postures of man in automated working systems in order to reduce his fatigue while increasing efficiency.
(4) The second development age for hardware ergonomics research (1971-1980)
   Fundamental ergonomics data were stored to some extent. Ergonomics knowledge or an ergonomics way of thinking was introduced and established among industrial people. New ergonomics research activities were carried out on monotonous work and the quantitative modernization of automated systems, machines and equipment, together with the further accumulation of data.
(5) The first development age of software ergonomics research (1981-1990)
   The dawn of the information technology era directed most ergonomics research towards the use and evaluation of information in industries, namely, to initiate fundamental or applied research to build up a comfortable relationship between man and machines where individual workers can work with high motivation and less fatigue. These activities have been based mainly on medicine, psychology, and industrial design
(6) The second development age of software ergonomics research (1991-)
   Owing to machines becoming more complex and more advanced, e.g. intelligent robots, and owing to the emphasis on individual preference, ergonomics research will be more detailed and will go into greater depth.
   Some examples are: Fatigue and comfort at work, analysis of human judgement and understanding, storing of data or knowledge applicable to ergonomics problem solving, development of research on ethical problems, development of methods enabling man to evaluate information correctly and design machines, equipment and environments that can be fitted to human behaviour, development of applicable ergonomics methods for insurance, banking, hospitals, education etc.
---

The tremendous amount of useful results, obtained by the Institute, greatly contributed to the advancement in production techniques, the improvement of the workers' physical working environments and working conditions from the viewpoints of physiology, psychology and hygiene.

At the same time, the book entitled "Ningen Kogaku" written by Kanichi Tanaka was first published in Japan. The contents of the book covered human behaviour, fatigue, working hours, rest and sleep, and training and work efficiency.

Between 1941 and 1945, Masamitsu Oshima, Fushiro Motobayashi and Kanichi Takagi rendered a distinguished contribution through considerable research activity into the improvement and development of the man-machine interface in aircraft. As a point of interest, Fitts and Chapanis's period of research coincided with that of Oshima and others.

In 1952, a study of the relationship between rest chairs in offices and body functions was carried out at the Research Institute of Industrial Products. This was the first piece of fundamental research work in the design field in Japan.

In 1957, the Institute of Experimental and Clinical Aeronautical Medicine was established with Oshima appointed as the first head. His considerable research work was applied to such problems as pilots' vision while flying, motion and time analysis of jet pilots, illusions and so on.

In 1957, an ergonomics preparation committee was established to study the man-machine interface and construction system for the super express train and its related facilities. Four years later, this committee changed its name to the Ergonomics Research Committee. It was concerned with the building of entirely newly designed equipment and facilities for train operators and passengers. The research activities of this committee were connected with the successful development of the ACT and CTC and the maintenance of the safe operation of the bullet trains.

In 1959, a research committee for ship ergonomics was organized by the seven biggest ship industries in Japan and this initiated joint research on problems concerning environmental conditions inside ships, protecting the human body, relationships between physical strength or muscular power, and the operation of machines and equipment, etc.

One other aspect of transport, the construction of the Tomei-Meishin highway between Tokyo and Osaka began in 1957, following the construction of the Autobahn in Germany (1942), the Interstate in the U.S.A. in 1950, and the Autostrada del Sol in Italy in 1956. The Tomei-Meishin highway opened for traffic after eight years and Japan moved for the first time into the era of high-speed long duration driving with speeds above 100 kilometers/hour. Under such circumstances, the vehicle ergonomics research committee belonging to the Vehicle Techniques Society was established. The committee had three research groups dealing with the problems of drivers' psycho-physiological reactions, static features of vehicles such as driving facilities, and dynamic features such as information transmitting functions, in order to design and develop safe vehicles for drivers.

More recently, the research committee has become the Japan Vehicle Institute and has been doing much research on the ergonomics and other aspects of accidents, exhaust fumes, noise, dust caused by tyres, and drivers' psychological and physiological responses.

In 1961, eight camera manufacturers belonging to the Technical Research Union of the Optical Industry organized a camera ergonomics research committee and began research to stimulate consumer-oriented design of general consumer products.

All of these social and research activities formed the basis of an association for Japanese ergonomics research in Tokyo on 1st December 1963.

Although we cannot ignore the fact that the research record of the Kurashiki Institute of Labour Science had some indirect influences on the initiation of ergonomics research activities, it can clearly be said that physiological and psychological research was also necessary for the improvement or development of man-machine systems in aircraft equipment or pilot training during the war and this also directly motivated the new society. Further, the impetus to organize the society was accelerated by research activities in Europe.

In Europe, ergonomics took root in the 1960's, while at that time in Japan ergonomics research activities were only just beginning to take off and gradually to become involved in the industrial field.

The features of ergonomics research in Japan can be defined, therefore, as a combination of industrial fatigue-orientated or work-physiology-orientated research based on ergonomics philosophy, mainly carried out by physicians and physiologists in Europe, and human factors-orientated research aimed at the design of machines and equipment for operators, mainly carried out by engineers and psychologists in the United States of America.

## General activities of the Japanese Ergonomics Research Society

As mentioned previously, the Japanese Ergonomics Research Society, established on September 1st 1963, was based on the vast and different kinds of ergonomics research from various industrial fields and began its activities on 1st December, with a membership of about 530 individuals and twenty sponsoring companies, and holding its first general assembly in Tokyo. The societies whose members were invited to join the JERS were the Japan Psychology Society, Society of Industrial Medicine, Society of Physical Medicine, Society of Japan Mechanical Engineering, Society of Japan Architectural Engineering, Society of Japan Industrial Engineering, Society of Japan Industrial Design, Society of Japan Household, and Society of Japan Medical Electronics.

Among those participants attending the first conference, were representatives from 79 universities including Tokyo, Chiba, Waseda, Keio, Nihon, and so forth, from all over the country, the main 18 Institutes of Industrial Products Testing, Railway Labour Science, Aeronautics etc., 102 large companies ranging from the automobile, transportation, electrical, machinery, iron and steel, camera, shipbuilding to the chemical industry, and 23 hospitals and offices.

Lecture courses and laboratories in ergonomics were established and opened in most universities and institutes in Japan and the development of ergonomics research sections was observed between 1962 and 1964. The results from a survey carried out by Kumashiro et al. (1983) with 317 academics belonging to 142 universities show that 42.3% of them had a knowledge of ergonomics before 1964 and 24.8% of them had begun research activities. This corresponds well with the initial period of the association.

The changes in the number of society members are also closely related to the research activities of the society. About 300 members belonged to the society when it started, with membership rapidly increasing to about 1100 between 1966 and 1967. This coincided with the onset of both the various committee activities and the research groups belonging to the society. During 1972 and 1973, the number of members again rapidly increased to about 1400. Part of this increase can be explained by the fact that research committees into safety ergonomics, rehabilitation and welfare were set up and there was a rapid increase in the participation of people from industry. Currently, there are around 1600 members of the JERS.

Figure 1 indicates the results of the classification of all the technical presentations for the 31 conferences into four categories, namely, basic and applied study based on theory or on experiment. Most research objectives obtained from the 1st to the 10th domestic conferences were basic research depending on a theoretical or experimental approach, while the objective has gradually been shifting to the field of application since the 11th conference.

Figure 2 shows the results of the research classified into speciality areas. Research into the medical, engineering and psychological fields was seen frequently during the 1960's, while research into systems engineering increased significantly in the 1980's.

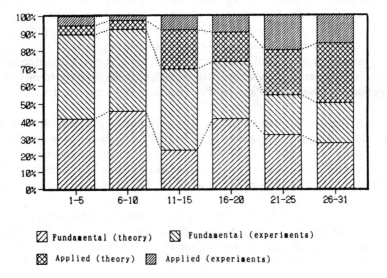

Figure 1. The proportions of basic and applied research reported at each JERS conference (from the 1st to the 31st conference)

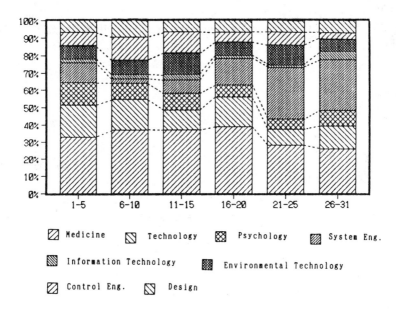

Figure 2. The proportion of research presentation themes to the JERS (from the 1st to the 31st conference)

## Conclusion

It goes without saying that ergonomics has achieved great success, not only in day to day activities but also in many vast industrial situations, with regard to comfort, safety and efficiency, owing to the efforts of ergonomists or human factors engineers worldwide. As shown in Table 1, progress in machine design has made it possible for people to reduce their muscular workload. Japanese ergonomics research activities have greatly contributed to solving hardware ergonomics problems to reduce fatigue, increase work output and to ensure safety by optimizing the design of machines, equipment, tools, facilities, etc. We are now at the beginning of the information technology era, as elsewhere in the world, with the research trend towards software ergonomics. One good example of this research is the universal VDT-related research.

As elsewhere in the world, research into new types of human problems at work have just begun. More emphasis is placed on individual preferences. New methods of production, shifting from large to small batch systems, need more investigation. We must challenge entirely new or relatively less researched situations, such as work for the disabled and elderly as well as overseas workers. So, many unsolved ergonomics problems still exist.

Fortunately, it is recognized that the importance of evaluating industrial information from an ergonomics point of view has been gradually increasing among Japanese industries. Research on problems of product liability or international standardization is keeping pace with countries abroad. If we look at the proportion of Japanese ergonomists and research activities, they are suitably distributed and well-organized. I believe we can expect further beneficial developments of ergonomics research in Japan.

*Reference*

Kumashiro, M. (1983) Ergonomics education in Japan. NINGEN KOGAKU 19(5) 278-281

# 4. THE APPLICATION AND DEVELOPMENT OF ERGONOMICS IN CHINA

**Demao Lu and Guoming Feng**

*Chinese Ergonomics Society (CES), SEPRI, Renjia Road
Qingshan District, Wuhan 430081, China*

**Abstract**

The inauguration of the Chinese Ergonomics Society (CES) in Shanghai, in June 1989, marked the change to organized, systematic and nation-wide ergonomics research and applications activities in China from previously scattered, local and sectorial activities. The development of ergonomics in China has only a ten year history. However, many Chinese ergonomists, specialists and engineers had previously carried out a great deal of successful ergonomics research, development, teaching and training etc. before the CES came into being. This paper gives a detailed presentation of standards concerning ergonomics in China, a description of ergonomics curricula for university and vocational teaching and training, and a review of research and application in production. Also the paper provides a comprehensive overview of ergonomics in China to ergonomists outside China to enhance the exchange of ergonomics activities with foreign countries.

**Keywords:** *history of ergonomics; education in ergonomics; developing countries; research in ergonomics; standards*

## *General description of the development of ergonomics in China*

The history of ergonomics in China spans only ten years. In the late 70s, research work began in a few research institutes and universities, mainly studying the relationship between workers' health and the working environment, the effect of work conditions on productivity, unsafe factors in the working process, etc. Since the implementation of the open door policy in the 80s, visiting scholars and engineers from the field of ergonomics were sent to USA, FRG, Sweden, UK and Australia for study, training and meetings. Meanwhile ergonomists from Japan, Australia, Canada, FRG, USA, Sweden and Singapore were invited to give talks and lectures in China. All these have promoted the development of ergonomics in China.

The International Conference on Ergonomics, Occupational Safety and Health, and the Environment was jointly organised by the Chinese Society of Metals, Safety and Environmental Protection Research Institute (SEPRI) of the Ministry of Metallurgical Industry, China and the Darling Downs Institute of Advanced Education, Australia in Beijing in October 1988, and this has further publicised ergonomics within China.

The Chinese Ergonomics Society was formally established in Shanghai in June 1989 with the approval of the China State Commission of Science and Technology and the State Commission of Education. Since then, Chinese ergonomists, experts and engineers have been working on research and development in ergonomics under the guidance of CES in an organized manner. Research and application of ergonomics in China has entered a new era.

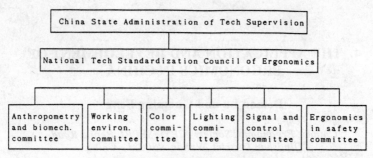

Figure 1. NTSCE organization

## Preparation of ergonomics standards in China

In 1982, the China State Administration of Technical Supervision (formerly, China National Bureau of Standards) set up the National Technical Standardization Council of Ergonomics (NTSCE) which has six technical committees including those concerned with anthropometry and biomechanics, working environment, colours, lighting, signals and control, and ergonomics in safety, as shown in Figure 1.

They are engaged mainly in the research and development of ergonomics standards. The guiding principles and requirements set by the Council include ergonomics principles of working system design, safe distances and measures, living space design, the position and range of machine operators, general ergonomics requirements for the design and arrangement of workstations for operators, ergonomics index of product quality and ergonomics terms. Over the last eight years, twenty-nine national standards relating to ergonomics have been prepared, twenty of which have been issued (Figure 2), and the rest are still in preparation. The development and implementation of those standards have provided a technical basis and legal criteria for achieving optimized design of the man-machine-environment.

## Higher education in ergonomics

At present about fifty Chinese universities or colleges include ergonomics in their teaching programmes as a compulsory part of a course. Ergonomics is taught in all departments of mechanical engineering, transportation, space and aeronautics, building, metallurgy, industrial safety and environmental protection, etc. Teaching programmes are arranged at two levels for students. For those students on 4-5-year courses the teaching programme considers in detail the theory and practice, so that students will be able to carry out research and design; for students on 2-3-year courses the focus is on teaching the applied theory of ergonomics, turning the students into more practical technicians.

The teaching programme in ergonomics usually covers the following four areas:
1. Human factors relating to engineering, mainly referring to human ability to receive and process information and to take action. The measurement of human body sizes, investigation of work physiological conditions and psychological factors.
2. Man-machine interfaces, mainly referring to those parts of equipment which are manipulated directly by workers. For example, the arrangement and design of displays should be fit to human psychological, physiological and anthropometric characteristics, making operation simple, easy, safe and reliable.

| Sr No. | Standard titles | (Year.Issue) |
|---|---|---|
| 1. | Signal danger at workplace and audio signal to danger | (1983. 3) |
| 2. | Human dimensions of Chinese adults | (1988. 12) |
| 3. | Method for assessing the quality of daylight simulators | (1987. 12) |
| 4. | Surface colors for visual signaling | (1987. 12) |
| 5. | Methods of measuring the color of light source | (1987. 6) |
| 6. | Uniform color space and color difference formula | (1987. 6) |
| 7. | Methods of anthropometry | (1985. 12) |
| 8. | Measuring instruments for anthropometry | (1985. 12) |
| 9. | Optimum temperature in air-conditioning room | (1985. 12) |
| 10. | Methods of daylighting measurements | (1985. 12) |
| 11. | Measurement methods for interior lighting | (1985. 12) |
| 12. | Terms of lighting for ergonomics | (1985. 12) |
| 13. | Glossary of color terms | (1985. 12) |
| 14. | Method of evaluating color rendering properties of light sources | (1985. 12) |
| 15. | Terms of anthropometry | (1983. 12) |
| 16. | The functional sizes of chairs and desks for school | (1983. 12) |
| 17. | Methods of measuring the color of materials | (1983. 12) |
| 18. | Methods of color specification | (1983. 12) |
| 19. | Standard illuminants and illuminating-viewing conditions | (1983. 12) |
| 20. | Colors of light signals | (1987. 12) |

Figure 2. *National standards prepared by CES*

3. Working environment, referring to workspace, ambient noise, colour, lighting, temperature, humidity, air quality, radiation, electromagnetic waves, positioning of machines and safe walkways, creating a highly efficient and comfortable working environment for workers.
4. Analysis and assessment of man-machine systems, referring to coordination, rationality, reliability and comfort among man, machine and environment, achieving an optimization of the whole system.

## The research application of ergonomics in China

So far, ergonomics research and application work has mainly been conducted by research institutes and universities. In 1986, a research institute under the Ministry of Metallurgical Industry initiated an investigation of physical workload for 2,150 workers engaged in 536 different kinds of jobs in 18 large-scale metal enterprises, and this has provided an ergonomic basis for improving work conditions and optimizing work organization.

SEPRI, applying ergonomic principles, has succeeded in developing an air quality controller which has been in operation on electric shovels and drills in open-pit mines, and in the driver cabins of large-scale scrapers in underground mines. As a result, working conditions have been improved and productivity has been raised. At present, field measurements and analysis are being conducted to study the occupational adaptability of crane operators in steel mills.

The China Research Institute of Palaeoanthropology and other research institutes have conducted measurements and statistics of Chinese head shapes, working out standard head shapes for Chinese adults. The Chinese military departments have employed ergonomics principles to design the man-machine interfaces for arms and equipment. Research has also been carried out under various simulated conditions (e.g. overweight, weight loss, high/low temperatures, vibration) to investigate physiological and psychological reactions of the human body in order to provide parameters for engineering design.

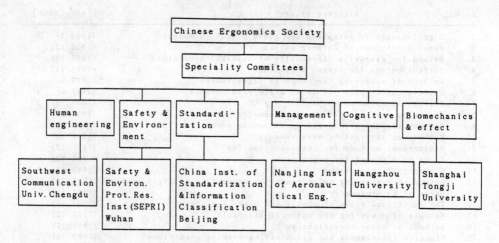

*Figure 3. CES speciality committees and their locations*

Specific examples of research and application by universities or colleges are as follows: Hangzhou University has conducted experiments in visual ergonomics and fatigue; Wuhan Tongji Medical University is engaged in ergonomics issues relating to high-temperature working environments; Southwest Communication University and Chongqing University have started a research project on the application of ergonomics to the design of machinery; Human University has been engaged in the design of mechanical products combining ergonomics with aesthetics, and Northeast University of Technology mainly specializes in the study of ergonomics in work management and organization.

## Conclusion and prospects

More attention has been paid to ergonomics in the fields of science, education and engineering in China. More and more people have become aware of the importance of ergonomics in management. The number of CES members is nearly 1,000. Efforts are being made in research, development and application of ergonomics under six CES committees, i.e. committees for the safety environment, biomechanics, human engineering, standardization, management and cognition (Figure 3).

More extensive education and training will be provided. Postgraduate Master of Science courses have been initiated in some universities and research institutes. The ergonomics teaching programmes and equipment facilities in universities will be further rationalized.

International connections will be further expanded between CES and overseas ergonomics societies. CES has applied for membership of the International Ergonomics Association. It is proposed that a liaison committee be set up in the Asia-Pacific region to promote ergonomics development and application in the region.

Chinese ergonomists will be happy to work with their counterparts abroad towards solutions to a better working environment and more safe, efficient and comfortable work conditions.

# 5. TRENDS IN OCCUPATIONAL ERGONOMICS IN KOREA

Sang-Do Lee

*Department of Industrial Engineering
Dong-A University, Korea*

**Abstract**

The historical development and the current research trends in ergonomics in Korea are reviewed in this paper. The recommended research directions for the near future are also discussed based on analyses of potential major key industries in Korea in the 1990's.

**Keywords:** *history of ergonomics; developing countries; research in ergonomics; future of ergonomics*

## Introduction

The complexity and number of machines used in modern manufacturing companies have increased rapidly along with developing industrial technologies. These machines require not only simple muscle power, but also judgemental, sensory and perceptual abilities. Sometimes the requirements to operate a machine are beyond the capabilities of well-trained operators. Therefore, it is necessary to apply human factors engineering to improve the performance of integrated man-machine systems. The need for ergonomics to improve performance in integrated man-machine systems will increase with time.

Korean industries have developed quickly since the 1960's by adopting modern technologies from developed countries and exporting the products to them. Necessarily, this process has caused several serious problems such as air pollution and weak industrial structures. In addition, high wages, welfare and better working conditions are other vital issues. Many key technologies are required to become a developed country, but human factors engineering incorporating both human values and high technology should play a key role. In this paper the historical background of ergonomics in Korea and current research trends are described. In addition, future research directions and areas to focus on in ergonomics are presented.

## Historical review and current research trends

### Ergonomics in ancient Korea

Ergonomics has been studied for only about 10 years in Korea. But, the origin of ergonomics in Korea can be traced back to ancient time. About a thousand years ago, Koreans constructed their living area according to a law relating to the average height of adults. A typical example is 'Seoggul-am' in Kyungjoo which was built in those days. The concept of applying anthropometric scale had been transferred to the period of the 'Korean-dynasty' and then the 'Yi-dynasty'. It was used not only for building living areas, but also in various industries in the Yi-dynasty. In about 1760, a scholar named Seong Ho (1681-1763) studied various proportions of human body dimensions to height, just as Woodson and Conover (1964) studied the relationship of height and chest circumference to other body dimensions. Seong Ho's proportions of human body dimensions are provided in Table 1. Seong Ho suggested using the proportions in the design of clothes and containers, the construction of castles, and the standardization of building materials.

Table 1. Seong Ho's proportions of human body dimensions

| Classification of proportions | Body dimensions |
| --- | --- |
| h/24 module (h = height) | Ring finger length; Basic module |
| h/4  module | Middle fingertip to elbow length |
| h/6  module | Chest breadth |
| h/10 module | Chest depth (thickness) |
| h/12 module | Thumb tip to wrist length |

## Current research activities and trends

### Research activities in ergonomics

<u>Work Standards and Environment</u>. The Korea Standards Research Institute was founded with governmental support in 1975 for the purpose of establishing national standards and improving the technology of precision measurement. This Institute began to study noise problems, environmental pollution and other living environments in the late 1970's, and now has the Ergonomics Center to lead research activities in ergonomics. Standards for safety signs in KS (Korean Industrial Standard) were announced in 1977, and in 1981 the law on 'Industrial Safety and Health' was promulgated by the government to promote industrial safety and health of workers. Also, the Korea Industrial Safety Cooperation for the prevention of industrial accidents was founded in 1987, with much research conducted into the prevention of occupational accidents and diseases, and into the development of safety facilities and equipment for workers.

On the other hand, non-governmental organizations such as the Korea Industrial Safety Association, and the Korea Occupational Health Association are working for industrial safety and occupational health. Recently, many companies have been working together with research centres, hospitals, and universities on the measurement of occupational and work environments.

<u>Anthropometry</u>. Studies in anthropometry in Korea began in the 1950's, but were limited to soldiers and students. More general surveys can be found a decade later. Sang-Do Lee (1978) provided 61 body dimensions based upon the measurements of 500 male and female workers. It was the first organized survey on a large scale into anthropometry in Korea. Another large scale anthropometric survey for 18,000 subjects was conducted in 1979 by KIST (Korea Institute of Science and Technology) with government sponsorship. This survey has now become an important project for the Korea Standards Research Institute with the statistics updated periodically.

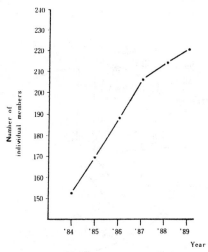

Figure 1. *Number of registered members of the HESK*

The design of shoes, automobile seats, and clothes are typical examples of applications of results in Korea. However, other data on anthropometry must be collected for working area design based upon human movement, physical functions and human body performance.

Communication and Automation. The standard codes of the Korean alphabet, 'Hangul' (consisting of 10 vowels and 14 consonants) were developed with the co-operation of several companies under government control. The codes are now in use and have become the basis for communication technologies in Korea. The study of ergonomics in office and manufacturing automation, VDTs (Visual Display Terminals), and expert system development are expected to be popular in the near future.

Sports Science. Sports science has developed rapidly since the '88 Seoul Olympic Games in Korea. Ergonomics research into sports science based on medicine, physiology, and social science is conducted actively for the scientific selection and training of players, and for the promotion of sport all over the country. With the support of the Human Engineering Society in Korea, many of the related organizations and research centres are participating in these studies.

Academy. The Korean Institute of Industrial Engineers, founded in November 1974, has a subcommittee for human factors. It is not too much to say that this committee has provided a base for ergonomics research in Korea. Currently, many organizations in various fields such as societies for home management, architecture, industrial design, and sports science are working closely with the Human Engineering Society of Korea for ergonomics applications.

Others. Ergonomics studies are also being applied to automotive industries to promote safety and control performances. Allocation of functions between man and machine has become an important area for study due to rapid industrial automation. Many companies already have their own work standards, but they are realizing that extensive research is also necessary for the modification and re-establishment of work standards due to quickly changing automation technology.

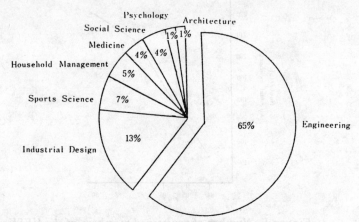

Figure 2. Educational backgrounds of the HESK members

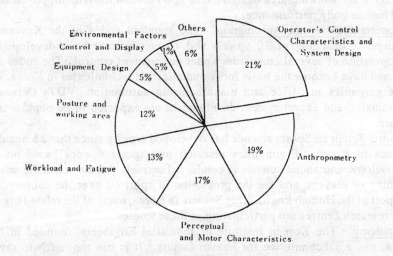

Figure 3. Classification of article topics published in 1982-1989 journals

## Human Engineering Society of Korea (HESK)

The Human Engineering Society of Korea (chairman: Kyung S. Park, Professor of KAIST) was founded in March 1982, with the objectives of overcoming environmental pollution and human isolation in a system resulting from rapid industrialization, and applying its technologies to industries based on humanism by considering human factors and system performance simultaneously. Currently, the Society has about 250 individual members and about 30 organizations such as research centres and companies (see Figure 1). The educational backgrounds of the registered members in 1989 are classified in Figure 2. More than half of them are engineers and the rest are specialists in industrial design, home management, and medicine. The society holds 2 conferences and publishes 2 journals every year. The classified article topics published in the journals in the period 1982-1989 are summarized in Figure 3. This society is a member of the United Society of Science and Technology of Korea and so is eligible for governmental support. The society particularly wishes to work with ergonomics societies in other countries, especially in Asia.

*Ergonomics education*

There are 120 universities in Korea with 45 of them offering engineering programmes. Among those 45, 41 have ergonomics programmes in the department of industrial engineering and about 70% of them offer a course as required. Most of the industrial design departments offer ergonomics as an elective course. In addition, almost all the 25 industrial management departments and about 50 of the departments related to industrial design from 117 two-year colleges offer the subject as an elective or a required course. Further, the subject is popular in several other fields such as architecture, home management, fashion design, and sports science. Currently, about 7 universities offer the programme at graduate level and have awarded several doctoral degrees in ergonomics. In the near future, there will be many human factors engineering specialists who will contribute to all areas.

## Future research areas in Korea

The next decade is going to be very important to Koreans since it is going to be the turning point to the 21st century and also the outpost in all areas such as politics, economy, culture, industry, and so on. It is possible to forecast the economy and industries in Korea in the 21st century by analyzing the current status of industries. The growth of industries in the 21st century possibly depends upon the growth of the industries in the next decade. Major industries for the next decade in Korea are discussed in this section. Ergonomists are strongly recommended to work in all these areas.

### Aviation

The Ministry of National Defense confirmed their plan to incorporate new technologies and to have assembly lines for 'FX' (Fighter Experimental), with private companies in Korea. This has been made possible with the development of assembly technologies for helicopters in the 1980's. The FX project greatly influences other technologies and industries such as the machine tool and automotive industries. It is not difficult to conclude that the aviation industry will be one of the important key industries in the 1990's.

### Information and communication

Information management could be vital to all industries. The importance of communication technologies has been realized by industries, universities and research centres. The Korean government plans to construct computer networks among all the national organizations, and to launch satellites for communications and scientific research.

### Industrial automation

High income and better welfare for workers can be achieved only by automation in both office and manufacturing facilities. For the last few years, many companies have incorporated automation and have installed robots on their production lines. Just as the Japanese economy has grown on the basis of automation, so it is hoped the Korean economy will also grow. Human factors research and applications in automation technology in Korea will be popular and the future looks optimistic.

**Traffic science**

Automotive industries in Korea have developed rapidly for the last five years. They now form an important part of the Korean economy. On the other hand, the government also plans to construct a rapid transit railway to run at 300-400 km/h in the year 2000, and the 1990's will see the preliminary stages of the project in which great investment will be made.

**Medical care and environment**

Environmental pollution is a serious problem in Korea. The government announced 1990 as the first year for solving this environmental pollution problem and took steps to strengthen the administrative organization. In a sense, medical care is the key of social welfare system. This area is going to be continuously supported by the government and expanded all over the country. Therefore, the application of ergonomics to medical care is strongly recommended.

## Conclusions

Ergonomics technologies from developed countries have been introduced, without modification, into Korea for the last 10 years. However, ergonomics research in Korea will increase rapidly for the next 10 years, both in quality and quantity. The next decade will be the growth period for Korean ergonomics. Several issues in the field can be summarized as following:
1. Ergonomics research should pay more attention to the development of application technologies than to theory.
2. Research topics should be more specific.
3. Well organized and more systematic approaches are required.
4. More comprehensive and abundant information related to human factors is required.
5. Ergonomists should be able to present the concepts of ergonomics and to apply the data to system designs. They also should be able to work well with equipment designers, other engineers, the specialists in medical science and psychology, and so on.
6. More international interchange programmes for scientific research are necessary.

## References

Brown, I.D., 1985, Ergonomics and technological change. Ergonomics, Vol.28, No.9, 1303-1309.

Human Engineering Society of Korea, 1982-1989, Journal of the Human Engineering of Korea.

Lee, S.D., 1978, A study of the differences and characteristics between the nations in the measured values of the human body. International Journal of Production Research, Vol.19, No.4, 335-347.

Lee, S.D. and Chung, J.H., 1976, A study of the standard working area by somatometria. Journal of the Korean Institute of Industrial Engineers, Vol.2, No.1, 61-78.

McFarland, R.A., 1971, Ergonomics around the world - The United States of America. Applied Ergonomics, Vol.2, No.1, 19-25.

Meister, D., 1982, The present and future of human factors. Applied Ergonomics, Vol.13, No.4, 281-287.

Oshima, M., 1979, Ergonomics around the world - Japan. Applied Ergonomics, Vol.1, No.2, 70-72.

Oshima, M., Hayashi, Y. and Noro, K., 1980, Human factors which have helped Japanese industrialization. Human Factors, Vol.22, No.1, 3-13.

Seong Ho, 1760, Seong Ho Sa Sul, Vol. 5, 49.

Woodson, W.E. and Conover, D.W., 1964, Human Engineering Guide for Equipment Designers. University of California Press, Berkeley, California, 483pp.

# 6. ERGONOMICS REACHES THOSE OCCUPATIONS THAT SOME BEERS SHOULD NOT REACH?

Ted Megaw

*Ergonomics Information Analysis Centre*
*University of Birmingham*
*Birmingham B15 2TT, England*

**Abstract**

This paper reviews the applications of ergonomics to a number of occupations which have been relatively unaffected by new technology. The nine occupations selected are brewery work, abattoir work, hairdressing, chimney sweeping, public speaking, circus performing, veterinary surgery, religious work and funeral undertaking. The review highlights a number of ergonomics factors common to several of the occupations including manual materials handling, posture, stress, and environmental and chemical hazards.

**Keywords:** *manual materials handling; posture; stress; environment; chemical hazards*

## Introduction

This paper owes its lighthearted title to the photograph of shape coding for some controls in a nuclear power plant taken by Seminara, Gonzalez and Parsons (1977) and shown in Figure 1. A more serious reason behind the title arises from the general impression given that ergonomics is primarily driven by technological developments and concerned mainly with human-computer interaction, VDU design, office automation, telecommunications, complex system reliability, expert systems, decision support systems and military systems. All these areas reflect the growing interest in knowledge-based tasks and cognitive ergonomics not only by professional

*Figure 1. An example of 'beer' coding (from Seminara et al., 1977)*

ergonomists but by cognitive psychologists, computer scientists and human engineers. This has not happened to the exclusion of other areas and the continuing interest in ergonomics related to manual materials handling, musculoskeletal disorders, and seating and posture reflects this. On the other hand, one could be forgiven for sometimes thinking that most people spend most of their working and leisure time in front of a video display terminal or in the control room of a nuclear power plant. Quite obviously this is not the case. For example, about 75% of the world's population does not live in what are often referred to as industrially developed countries and between 80% and 90% of those people work in agriculture. Even in the industrially developed countries, many workers are unaffected by new technology and it is this point which is taken up in this paper.

It is not possible to be anywhere near comprehensive in reviewing the ergonomics contribution, or as is often the case, non-contribution, to non-technology based occupations. Of course, many occupations share several task characteristics so that the possibility exists for transferring ergonomics methods and data from one occupation to another in order to avoid duplication of effort. Duplication of effort could also be reduced by increasing awareness of the ergonomics studies being conducted in different countries. Even in the limited review presented in this paper, studies are referenced from over 15 countries and it would seem unlikely that ethnic and cultural factors can significantly confound the interpretation of the findings.

## Nine selected occupations

### Brewery work

Given the title of the paper, it is only right to begin by looking at the contribution of ergonomics to the brewing industry. The potential occupational problems associated with drinking at work were described by Baker as early as 1767 where he identified the cause of Devonshire colic as the presence of lead in the apples during the course of cider making. A more recent survey in France has suggested that diseases such as lead poisoning and silicosis which were once quite commonly associated with the production of brandy are now rarely encountered (LaFargue, 1978). Carbon dioxide exposure of cellar workers has been investigated by Riley and Bromberger-Barnea (1979). Although the recorded blood bicarbonate levels were higher for the exposed group compared to a control group, the differences were not significant. The levels were found to increase during the course of the week's work but this was probably attributable to the large consumption of beer rather than the continuous exposure to the carbon dioxide. Noise, especially in bottling plants, is a common occupational hazard (Hetu and Parrot, 1978; Rosborg, 1979; Malefant-Pierre, 1984). A considerable amount of manual materials handling is found in brewery work and this has been studied in the laboratory by Ljungberg, Gamberale and Kilbom (1982) and by Labonte et al. (1982), the latter authors examining the contribution of breathing control during the lifting cycle. Very heavy workloads have been identified (Blazek, Chaloupkova and Joachimsthaler, 1975) from people working in the vicinity of a boiling plant in a brewery, with heart rate levels reaching as high as 176 beats per minute at the termination of work. Excessive workloads were identified by Rohmert and Luczak (1974) in their study of bottle inspectors as a result of which the workplace was redesigned. They also identified high noise levels. The control and supervisory roles of operators of an automated distillation plant were investigated under simulation conditions by Paternotte and Verhagen (1979). The authors were particularly interested in learning effects.

Some attempts at improving work organization have been made, though with little success. These include the introduction of Quality Circles in a whiskey bottling plant (Weir, 1986) and participative management style in the Guinness brewery in Dublin (Lawlor, 1984). A more successful innovation has been the introduction of accident investigation groups within safety committees in some Swedish breweries producing significant decreases in the number of reported accidents and in the severity of those accidents reported (Carter and Menckel, 1988).

**Abattoir work**

Injuries and accidents leading to high absenteeism rates are common in abattoir work, the rate being nearly twice as high as that for manufacturing industries (NIOSH, 1981). Particularly common are neck and upper limb disorders (Viikari-Juntura, 1983), often associated with the back and spine (Biering- Sorensen, 1985) and injuries to the hands (Velicic and Mikov, 1985; Temmyo et al., 1986). Severe hearing loss has been identified (Fontaine, Furon and Levent, 1984).

The occurrence of chronic cold discomfort in a slaughterhouse has been reported by Vezina and Laville (1986) with minimum finger temperatures reaching as low as $6^{\circ}C$. Nielson (1986) concluded from her field studies that the clothing commonly used in cold environments in abattoirs was unsuitable and confirmed the widespread cold discomfort experienced by workers particularly in the feet and hands. Apart from low temperatures, noise levels as high as between 85 and 95 dB and excessive humidity have been recorded by Olivier-Roussel et al. (1985). Based on heart rate recording (Blazek, Chaloupkova and Joachimsthaler, 1980) and on electromyography (Temmyo and Sakai, 1987) excessive static and dynamic workloads have been identified. In addition to redesigning the workplace, these results have lead to the more effective use of group work in order to spread the workload more evenly.

**Hairdressing**

A general survey of the working conditions of hairdressers was carried out by Hekala, Jarvenpaa and Hassi (1979). Nearly a third of the sample showed skin rashes resulting from the chemicals used. Work posture problems were identified as well as deficiencies in the design of hand-held hair dryers. More recent studies by Kumaki (1988) and Kumaki, Kurosaki and Yunoki (1985) have confirmed the frequent occurrence of postural fatigue and acute lumbago. A study of the use of domestic hair dryers, including an engineering safety evaluation, was made by Stefl (undated). There were no reported hazardous patterns of usage.

**Chimney sweeping**

This occupation has a long history of accidents and illness, including cancer which was recorded as early as 1775 (see Wade, 1964). Although the occupation is in decline, insurance companies in Finland have been concerned about the claims from some of the 1200 sweeps working there. In response to this concern, Tiitta (1988) generated an accident model with a view to improving occupational safety. In his analysis, Tiitta found that over 60% of accidents were due to slipping, tripping and falling.

**Public speaking**

Two quite different aspects of this activity have been investigated. Quite a few studies have been concerned with the stress experienced by speakers and a recent study by Bolm-Audorff et al. (1989) has demonstrated a rise in plasma catecholamines and total cholesterol during the course of public speaking indicative of acute emotional stress. The other aspect that has been studied is the arrangement of lighting on the speaker. Ironically, the arrangement of point sources of illumination preferred by audience subjects in a study by Golden (1985) was found to be the most generally unpleasant for the speaker providing the least clear view of the audience.

### Circus performing

An overview of ergonomics in the circus has been provided by Shaw (1985). Unfortunately, Shaw was able to do little more than indicate where the application of ergonomics could be beneficial. Copley (1986) recorded heart rate from a variety of circus artists and found the highest mean exercise rate (175 beats per minute) from a tight rope artist. He concluded that in general the rates were lower than those reported for ballet dancers, runners and swimmers. Moreover, much of the work was of short duration and probably the high heart rates attained reflected the severe risks involved in the acts.

### Veterinary surgery

While medicine in general has received a considerable amount of attention from ergonomists, veterinary science has been ignored. A study by Earley (1985) investigated the effects of the participation of animal care-givers in setting goals in performance as a function of the amount of task-relevant information provided and task complexity. He found that choice in establishing goals was only effective in improving performance if sufficient information was provided. On a more practical level, McCrobie (1985) evaluated a newly designed animal euthanasia workstation incorporating human factors standards. As measured by questionnaire responses, there was no difference in reported stress between the new and original designs, but surprisingly, the new design resulted in lower job satisfaction. A possible explanation for this result might have been a result of confounding effects. For example, the new facility developed an odour problem from the nearby storage of dead animals which could not easily be eliminated.

### Religious work

Church lighting has been discussed by Wood-Robinson (1986) who made recommendations on illumination levels, illumination quality, distribution of illuminance and general arrangement of illuminaires. Three sources of job stress for New Zealand ministers were identified by Dewe (1987). These were parish conflicts and church conservatism, difficulties involving parish commitment and development, and finally emotional and time difficulties surrounding crisis work. Dewe goes on to identify the various coping strategies. Heart rates were recorded from Zen priests by Takeo et al. (1984). While average rates were generally very low, the highest rates accompanied religious mendicancy (begging), but even these rarely exceeded 100 beats per minute.

### Funeral undertaking

It is appropriate that this review should end with an occupation associated with the end of our lives. A biomechanical analysis of the loads involved in moving coffins was made by Occhipinti et al. (1988) which demonstrated that the loads were often higher than those recommended. This was confirmed by X-ray examinations which revealed a chronic degeneration of the whole spine. Let us hope these results will enable our bodies to be put to rest without accelerating the death of those who bear us to the grave.

## References

Baker, G. (1767). An Essay Concerning the Cause of Endemial Colic of Devonshire. London: J. Hages.

Biering-Sorensen, F. (1985). Risk of back trouble in individual occupations in Denmark. Ergonomics, 28, 51-60.

Blazek, K., Chaloupkova, E. and Joachimsthaler, J. (1975). A contribution to the problems of the working heat load at some brewery workplaces. Pracovni Lekarstvi, 27, 37-40.

Blazek, K., Chaloupkova, E. and Joachimsthaler, J. (1980). Work load of employees in processing to cattle. Pracovni Lekarstvi, 32, 177-181.

Bolm-Audorff, U., Schwammle, J., Ehlenz, K. and Kaffarnik, H. (1989). Plasma level of catecholamine and lipids when speaking before an audience. Work & Stress, 3, 249-253.

Carter, N. and Menckel, E. (1988). Effective group routines for improving accident prevention activities and accident statistics. In Trends in Ergonomics/Human Factors V (Edited by F. Aghazadeh) Amsterdam: North-Holland, pp.567-571.

Copley, B.B. (1986). Telemetered cardiac response to selected circus acts. New Zealand Journal of Sports Medicine, 14, 63-65.

Dewe, P. J. (1987). New Zealand ministers of religion: identifying sources of stress and coping strategies. Work & Stress, 1, 351-363.

Earley, P.C. (1985). Influence of information, choice and task complexity upon goal acceptance, performance, and personal goals. Journal of Applied Psychology, 70, 481-491.

Fontaine, B., Furon, D. and Levent, D. (1984). Deafness in swine slaughter-houses: a fact. Archives des Maladies Professionnelles de Medecine du Travail et de Securite Sociale, 45, 594-598.

Golden, P.J. (1985). The effect of lighting a public speaker upon impression. Lighting Design and Application, 15, 37-43.

Hekala, E., Jarvenpaa, I. and Hassi, J. (1979). Hairdresser working condition - a study on the chemicals used and the ergonomic factors involved. Tyoolosuhteet 17, TTL 26-7979, Vantaa, Finland: Institute of Occupational Health, 62pp.

Hetu, R. and Parrot, J. (1978). A field evaluation of noise-induced temporary threshold shift. American Hygiene Association Journal, 39, 310-311.

Kumaki, T. (1988). A questionnaire study of the physical fatigue feelings of barbers and beauticians. Journal of Science of Labour, 64, 471-490.

Kumaki, T., Kurosaki, S. and Yunoki, H. (1985). An occupational health survey on beauticians in Japan. III. Survey of lumbago. Journal of Nippon Medical School, 52, 111-116.

Labonte, J.P., Gilbert, R., April, G.E., Toulouse, G. and Warmoes, J.C. (1982). Impulsive forces during manual truck unloading operations. In Proceedings of the 15th Annual Meeting of the Human Factors Association of Canada, Ontario, Canada: Human Factors Association of Canada, pp.109-111.

LaFargue, M. (1978). Workplaces and occupational health damage in glass works - a case study of a bottle manufacturing plant in Cognac. MD Thesis, University of Bordeaux.

Lawlor, M. (1984). Participation - involvement: the Guinness (Dublin) experience. In Proceedings of the 1st International Conference on Human Factors in Manufacturing (Edited by T. Lupton) Amsterdam: North-Holland, pp.17-28.

Ljungberg, A.S., Gamberale, F. and Kilbom, A. (1982). Horizontal lighting - physiological and psychological responses. Ergonomics, **25**, 741-757.

Malefant-Pierre, E. (1984) Occupational hazards on the bottling lines of breweries. Report from Universite Pierre et Marie Curie, Paris.

McCrobie, D. (1985). A human factors approach to the design of an animal euthanasia workstation. In Proceedings of the Human Factors Society 29th Annual Meeting (Edited by R.W. Swezey) Santa Monica, California: Human Factors Society, pp.580- 583.

Nielson, R. (1986). Clothing and thermal environment on industrial work in cool conditions. Applied Ergonomics, **17**, 45-57.

NIOSH (1981). Occupational hazard assessment: criteria for controlling occupational hazards in animal rendering processes. National Institute of Occupational Safety and Health, Publication No. 81-132, Cincinnati, Ohio: The Institute.

Occhipinti, E., Colombini, D., Cattaneo, G., Cervi, E. and Grieco, A. (1988). Work postures and spinal alterations in grave-diggers. La Medicina del Lavoro, **79**, 452-459.

Olivier-Roussel, B. et al. (1985). Study of the workload of four workposts in a slaughter-house in Grenoble. Archives des Maladies Professionnelles de Medicine du Travail et de Securite Sociale, **46**, 477-482.

Paternotte, P.H. and Verhagen, L.H.J.M. (1979). Human operator research with a simulated distillation process. Ergonomics, **22**, 19-28.

Riley, R.L. and Bromberger-Barnea, B. (1979). Monitoring exposure of brewery workers to $CO_2$: a study of cellar workers and controls. Archives of Environmental Health, **34**, 92-96.

Rohmert, W. and Luczak, H. (1974). Determination of work load in field studies: evaluation and design of an inspection task. Travail Humain, **37**, 147-164.

Rosborg, J. (1979). Noise-induced hearing loss in Danish brewery workers. Acta Otolaryngologica, Supplement **360**, 102-104.

Seminara, J. L., Gonzalez, W.R. and Parsons, S.O. (1977). Human review of nuclear power plant control room design. Electric Power Research Institute, Report No. EPRI NP-309.

Shaw, I. (1985). The ergonomics of the circus. In Proceedings of the 22nd Annual Conference of the Ergonomics Society of Australia and New Zealand (Edited by M.I. Bullock and N. Eddington) Carlton South: Ergonomics Society of Australia and New Zealand, pp.57-63.

Stefl, M. (undated). Human use of hand-held hair dryers. National Bureau of Standards, USA, Report on CPSC Project No. 132, 32pp.

Takeo, K., Minamisawa, H., Kanda, K. and Hasegawa, S. (1984). Heart rates during daily activity of Zen priests. Journal of Human Ergology, **13**, 83-87.

Temmyo, Y. and Sakai, K. (1987). An ergonomic study of workload in a slaughterhouse. In Proceedings of the International Symposium on Ergonomics in Developing Countries, Geneva: International Labour Office, pp.375-384.

Temmyo, Y. et al. (1986). An ergonomic study of workload in a slaughterhouse. Journal of Science and Labour, **62**, 1-6.

Tiitta, P. (1988). Occupational hazards involved with chimney sweeping. In Trends in Ergonomics/Human Factors V (Edited by F. Aghazadeh) Amsterdam: North-Holland, pp.663-667.
Velicic, M. and Mikov, M. (1985). Presentation and analysis of occupational traumas on the Carnex slaughter line. Ergonomija, 12, 23-28.
Vezina, N. and Laville, A. (1986). Work in the cold in a poultry slaughterhouse. In Proceedings of the 19th Annual Meeting of the Human Factors Association of Canada, Rexdale, Ontario: Human Factors Association of Canada, pp.107-110.
Viikari-Juntura, E. (1983). Neck and upper limb disorders among slaughterhouse workers - an epidemiologic and clinical study. Scandanavian Journal of Work, Environment and Health, 9, 283-290.
Wade, L. (1964). Occupation and cancer. Archives of Environmental Health, 9, 364-374.
Weir, D.T.H. (1986). Ascribed and achieved philosophies of quality in Scottish manufacturing companies. In Proceedings of the 3rd International Conference on Human Factors in Manufacturing (Edited by T. Lupton) Berlin: Springer-Verlag, pp.145-160.
Wood-Robinson, M. (1986). An approach to the interior lighting of churches. In Proceedings of the National Lighting Conference, London: The Chartered Institution of Building Services Engineers, pp.16-25.

# 7. THE CORRECT USE OF ERGONOMICS AND THE TREND OF SOCIAL CHANGE AND DEVELOPMENT

## Sadao Sugiyama

*School of Sociology, Kwansei Gakuin University*
*Nishinomiya, Hyogo, Japan*

**Keywords:** *history of ergonomics; future of ergonomics; ergonomics intervention*

## Introduction

Ergonomics concerns in human society began, it seems, about the time of the Industrial Revolution, that is, from the second half of the 18th century to the mid-19th century, when, with the changes in the manufacturing process, the social system underwent drastic changes in Europe, especially in England. Although this has been investigated in social sciences, very few studies have been made in the field of ergonomics. During the period of the revolution, the role of production was entrusted mainly to factories as well as shop owners who made goods for sale. The Industrial Revolution was the first time in the history of mankind, if my speculation is correct, when the social structure underwent changes owing to the impact of science and technology. The changes that took place were:

(1) Within the process of manufacturing, the role of the machine was established so that the role of the worker changed from direct handling of the products to indirect handling. Machine operation became the worker's role.
(2) Machine-driven power changed from hydraulic to steam, consequently, the operability of the machine changed.
(3) The labour system changed from previously self-paced work to machine-paced work, so that the management of workers became important.

If you look at the social changes in such a way, you may notice that the impact of the manufacturing process on our modern age is essentially the same as in the time of the Industrial Revolution. However, one important fact is that, since then, the style of human working had changed and, with the development of science and technology, the structure of the machine system has been hybridizing and its function has become more complex. The human role has been changing accordingly, and the division of labour has become more common.

In recent years, which we call the information era, people have become involved in social changes due to the rapid development of computer science and technology. The issues have to do with computers, automation, robotics and data communication. Again the development of those sciences and technologies are not independent of the rapidly changing human society.

## Ergonomics issues, past, present and future

Certainly, our ergonomics concerns have been changing as the evolution of science and technology goes on. I assume that, at the time of the Industrial Revolution, the issues were mostly concerned with health, illness, management of workers, and people's survival in the changing society. However, scientists' efforts to cure workers were insufficient or disregarded at that time because those sciences were still immature. In most cases, medical care must have been beyond their reach. Our technological society has developed, indeed, through many decades of confusion.

Thom (1963) classified the six stages of technology. According to him, the first technological revolution began with the discovery and use of the wheel, and the second one with the discovery of methods for melting ores and making alloys for forged tools and weapons. The Industrial Revolution I mentioned above was the third. Then, the fourth revolution was caused by chemical engineering, the fifth by electronics, and the sixth by the advancement of transportation. At present, and in the near future, computer science, automation by robotics, bio-technology with molecular biology, use of nuclear energy, space science and technology, etc. are the current issues which have changed not only society itself but our way of behaviour, of thinking, of feeling and our life style drastically. Ergonomics has always to be alert to such social trends with a view to man in the future.

## A tendency to "be too late" in ergonomics countermeasures

In a sense, the social issue, at any age, represents a cluster of various socio-emotional elements floating freely in the stream of social consciousness. It always creates a social tension which has to be resolved either in the long run or even in the short run so that a social balance can be obtained. For resolving the tension, again science and technology play an important role, as you see in the case of the safety issue in air transportation which has to be solved by efforts of science and technology. However, resolving social tension can not only be made simply by technological countermeasures, but can be resolved by proper social countermeasures at the same time. Ergonomics is always one of the countermeasures. It is not a countermeasure based on some absolute theory. Thus, I feel that if prediction is required for next generation products to be accepted by a future society, a much wider framework of ergonomics has to be necessarily reconsidered. Always such values as efficiency, safety, reliability, etc. are dependent on other unknown factors. The only way to predict about values taking place in the future is to speculate about various causal relationships, so that we can keep a proper balance among them.

## Diversification and unification of values

It has been said that, as science and technology develop, we will have more freedom to choose any values we desire. Now we can enjoy such a wide range of freedom as to allow us even to experience unrealistic matters by means of computer simulation. In daily life, we can enjoy choosing a variety of foods, travelling by making use of the vast transportation network, etc. However, the fact is that we should keep a balance among those given values when selecting the course of our own life. Once a man has decided on the course of his life, he has to choose certain values from many. The life style developed like this has much to do with ergonomics in a particular culture. Thus, in order to promote a valuable life style, I think man has to manage skilfully the balance of diversification and unification. This might be a matter of future investigation of so-called value technology which is deeply concerned with the purpose of ergonomics research.

## A tentative conclusion

In this short paper, I have described what I regard as a necessary vision for the future development of ergonomic science and technology.

Along the line of man's history, there have been several occasions for society to change drastically due to the impacts of technological development. The third revolution, which we call the Industrial Revolution, taught us to use machines in the manufacturing process. Ever since then, we have come through different stages of mechanization, with which labour and health concerns have become social questions. Man has come to handle materials more and more indirectly via operating machines. Accordingly, control of machines has become more and more the main task of workers. In addition to such a change in tasks, the labour system, presently the organizational concerns of ergonomics, had to change. At that time, however, none of the human efforts had materialized enough and every effort was too late to cure the workers. This might suggest to us that all the ergonomics countermeasures have to be considered and applied beforehand. In this respect, it is evident that we must predict what is going to happen along with the rapid changes in technology, as well as in society.

Man belongs to what we call human organization or human society. Within this framework of civilization, man can live safely and comfortably. Technology is also a matter within the framework. Therefore, once they interact with each other inadequately, confusion easily arises. This means that ergonomics has to be with and for the human society. Once confusion takes place, it is easily speculated that the organizational and social system, as well as the impacts of science and technology, will be criticized by the public. Therefore, in the development of ergonomic science and technology, consideration of socio-emotional factors is urged.

The tidal stream towards future sciences and technologies seems directed towards matters related to human survival on earth on the one hand and matters related to the better quality of human life on the other. These matters will include the technological ones of medical treatment, of health maintenance, of ageing, of working environment, of total life style, of bio-technology and life science, of energy science and technology, of new transportation and computer-based communication, of space exploration, etc. within the framework of the ecological balance on earth.

Conceptually speaking, these matters will require an adequate balance among them if the well-balanced advancement of science and technology is aimed for, especially now that we understand about the limitation of our physical environment. Also it is evident that ergonomists have to observe human beings who are not only complex conceptual outcomes of the sciences of physiology, psychology, sociology, etc., but also the complex result of the interrelation between man and existing numerous natural sciences, and existing available technologies, and even values created in areas of humanities. In this respect, I believe every ergonomist has to re-consider what a human being is in this age of ours.

## Reference

Thom, W. T., 1963, Science and Engineering and the Future of Man, Science and the Future of Mankind, W.A.A.S

# Part II
# Health and Safety

# 8. ERGONOMICS AND DESIGN FACTORS IN THE PROMOTION OF HEALTH AND SAFETY

### Ilkka Kuorinka

*Institut de Recherche en Santé et en Sécurité
du Travail du Québec
505, de Maisonneuve ouest
Montréal (Québec), Canada H3A 3C2*

**Abstract**

The incorporation of health and safety factors into the design process encounters many difficulties. A special approach is needed if one wants to go beyond the application of standards and normative regulations. On the one hand occupational safety and health personnel must understand the nature of the design process itself in order to be able to interact appropriately at the right moment. On the other hand, health and safety criteria should be formulated in terms which are operational and usable in the design process. The first step would be to set target values for the working environment, which could then be converted into concrete terms for the design process itself. The participation of the employee is an important element in the successful design of work places.

**Keywords:** *health and safety; design methods; risk; fatigue; discomfort*

## Introduction

Design factors, in the wide sense of the word, have definitely changed our work environment, mostly in an advantageous way. A question that can be asked is whether ergonomics has had a specific effect on our everyday work environment and whether the effect has also been long-term. If the answer is not totally positive, we should analyse why ergonomics factors have been neglected and what we could do to improve the situation.

This paper analyses the design process and its models and the role of ergonomics in it. Health and safety factors in ergonomic design are then analysed and a model is proposed to allow these factors to be incorporated into the design process.

## Ergonomics in the design process

Both "ergonomics" and the "design process" are far from being monolithic concepts thus making generalizations difficult. What is typical in a plant design may not be typical in a machine design. The scope here is limited to workplace design in the sense in which "occupational ergonomics" treats the subject.

Ergonomics in the design process is understood mostly as a limiting condition, a constraint. The designer sets the goals and identifies means to reach them. In this process he/she is limited by a number of constraints, which do not form an essential part of the design process itself. For example, government regulations must be taken into account, and standards may limit free design. Ergonomic design criteria often belong in the same category. Constraints obviously are not the main issue in designers' innovation and professional ambition.

There are other factors that affect the adoption of ergonomics in workplace design. Some older as well as recent reports have enumerated some of them (e.g. Meister and Farr, 1967). These include: ergonomics knowledge is presented to designers in an unsuitable or complicated form; there is an information gap between the designers and ergonomists or health and safety professionals; there are no convenient means of incorporating ergonomic design criteria or user experience into the beginning of the design process. The designer also has a tendency to copy existing and proven solutions, which also means that the errors are often copied. Perrow (1983) has analysed human factors design practice in the organizational context.

In the general design process, there are various stages (e.g., system level, workstation level and installation), which allow for certain types of ergonomics contributions and inhibit others. Consequently, a certain type of ergonomics knowledge is useful only in a specific phase. For example, the criterion of individual well-being fits most closely with the workstation design phase (Shackel, 1986).

## Models for ergonomic design

There may not be a specific model for ergonomic design but there are specific features for sound ergonomic design. Several authors have deviated from the usual linear design model from goal setting to overall design, and then on to detailed design by introducing the iterativity of design. In the iterative model, the designer repeatedly evaluates the design and makes the necessary corrections, each time getting closer to the design goal. Actually, according to many opinions (e.g. Powrie, 1987) designers may not use any fixed, logical model in their work. They may jump to far-reaching conclusions, reject them, try another approach, turn back and put together pieces that do not seem to fit, and the result can be perfect. No designer can however, work without specifications of the expected outcome.

Iterativity, evolutionary design (Eason, 1982) and participation seem to be essential elements in ergonomic design for occupational safety and health.

## Ergonomic design criteria for occupational safety and health

An obvious point of departure for considering ergonomic design is that ergonomic design as such promotes health and safety. Although this may be true in most cases, some reservations are warranted. The organizational or production context may finally decide whether application of an ergonomic criterion is beneficial from the health and safety point of view. For example, a workplace redesign which diminishes postural stress may allow an increased work pace, and consequently increased strain from repetitive movements.

It should also be stressed, that the objectives of ergonomics do not take into account only health and safety criteria, but also system and productivity criteria.

## Short- and long-term criteria for occupational safety and health

Short- and long-term effects form a watershed in the acceptance of safety and health criteria in design. Short-term effects causing, for example, accidents, back injuries or acute toxic effects are more easily accepted as a design element. Government regulations and other normative design criteria re-inforce the acceptance.

Long-term effects, which are usually less formal, tend to be neglected. Often the elimination of negative long-term effects is expressed in terms of "better working environment", which has no operational content for the designer. It takes visionary enterprise management to define a "better working environment".

## Operationalizing ergonomic health and safety criteria for design

When combining the various goals, means and resources into an integrated and balanced whole, the designer faces a tough task. If ergonomic design criteria are to be taken as a factor in this equation, they must be operationalized (see for example, Sen, 1987 for participatory design) in order to form clear, conscious and deliberate specifications (Meister, 1984).

It is surprising that operationalizing often meets resistance from the medical professions: "Human criteria cannot be put in simple numbers". The lack of understanding of the design process, which the above quote demonstrates may explain why there are significant gaps in ergonomic criteria for health.

## An attempt at operationalizing the health and safety criteria for ergonomic design

The first task in operationalizing would be to outline the *framework* in which health and safety factors can be dealt with in ergonomic design. The framework would exclude general ill health conditions, e.g., the spread of schistosomiasis (a worm disease) or AIDS, which may have occupational components, but which are for the time being beyond the scope of ergonomic design.

Short- and long-term effects serve as a starting point for our exercise. The list is a tentative one. It should be the task of the ergonomics community to develop a more complete and well analysed framework proposal.

- Short-term effects

    Accidents
    Discomfort
    Fatigue

- Long-term effects

    Chronic Stress
    Accelerated Degenerative Disease
    Accelerated Ageing.

The second stage would be to define the descriptors of the items in the framework. The task is an unrewarding one because of the conceptual confusion about the individual items and the lack of even elementary data. The completed table could be as follows:

- Short-term effects

    Accidents: Risk, frequency and gravity
    Discomfort: % of population experiencing discomfort after work
    Fatigue: % of population not recovering after the weekend

- Long-term effects

>   Chronic stress: Identification of stress-producing factors in
>   the work environment
>   Degenerative diseases: Identification of biomechanical load
>   factors in the work environment
>   More rapid ageing: Identification of negative physical and
>   psychological workload factors in the work.

The list of descriptors shows differences in dealing with short- and long-term factors. We may have an opportunity to measure some short-term effects in an actual population, but for long-term effects, the only way is to analyse workplace factors. The relation of the latter to the actual phenomenon we are interested in (e.g., chronic stress or degenerative diseases) has usually not been proven by epidemiological studies.

The next stage is to find operational parameters for each descriptor. Here we step outside the analysis-deduction logic and actively set goals for the end state of the design. There is little guidance in this task except from knowledge related to general ergonomics and safety and health, and experience.

The operational parameters could take the following form:

- Short-term effects:

>   Accidents: Low risk, 0% frequency and 0% gravity
>   Discomfort: No more than 10% of the population experience
>   discomfort after work
>   Fatigue: No more than 5% of the population say they have not
>   recovered over the weekend

- Long-term effects:

>   Chronic stress: A formal work analysis should not reveal other
>   than low level stressors in the task (FIOH, 1989)
>   Degenerative diseases: For back load, only occasionally exceeding
>   the "maximum permissible limit" (NIOSH, 1981) and less than
>   10% high-risk repetitive tasks (Silverstein et al., 1987)
>   More rapid ageing: Only low level stressors as revealed by a formal
>   work analysis (e.g. Landau and Rohmert, 1981).

The operationalizing of the parameters should go into further detail by specifying the target populations, methods, the work analysis scheme, etc. This task is easier than the previous ones, because ergonomics literature gives ample guidance for that.

## Conclusion

There is a large area that has been left untouched in closing the gap between ergonomic safety and health factors and design. The problem is not only the lack of adequate data but also a lack of a conceptual framework and tools. A better understanding of the design process by health and safety professionals and an improved readiness on the part of the designers would pave the way to better cooperation and better results. The participation of workers and management in the design process is an important element in enriching the design and validating its results.

## References

Eason, K.D. The process of introducing information technology. Behaviour and Information Technology, 1(1982): 197-213.

Finnish Institute of Occupational Health (FIOH). Ergonomic work place analysis, FIOH, Helsinki, 1989, 33pp.

Landau, K., Rohmert, W. AET - A new job analysis method. AIIE Annual Conference, Detroit, Michigan, USA, 1981.

Meister, D. A catalogue of ergonomic design methods. Proceedings of the 1984 International Conference on Occupational Ergonomics, Toronto, Ontario, Canada, May 7-9 1984, pp 17-25.

Meister, D., Farr, D.E. The utilization of human factors information by designers. Human Factors, 9(1967): 71-77.

National Institute for Occupational Safety and Health. A work practice guide for manual lifting. Technical Report No. 81-122, US Dept. of Health and Human Services (NIOSH), Cincinnati, Ohio, USA, 1981.

Perrow, C. The organizational context of human factors engineering. Administrative Science Quarterly, 28(1983). 521-541.

Powrie, S.E. Design models and design practice: An overview. In: Megaw, E.D. (Ed), Contemporary Ergonomics 1987, Taylor and Francis, London, 1987, pp 123-128.

Sen, T. Participative group techniques. In: Salvendy, G. (Ed), Handbook of Human Factors, John Wiley & Sons Inc., New York, 1987, pp 453-469.

Shackel, B. Ergonomics in designing for usability. In: Harrison, M.D., Monk, A.F. (Eds), People and Computers: Designing for Usability, Cambridge University Press, Cambridge, 1986, pp 44-64.

Silverstein, B., Fine, L.J., Armstrong, T. Occupational factors and carpal tunnel syndrome. American Journal of Industrial Medicine, 11(1987): 343-358.

# 9. THE ORGANIZATION OF COMPREHENSIVE HEALTH CARE FROM THE POINT OF VIEW OF STRESS MANAGEMENT AND MENTAL HEALTH

### Hiroshi Sakamoto

*Department of Hygiene*
*Mie University School of Medicine*
*Edobashi - 2, Tsu, Mie, Japan*

**Keywords:** *mental health; health and safety; health education; stress management; health care*

## Era of mental hygiene

Activities under the name of mental hygiene were started after World War II. The Committee of Mental Hygiene was established in 1966 by the Japan Association of Industrial Health, who issued a report on the concepts of occupational mental hygiene, contents and procedures of the activities of mental hygiene, and the understanding and conceptual readiness of society and corporations for such activities (Konuma, 1968). In this report, how the guidelines should be formulated for detection, treatment, and rehabilitation of the mentally ill, the criteria for their employment, and their counselling are described as urgent problems. As these developments suggest, the period up to the 1970's may be described as a period of developing and applying measures for the mentally ill and education in mental hygiene.

## Rush of technical innovations

In Japanese industry, technical innovations have been introduced into manufacturing processes since 1955 progressing rapidly in the 1960's. This has caused drastic changes in the working environment and working procedures, and so the mental strain associated with the reassignment of workers, the adaptation to new techniques, and the loss of the value of skilled labour began to be reported from various sections of industry. However, most Japanese corporations tried to cope with these complaints by means of labour administration. Academic research also focused primarily on strains due to a single factor, and the results were often inapplicable to actual work situations. Moreover, this mental strain was mostly overlooked by mental hygiene.

## Mental stress caused by technical innovations, especially those related to human relations (Sakamoto, 1990)

### Labour-management relationship

New manufacturing processes brought about by technical innovations relieved workers from heavy manual labour. They also improved the working environment and drastically reduced occupational diseases due to physical and chemical factors. In the mean time, a remarkable increase in output led to improvement in the workers' standard of living.

Since the second half of the 1970's, computerization has emerged as a leading feature of technical innovation, and heavy manual labour has been replaced by the key punching operation. In the times of manual labour, a heavy workload made the workers pant and perspire. Seeing these signs of exertion, their bosses appreciated their labour and rewarded them with cordial words such as "Thanks. That's good enough. Why don't you take a break?". Workers, for their part, could see from this attitude of the bosses that their efforts had been acknowledged. Thus, heavy physical labour itself was a medium that bound the worker with his boss. On the other hand, as computerization has progressed, workers may have become frustrated by or exhausted from data processing. However, they do not pant or perspire so that their boss does not know when he should thank or comfort them. Thus, the intermissions instituted by working rules ceased to be opportunities for exchanges of humane feelings.

**Relationship among workers**

The entire process of manufacturing integrated circuits, which are essential parts of a computer, is carried out in a cleanroom. Workers engaged in this process are required to wear clean suits and are forbidden to hold private conversations. Therefore, human contact or conversation with the boss or colleagues is markedly reduced in these workers.

We evaluated stress responses in rats by measuring adrenal 11-OHCS concentration while changing the number of animals housed together. No significant difference in the adrenal 11-OHCS concentration was observed between male rats housed alone and those housed with another rat. In females, however, the concentration was significantly higher in those housed alone than in those housed with another animal, indicating an isolation syndrome in the first group. These results may not be applied directly to humans, but there are also other interesting reports. From the results of the 1988 national survey on life style, a clear gender difference was observed in the manner by which individuals coped with stress. Significantly more males tried to escape from stress by drinking than females, but significantly more females did so by talking. The frequency of talking as a means to resolve stress was also comparable to that of drinking in males.

## From mental hygiene to mental health

Computerization has eroded appreciation, comfort, and confidence from the labour-management relationship and deprived workers of opportunities for conversation with their colleagues. Mental stress due to these causes combined with ergonomics factors such as the man-machine interface and the load on data processors has increased the intensity of mental strain. Demand for total health care designated to cover such occupational stress has grown since 1980. The term "mental hygiene", the primary activity of which used to be taking measures for individuals with psychiatric disorders, has also been substituted by "mental health" to illustrate this conceptual change.

## The survey on worker's health (by the Ministry of Labor)

Part of the results of a survey conducted by the Japanese Ministry of Labor in 1982 is cited as it relates to my presentation (Japanese Ministry of Labor, 1986). About 50% of all workers experienced intense anxiety, troubles and stress in their life as workers. According to the size of the corporation, the percentage of workers

| Flow | Care level | Objects | Person in charge | Technique |
|---|---|---|---|---|
| → | IV | Workers needed cooperation with resources of extra-industry | Medical & technical specialist | Treatment & consultation |
| | | | Case-worker | Welfare work |
| | | | Lawyer | Legal steps |
| → | III | Workers with complicated background | Physician & nurse | Medical service |
| | | | Ergonomist | Improvement of work design |
| | | | Hygienist | Environmental control |
| → | II | Workers with mental stress | Counseler | Counseling |
| → | I | Noticed workers All workers | Foreman | Active listening Observation |

*Figure 1. Comprehensive care system*

complaining of mental stress was higher in a middle-sized corporation with 100 or more employees than in small corporations. According to the industrial types, the frequency of stress was high in specialists, technicians, communication workers, and workers of the service industry. These high frequencies of stress in the middle-sized corporations and in workers of particular industrial divisions suggest a relationship of the stress with the degree of computerization.

## Degree of spread of mental health

Means for effective management of stress are urgently needed in the health care of workers. However, our survey carried out in 1987 yielded the following results (Matsumura and Sakamoto, 1989). About 27% of workers had not heard the term "mental health", and no mental health care was available in about 35% of corporations.

## Organization of care system

The care system that I advocated is outlined in Figure 1 (Sakamoto, 1989). It consists of four levels. Care at the primary level covers all workers and is done by observation by the foreman. If he notes any changes in complexion, expression, or movements during work of workers under his supervision, he uses active listening with the workers and attempts to understand their situation and mental state. This provides catharsis, and most of their problems are resolved at this level. However, workers with more complex problems must be treated at the secondary or tertiary care level.

Counselling is done at the secondary care level. This counselling is provided not only for individuals referred by foremen but also for those who voluntarily seek help. However, some workers need additional assistance. An occupational medical service by industrial physicians and nurses, environmental control by industrial hygienists, and assessment and improvement of the work design by ergonomists are examples of the aids provided at the tertiary level.

However, some problems may remain even with the help of corporate specialists. In such cases, support from outside the corporation is required (fourth level). Cooperation may be obtained from medical, environmental, and ergonomics specialists as well as case-workers and lawyers in the community.

*Table 1. Training programme in active listening for foremen*

| Lesson | Hours | Method | Content |
|---|---|---|---|
| I | 4 | Lecture | Man-oriented supervision<br>Physiology and psychology of burden<br>Adjustment<br>Human relations<br>Active listening |
| II | 4 | Case study | Insight into the cause of stress<br>Coping behavior of the foremen |
| III | 4 | Role play | Communicating with the workers<br>Way of making the relations |
| IV | 2<br>2 | Case study<br>Role play | Present a different case from lesson II<br>Technique on active listening |
| V | 4 | Role play | Technique on active listening |
| VI | 1<br>2<br>1 | Demonstration<br>Group discussion<br>Role play | Active listening by instructor<br>Doubt solving<br><br>Technique on active listening |
| VII | 2<br>2 | Group discussion<br>Role play | Doubt solving<br>Technical training |

In organizing such a care system, it is essential that the management provides a man-oriented policy. Also, the function of active listening by the foreman at the primary level of this system is the key to its successful operation.

## Education in listening

An educational programme for the preparation of foremen for active listening is described (Sakamoto, 1989). As shown is Table 1, the programme is made up of 7 lessons.

The first lesson is given as a lecture. The contents of the lecture are, for example, man-oriented supervision, physiology and psychology of burdens, human relations, and active listening. The second lesson consists of a case study, in which training for developing insight into the causes of stress is given. The third lesson involves role play. Prospective foremen are trained in communicating with workers and acquire the art of active listening.

In the fourth and fifth lessons, training as in the second and third lessons is given by using other cases. The sixth lesson consists of a demonstration of active listening by an instructor and group discussion. Each lesson is separated from the next by an interval of one week, during which homework is assigned.

The trainees have exercises in listening at their own place of work for about three months after the sixth lesson. Then, they attend the seventh lesson, in which they ask questions and have group discussion and role play the problems raised.

## Conclusions

I have described a tentative programme for the management of mental stress associated with technical innovation. In conclusion, I would like to emphasize the following two points -

One is that ergonomics approaches and a medical service should not be provided separately, but they must be performed so that they can function as parts of a total health and safety system.

The other is that the foreman plays the supervisory role by listening to complaints of individual workers. The fruit of academic research is considered to contribute to improvements in industry only when information collected by foremen sets the total system into motion.

## References

Japanese Ministry of Labor. Mental health in the administration of occupational hygiene, Foundation of Occupational Medicine, Tokyo, 1986. (in Japanese)

Konuma, M. Report on occupational mental hygiene, Japanese Journal of Industrial Health, 10, 559-596, 1968. (in Japanese)

Matsumura, Y. and Sakamoto, H. Present status of mental health care activity of some selected industries in Japan, Mie Medical Journal, 39, 243-247, 1989.

Sakamoto, H. Easy understanding mental health, Chusaibo, Tokyo, 1989. (in Japanese)

Sakamoto, H. Some opinions from the point of view on mental health, Proceedings of the 63rd Meeting of the Japanese Society of Industrial Health, 23-25, 1990. (in Japanese)

# 10. WHAT WE SHOULD DO FOR OCCUPATIONAL HEALTH AND SAFETY: SOME PERSONAL VIEWS FROM AN OCCUPATIONAL PHYSICIAN

### A. Ward Gardner

*Consultant Occupational Physician, U.K.*

**Abstract**

Prevention should be the main aim of occupational health, safety and environmental professionals. The management concepts of efficiency, economy and effectiveness should ensure that we think about what can, must and should be done - or not done. Questions such as why do it at all, why do it this way and why do it now must be asked. Levels of technology and skills should be related so that over-pursuit of the excellent is not the enemy of the good. Ergonomists and all professionals here must speak out loudly in the world about prevention in the widest possible context thus contributing massively to the prevention of war, over-population, damaged ecology, and ethically unacceptable practices. Without action we and our successors are probably doomed.

**Keywords:** *health and safety; health care; ethics*

An overview of the contributions which any occupational physician can make to the overall objectives of a healthy and safe workforce in a healthy and safe environment must put one key work into prominence: PREVENTION.

To carry out effective prevention in any workplace - whether an established workplace or a new one - requires us to **THINK** about what is desirable, what can and should be done, who will do it, what the costs and the cost-benefits may be and how the desirable environment both at the workplace and outside of the workplace in the world around can be maintained. These physical, engineering and ergonomics factors also require that people are properly **managed, supervised** and **motivated** to perform their tasks in the best possible ways.

To many these statements must be blinding glimpses of the obvious. Unfortunately, in the real world many people - from politicians to managers, from workers' representatives to workpeople - have little concept of prevention in its widest sense and rather sketchy views about the true worth of prevention *first* in regard to the health and safety of the workplace and *second* regarding the contribution that our disciplines can and should make to improving the environment.

Prevention is a **team effort** and presupposes both adequate training, knowledge and management-effectiveness of management in the setting and carrying out of tasks. Unfortunately again, there are many inadequacies - of management, of the will to do things properly, and of money - although I have many times known this to be used as an excuse to do nothing. Preventive solutions to problems must be <u>e</u>ffective, <u>e</u>fficient and <u>e</u>conomic. Dr. Kazutaka Kogi of ILO in the splendid ILO booklet on *Higher Productivity and a Better Place to Work* has shown that "practical solutions for problems relating to working conditions should be demonstrated using available materials and skills and should be related to productivity and quality". He made this statement mainly in relation to the less developed countries, but the ring of truth which it contains in the statement that **"level(s) of technology and skill(s) must be related"** is a lesson for everybody everywhere. It also demonstrates a healthy pragmatism and a desire to do what can be done in line with the view that the over-pursuit of excellence may lead to the frustration of the good. If you can manage only the good, **DO IT**.

Next a personal word to physicians and perhaps, if I may, also to occupational health nurses. I see far too many so-called occupational physicians - and nurses - who, in an occupational setting or as part of a workplace service shut themselves away in so-called health or medical centres playing out the role of healing the sick which they learned in hospitals. They seldom or never visit and inspect workplaces, perhaps because they have not been trained in the techniques and importance of this activity. They become preoccupied with undirected and general physical examinations of well people - a sure way to blunt clinical skills. Management by objectives is not a part of their training or philosophy. They don't ask **why** are we doing this **at all, why** are we doing it this **way** and **why** are we doing it **now**? They make no attempt to keep group records and to apply the disciplines of epidemiology. Cost-effectiveness appears to be an unknown concept. In short, living fossils still exist. Prevention is ignored and elementary management and cost-benefit concepts do not enter into their thinking. How can such people influence managers and stress the importance of prevention, of design, of foresight to eliminate hazards, of hazard identification, of safety, of ergonomics - and so on?

What should we do for occupational health and safety? Occupational health and safety is only one part of the wider world. We should spend much more time and effort on the problems of the wider world. We should try, as privileged and educated people, to raise the general level of awareness about the likely near-future problems of planet earth. The prospects are grim on present-trend extrapolations.

Everyone involved in health and safety should be a leader in education and should be aware of the many failures of prevention today. We should accept that prevention and health promotion should receive priority in public health policies everywhere.

I cannot do better here than to quote from a recent lecture given in Dublin by Professor Risteard Mulcahy:

> "The threat to the ecology of the world is of such magnitude that it is almost too big and too obtrusive to be seen in the limited perspective we have of ourselves and of our surroundings. The dangers should be obvious and urgent when we witness the evidence of a changing ecology, the increasing soil erosion, the rapid destruction of forests, the many species of animals and plants that have recently disappeared forever from the face of the earth, the increasing shortage of water in many areas, the pollution of our land, water and air, the changing climate, the increasing refugee problem, and **the exponential and uncontrolled rise in the world population**.
>
> It is tempting to ignore the issue because, despite its gravity, we cannot think of possible solutions to reverse the trend. We are ignoring the issue because the solutions to our ecological problems conflict with those imperatives which are a part of our modern philosophy of living, the acquisitive philosphy that induces unlimited expectations of personal possessions and wealth, and of comfort and freedom from want, a philosophy devoted to the sacred cows of profit and economic growth. Inherent in this philosophy is conviction that the planet was assigned for the exclusive benefit of mankind and that its resources are unlimited. We forget our responsibility, imposed on us by our superior faculties of mind and body, to care for the more helpless flora and fauna which have an equal right to inhabit the earth and which ultimately are necessary for our own survival.

This issue remains largely ignored because we lack the courage to face up to the fundamental causes of this environmental crisis. Current attempts aimed at a solution are pitifully inadequate and are merely dealing with the symptoms. Policy-makers and politicians who appear to be outwardly caring are only paying lip-service to the problem.

**A major factor is the exponential rise in the world population. A world population of two and a half billion in 1950 had increased to five billion in 1985 and is projected to increase to eight and a half billion by 2020.**

If we are to avoid ecological disaster and to reverse present trends, political policy in all countries must undergo a radical change. This must include limitation of our exploitation of the earth's resources, a reduction in energy usage, the use of natural sources of energy and a reduction in fossil fuel consumption. It demands a more even distribution of wealth, a world authority to protect the flora and fauna that have survived, and to protect the environment.

**Above all, it demands effective population control.**

The ecological threat is a major challenge. It can only be effectively dealt with by world leadership and by enlightened political action but we cannot expect to have enlightened leaders unless we vote for them and we will not do so unless we are enlightened ourselves and properly motivated. It comes down to better public education and to an active public conscience, to the need to make fundamental changes in our philosophy of living and to shed our natural cupidity in face of caring for the world and future generations."

We must all see that our own discipline is but one amongst many which can contribute to health, safety and the environment. A **team approach,** better **management skills,** a background of **good ethics** and **real professionalism** could be our **individual contribution** both to **local** and to **global problems.**

It is said that good planets are hard to find: does anyone know another good planet?

## *Reference*

Mulcahy R. Of Medicine and the Millennium. Journal of the Irish College of Physicians and Surgeons, 1990, 19, 33-136.

# 11. HUMAN-ROBOT INTERACTION: AN OVERVIEW OF PERCEPTUAL ASPECTS OF WORKING WITH INDUSTRIAL ROBOTS

Waldemar Karwowski

*Center for Industrial Ergonomics*
*University of Louisville*
*Louisville, KY 40292, U.S.A.*

**Abstract**

The common cause of many robot-related accidents can be attributed to human physical, perceptual and psychological limitations. In order to prevent robot-related accidents in industry it is imperative to understand how humans perceive and respond to robots in the workplace. This paper discusses the current research efforts related to perceptual aspects of working with industrial robots. Specifically, the following key issues are discussed: 1) what maximum robot speeds are perceived by the workers as safe for monitoring and other tasks that need to be performed from outside, but in close proximity to the robot's working zone; 2) what is the minimum threshold of robot inaction (idle) time perceived by the workers as an indicator of the *safe-to-approach condition*, based on the belief that the robot stopped and is not going to move again; and 3) what is the perceived geometry of the robot's working space (reach zone for the robot's arm, and safety zone around the working robot), as a function of the robot's action dynamics. It is suggested that in order to provide for the operator's comfort and safety, the minimum separation between the worker and the robot's arm should be 43 cm or 17 inches.

**Keywords:** *accidents; robotics; safety; risk; risk perception; perception of movement; workplace*

## Introduction

The common cause of many robot-related accidents can be attributed to human physical, perceptual and psychological limitations (Carlsson, 1984). Worker misperception of the hazardous robotic workstations originates from the human inability to perceive changes in the robot's spatial envelopes and arm motions, which might result in collision and fatal injury. In order to prevent robot-related accidents in industry it is imperative to understand how humans perceive and respond to robots in the workplace, and evaluate the hazards associated with the operation and maintenance of robotic workstations (Sugimoto, 1985; Nagamachi, 1986; Karwowski et al., 1987; Karwowski et al., 1988). Parsons and Kearsley (1982) and Noro and Okada (1983) stressed that industrial robots are built to work with humans and, therefore, their design and operation call for ergonomics considerations.

According to the *Request for Assistance in Preventing the Injury of Workers by Robots* (NIOSH Alert, 1984), both safety training and supervision of workers who are involved with programming, operating, or maintaining robots should be provided in order to reduce the risk of injury. The training programme should be specific to the particular robot, and should include "refresher courses which re-emphasize safety" and discussions of "new technological developments". Such training should be available to experienced programmers, operators, and maintenance workers. The NIOSH (1984) document also outlined the following safety procedures for any personnel interacting with industrial robots: 1) workers must be familiar with all working aspects of the robot, including full range of motion, known hazards, how the robot is programmed, emergency stop buttons, and safety barriers, before operating or performing maintenance work at robotic workstations, 2) operators should never be in reach of the robot while it is operating, and 3) programmers, operators, and maintenance workers should operate robots at reduced speeds consistent with adequate worker response to avoid hazards during programming and be aware of all conceivable pinch points, such as poles, walls, and other equipment, in the robots' operational areas.

## Human factors: implications for robot safety

Numerous safety features have been proposed and implemented to ensure safe operation of industrial robots. These include protection against software and hardware failures, fail-safe designs, intrusion monitoring, deadman switches and panic buttons, workplace design considerations, restriction of arm motions, warnings, and operator training (Bonney and Yong, 1985). Most of these features and devices (i.e., fences, enclosures, guards, sensing devices) are hardware-centered, and attempt to physically separate the robot from the human or warn operators about the potential danger.

Carlsson (1984) examined several robot-related accidents, the majority of which involved pick-and-place operations. About 39% of these accidents occurred while the operator was adjusting a tool or a workpiece within the operating envelope of a robot. Another 36% of the accidents occurred during repair, programming, or returning of an out-of-sequence robot to operation. Such procedures required the operators to bypass safety systems designed to prevent entry into the robot "danger zone".

The studies by the Japan Ministry of Labor (1985) and French investigators (Vautrin, 1985) gave similar conclusions with respect to accidents resulting within the robot's danger zone. They point out the inability of even the most sophisticated gate, fencing, and enclosure systems to prevent robot accidents due to operators bypassing these systems for many reasons.

Cox and Butler (1986) observe that the robot "thinks" with its computer, makes "decisions" through its programs, performs curiously human-like manoeuvres with its "arms" and "hands", and does all this with great precision. A human who miscalculates a robot's path and gets inside the painted lines (working envelope), or someone who inadvertently wanders there, stands a good chance of being injured or killed. Such a possibility is enhanced if the person is working in a manufacturing facility where the work envelope of the robot may quickly change.

Previous experience with robots also influences the safety behaviour of the human observers. There may be quite lengthy pauses to allow for processes to take place which could lead operators to think that it is safe to approach the robot. On the other hand, even those who work around the robot for some time begin to take its movements for granted, exposing themselves to hazards if, for example, due to power failure or other malfunction, the robot arm is brought to the resting position or suddenly takes an "unnatural" and unexpected path of motion. Even experienced operators may have difficulty in understanding why the robot stopped; whether it is safe to approach it; if the cause of stoppage is a malfunction due to abnormality, a runaway halt for machine failure, or a condition halt for machinery cycling (Sugimoto and Kawaguchi, 1983).

Slow operating speeds of a robot allow minimization of such hazards. For example, it is recommended that the teach pendants should have an automatic slow-speed facility. In the Japanese and Soviet standards, such speeds have been set at 14 and 30 cm/sec, respectively, while the German and U.S. drafts point to 25 cm/sec. The above speeds appear to be based on current practice rather than any criterion which implies that speeds below these figures are safe for all operations which require close interaction between the operator and the moving robot.

As reviewed above, many common causes of the robot-related accidents can be attributed to human physical, perceptual and psychological limitations. In addition, as pointed out by the *Robotic Industrial Association Subcommittee R15.06 on Safety*, (Strubhar, 1986) "Certain maintenance and programming tasks may require a person to be in the proximity of the robot while motive power to the robot is available". It was also recognized that the words safe and safety are not absolutes, and that the ultimate link in a safeguarding system is a person.

## Early studies in human perception of robotics

The pioneering work on human perception of robots was performed in Japan by Nagamachi (1986) and Sugimoto (1985). Sugimoto (1985) conducted a series of experiments in order to measure the reaction time for subjects who were to correct the mode of a control button on a teach pendant. The teach pendant was set to advance the robot arm in their direction. Based on the assumption that the minimum safety distance between a robot and the human operator should be 20 cm, it was found that at the speed of 14 cm/s all subjects were able to stop the robot at this or greater distance. It was, therefore, recommended that the speed of 14 cm/s should be used for teaching purposes.

Nagamachi (1986) used a 5-point psychological scale (with endpoint values marked from 'very easy' to 'completely impossible') to quantify subject responses in three laboratory experiments that investigated human-robot interactions. In the first experiment, subjects were asked to evaluate the degree of difficulty (with respect to safety) in a task that required adjustment of the object held in the robot's arm, as a function of robot speed and axis of motion. The speed was limited to the range from 10 cm/s to 50 cm/s, and the three axes were: 1) back and forth motion (X-axis), 2) up and down motion (Y-axis), and 3) right to left motion (Z-axis). The results showed that the X-axis (with subject facing the robot) was perceived as the easiest one to perform the adjustment task.

In the second experiment by Nagamachi (1986), subjects were asked to retrieve an object located on the floor (dropped by the robot), by entering the work envelope while the robot stopped for a short period of time (1, 2, 3, or 4 seconds). The speed of the robot ranged from 14 cm/s to 50 cm/s. It was concluded that the robot's idle time of 4 seconds was judged by the subjects as sufficient to "pass under the arm of the robot quite easily".

In the third experiment, subjects were asked to determine the minimum distance from the robot at which they would not fear that the robot may harm them. No information, however, was given regarding the relative position of the subject with respect to the robot's envelope and location of the robot's arm. Since the conditions of this experiment were reported to be the same as in the case of the first experiment, it can be assumed that the subject was facing the robot's arm (X-axis in the sagittal plane). At a very low speed (14 cm/s), the mean distance the subjects kept from the robot's arm was 1.5 cm, while at the speed of 25 cm/s the mean distance chosen was 22.5 cm.

Etherton et al. (1988) investigated human response to unexpected robot movements at selected slow speeds of a robot's arm. The subjects were asked to hit an emergency stop button when the robot arm moved linearly beyond an expected target point known to the subjects. The speed of robot motion was set to four levels between 15 to 45 cm/s, and the subject was positioned in one of the three angles (0, 45 or 90 degrees) with respect to the robot's line of motion. A linear relationship was found between robot arm speed and the overrun distance, defined as that exceeding the target point of stoppage. At a low speed of 25 cm/s the mean overrun distance was 7.77 cm, with a maximum value of 16 cm, while at the speed of 45 cm/s the mean overrun distance was 10.9 cm with a maximum value of 20.5 cm. It was stressed, however, that safety interpretation of the overrun distance was difficult.

## Worker perception of industrial robots

Recently, Karwowski et al. (1990) investigated human perceptual aspects of hazardous robotics workstations. Three laboratory experiments were designed to investigate workers' perception of the safe speed of a robot's movements, its motion patterns, reach capabilities and safe operating zones for two industrial robots of different physical configurations, and performance capabilities. The main objective of the study was to investigate worker perception of the operational characteristics of two industrial robots.

The following key problems were addressed: 1) what maximum values of the robot's arm speed are perceived by the workers as *comfortable and safe* for monitoring and other tasks that need to be performed from outside, but in close proximity to the robot's working zone? 2) what is the minimum threshold of robot inaction (idle) time perceived by the workers as an indicator of the *safe-to-approach condition*, based on the belief that the robot stopped and is not going to move again? The above time is the minimum time during an operating cycle in which the robot is temporarily immobile, but which is perceived by the operators as being an indicator of a stop due to system malfunction (versus a pre-programmed stop) that can only be corrected by entering the robot's work envelope; and 3) what is the perceived geometry of the robot's working space (reach zone for the robot's arm, and safety zone around the working robot), as a function of the robot's action dynamics?

Twenty-four industrial workers from the metropolitan Louisville area, who had previous exposure to automated systems and robotics, participated in the study. Two industrial robots, i.e. P50 and MH33, were used in three laboratory experiments. The large robot was an MH33 material handling robot made by Volkswagen Corporation. The small robot was a P50 made by General Electric Corporation. In Experiments 1 (perception of maximum safe speed of robot motions) and 2 (perception of robot's idle time), a two-factorial (2x2) design was used. The pool of 24 subjects was randomly divided into four groups (s=6), and assigned to each of the 4 treatment combinations. Experiment 3 (perception of robot's working space) utilized a three-factorial (2x2x2) design, with subjects randomly assigned to each of the eight treatment combinations (s=3).

In Experiments 1 and 2, the effects of speed of robot motions, the size of the robot, replication of the initial speed of the robot's arm, and the exposure to simulated accident on the respective response variables, were investigated. In Experiment 3, in addition to the above listed independent variables, the effects of the angle of approach towards the robot and instruction regarding assessment of the distance from the robot (maximum reach of the robot's arm versus safety), were also investigated. The simulated industrial accident involved a sequence of motions with the robot arm hitting the manikin placed inside the robot's work envelope. The simulated accident was presented to the subjects in order to illustrate the potential danger that industrial robots pose in the workplace when they collide with humans.

## Recommendations for safe operation of robotic workstations

Results of the three experiments conducted by Karwowski et al. (1990) point out the following conclusions regarding worker perception of the operational characteristics of industrial robots. Almost ninety percent of the workers in this study selected the speed of 70 cm/s or less as the maximum safe speed of robot motions for working in close proximity, but outside the robot's work envelope. In the workers' opinion, such a speed of robot motion was safe and comfortable, and did not cause them any stress. With respect to the robot size, the average preferred values of speed were about 64 cm/s and 51 cm/s for the small and large robot, respectively.

In view of the above, to assure operator's comfort and safety, the robot's speed should be restricted not only for teaching purposes, when the human operator may be inside the work envelope, but also during normal robot operation, even though this may influence effectiveness of the production process. It is, therefore, recommended that the maximum operational speed of industrial robots be limited to 70 cm/s. However, whenever feasible from the production point of view, the robot's speed should be 50 cm/s or less.

The pre-programmed idle times affect robotic cell performance in terms of efficiency and quality, as well as safety behaviour of the human operators. Ninety percent of the workers in this study waited for 24 s or less before assuming that the robot's idle condition was due to the system malfunction, and therefore, it would be *safe-to-approach* the robot by entering its work envelope. Therefore, it is recommended that the pre-programmed robot stops should be as short as possible, with the maximum value of 24 s, but preferably less than 16 s (50th percentile value for the MH33 robot), regardless of the robot size. Beyond the above time of 24 s, there is a strong possibility that most of the workers may perceive the pre-programmed stop as the system malfunction and, therefore, attempt to enter the robot's work envelope putting themselves at a high risk of injury. Another design recommendation is to provide the operators with a clear indication when, and if, the robot has finished its pre-programmed operating cycle and to identify systems stops.

The results of this study by Karwowski et al. (1988 and 1990) also revealed that workers misperceive the true robot's work envelope and may frequently invade the robot's working space, especially when workplace layout places them in the middle section (front) of the robot. Table 1 shows the comparison of recommended minimum safe distances from the robot's arm reported in the subject literature with the results of this experiment. According to the present study, the minimum separation distance between the working robot and the human operator should be greater than previously proposed. Nagamachi (1986) suggested that the minimum safe distance from the robot operating at a speed of 25 cm/s should be about 23 cm.

According to Sugimoto (1985), when the speed of robot motions is only 14 cm/s, the minimum separation between the human operator and the robot should be about 20 cm. Etherton et al. (1988) studied human ability to avoid collision with the approaching robot arm under the condition of simulated emergency, and reported average overrun distances of 16.0 cm and 20.5 cm with the robot's speed of 25 cm/s

Table 1. Comparison of recommended safe distances between a robot and the human operator

| Study | Robot Speed | Minimum safe distance |
|---|---|---|
| Nagamachi(1983)[1] | 25cm/s | 22.5 cm |
| Sugimoto et al. (1985)[2] | 14cm/s | 20.0 cm |
| Etherton et al. (1988)[3] | 25 cm/s | 16.0 cm |
|  | 45 cm/s | 20.5 cm |
| Karwowski et al (1988)[4] |  |  |
| Large robot (MH33) | 10cm/s | 47.9 cm |
|  | 90 cm/s | 50.5 cm |

| Angle of approach | Robot speed 10 cm/s | 90 cm/s |
|---|---|---|
| 1 (left side of the robot) | 58.4 | 62.7 |
| 2 (left side of the robot) | 50.3 | 54.1 |
| 3 (in front of the robot) | 36.3 | 37.7 |
| 4 (in front of the robot) | 35.5 | 34.4 |
| 5 (right side of the robot) | 52.7 | 55.1 |
| 6 (right side of the robot) | 54.0 | 59.1 |

| Study | Robot Speed | Minimum safe distance |
|---|---|---|
| Karwowski et al (1990)[6] [P50 robot speed of 90 cm/s, accident group] |  | 43.0 cm |
| Small robot (P50) | 25 cm/s | 31.1 cm |
|  | 90 cm/s | 35.7 cm |
| Large robot (MH33) | 25 cm/s | 17.1 cm |
|  | 90 cm/s | 29.1 cm |

1- Mean safe distance selected by the subjects
2- Recommended safe distance for teaching purposes
3- Mean safe distance based on the maximum overrun distance
4- Mean safe distance selected by college students (*no-accident group*)
5- Mean safe distance selected by industrial workers (averages across all conditions)
6- Mean safe distance selected by the *accident group*

and 45 cm/s, respectively. An earlier study by Karwowski et al. (1988) revealed the minimum safe distance from the robot's work envelope selected by the college students to be about 50 cm.

The results of the present study (Karwowski et al., 1990) confirm our previous findings (Karwowski et al., 1988), and suggest that in order to provide for the human operator's comfort and safety, the minimum separation between the operator and the robot's arm (including any extensions) should be at least 43 cm, regardless of the robot size and speed. This recommendation is based on the average value of safe distances from the P50 robot's work envelope selected by the workers who witnessed the simulated industrial accident. The above conclusion implies that even in the teaching mode, when the recommended slow speed of 25 cm/s is used, the path of robot motions should be designed (pre-programmed) in such a way as to allow for the minimum separation between the worker and the robot's arm of 43 cm (or 17 in.) or more. Also, the layout of the robotic workstation should be flexible enough to allow the operator to keep the distance of 17 in. (or more) from any part of the workstation that could come into contact with the moving robot's arm.

## Acknowledgements

This study was supported in part by a grant from the National Institute of Occupational Safety and Health (NIOSH), NO. 1 R01 OHO2568-01, 1988-89, Cincinnati, Ohio, U.S.A.

## References

Bonney, M.C. and Yong, Y.F. (Eds.), 1985. Robot Safety, Springer-Verlag, Berlin.
Carlsson, J., 1984. Robot Accidents in Sweden, National Board of Occupational Safety and Health, Solna, Sweden.
Cox, J.L. and Butler, J.K., 1986. Human factors issues in robotics: physical, mental, safety, legal. In: P.M.Strubhar (Ed.), Working Safely with Industrial Robots, Robotics International of SME, Dearborn, MI, pp.34-45.
Etherton, J., Beauchamp, J., Nunez, G. and Ahluwalia, R., 1988. Human response to unexpected robot movements at selected slow speeds. In: W.Karwowski et al. (Eds.), Ergonomics of Hybrid Automated Systems I, Elsevier Science Publishers, Amsterdam, pp.381-389.
Japan Ministry of Labour, 1985. Study on Accidents Involving Industrial Robots, Occupational Safety and Health Department, Labor Standards Bureau, Japan, Report No. 5.
Karwowski, W., Plank, T., Parsaei, M. and Rahimi, M., 1987. Human perception of the maximum safe speed of robot motions. In: Proceedings of the Human Factors Society - 31st Annual Meeting, Santa Monica, CA, pp.186-190.
Karwowski, W., Rahimi, M., Nash, D.L. and Parsaei, H.R., 1988. Perception of safety zone around an industrial robot. In: Proceedings of the Human Factors Society - 32nd Annual Meeting, Santa Monica, CA, pp.948-952.
Karwowski, W., Parsaei, H.R., Nash, D.L. and Rahimi, M., 1988. Human perception of the work envelope of an industrial robot. In: W.Karwowski et al. (Eds.), Ergonomics of Hybrid Automated Systems I, Elsevier Science Publishers, Amsterdam, pp.421-428
Nagamachi, M., 1986. Human factors of industrial robots and robot safety management in Japan, Applied Ergonomics, 17(1), 9-18.
Noro, K. and Okada, Y., 1983. Robotization and human factors, Ergonomics, 26, 985-1000.
Parsons, H.M. and Kearsley, G.P., 1982. Robotics and human factors: current status and future prospects. Human Factors, 24, 534-552.
Strubhar, P.M., 1986. Working Safely with Industrial Robots, Robotics International of SME, Dearborn, MI.
Sugimoto, N., 1985. Safety measures for automated machines. In: Safety Staff, 12, 4-18.
Sugimoto, N. and Kawaguchi, K., 1983. Fault tree analysis of hazards created by robots. In: M.C.Bonney and Y.F.Yong (Eds.), Robot Safety, Springer-Verlag, Berlin, pp.83-98.

# 12. THE APPLICATION OF ERGONOMICS TO THE PROTECTION OF FIREBRICK-MAKING WORKERS

## Xiaoxiong Liu

*Safety & Environmental Protection Res. Inst.*
*Ministry of Metallurgical Industry*
*Renjia Rd, Qingshan, Wuhan 430081, China*

**Abstract**

In China's refractory factories, friction presses are still widely used in the shaping process of refractory bricks. The machine can easily cause hand injuries for its operators. Therefore, preventing such accidents has become important for Chinese refractory manufacturers. For years, research in this area has been largely confined to designing and installing protective devices on presses. Little attention has so far been paid to the investigation and evaluation of human behaviour in production activities. An ergonomic analysis of the causes of hand injuries from friction presses is discussed in this paper, along with the ergonomics considerations for the design of protective devices with emphasis on two aspects: the match between man and machine, and information detection and feedback. The author designed a new protective device based on ergonomics principles. Its use in production has proved successful in protecting workers.

**Keywords:** *accidents; fault tree analysis; safety; human error; accident prevention; brick making; machine guarding*

## Introduction

The Double Disc Friction Press (DDFP) has been the main shaping machine for producing refractory bricks in China for decades. It is estimated that at present more than 90 percent of the firebricks used in the Chinese metal industry are made with DDFPs. The machine also causes accidents - it is easy to squash or cut off operators' fingers or sometimes even the hands.

Much research has been carried out by refractory manufacturers in China into safety protection for firebrick workers and many kinds of protective devices have been developed. These devices once prevailed in many of the refractory factories but failed to maintain a permanent application. The reason is that they were only functional from a safety point of view, and were in fact neglectful of the ergonomic relationship between man and machine in the production system. Some devices even succeeded in worsening the existing man-machine relationship, and so prevented them from functioning satisfactorily.

## An ergonomic analysis of hand injuries caused by DDFP

The operator-machine system consists of the firebrick workers and a DDFP and is illustrated in Figure 1. In this system, a group of workers acts as operators of the production process - the foreman manipulates the DDFP, and the others carry out the operations such as feeding the brick-die and fetching out the ready-made brick adobes, etc.

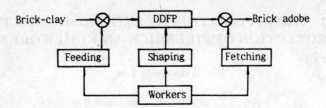

*Figure 1. The operator-machine system in firebrick production*

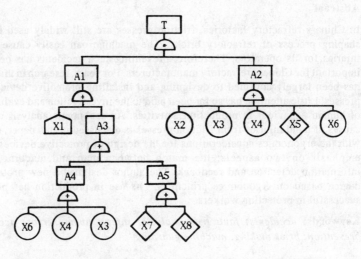

*Figure 2. A fault tree analysis of hand injuries*

The information channel in this system is one-way. The workers give a variety of operation orders to the machine, from which there is no information feedback. They learn the production situation only from their own observation and judgement. It is compulsory for workers to adapt their actions to the DDFP's specifications. Whether the man-machine relationship is harmonious or not fully depends on the workers' behaviour in the operation. Once a worker makes a mistake, the ergonomic relationship will be unbalanced, and this may lead to an accident.

Figure 2 shows a fault tree analysis of hand injuries. T stands for the top event 'Hand injury accident'; and A1, A2 'Hammer moves down' and 'Hands are beneath the hammer' respectively. X1 and A3 represent hammer moving down 'Wantedly' and 'Unwantedly'. X2, X3, X4, X5, X6, X7 and X8 are basic events 'Working habits', 'Misjudgement', 'Miscooperation', 'Brick-push-up failures', 'Low arousal', 'Break of spiral screws' and the 'Failures in motion systems'.

There exist seven 'minimum cuts' in the fault tree which indicate the main points leading to hand injuries:
$T = X3 + X4 + X6 + X1X2 + X1X5 + X2X7 + X2X8$

(1) Workers' misjudgement of information -

During the machine operating process, a worker must receive and respond to a large amount of information. For example, he must keep watch on the pressing pressure to maintain the quality of the products, and this often places him in a state of mental concentration. But due to the single channel mechanism of information processing, he can only be absorbed at one time in receiving and processing one sort of information. Therefore, when some dangerous situation happens abruptly, it is very difficult for him to avoid errors in judgement.

(2) Errors in cooperation among the workers -

Production with DDFPs requires great harmony in cooperation among workers, especially between the foreman and his workmates. To enhance the efficiency and ensure safety, each worker should make his operation quick, correct and convenient for the next procedure. But because of the personal differences among the group, the completion of each working procedure and the link between each one are very complicated. The workers cannot completely avoid any errors in information communication and in the cooperation among themselves, and this may cause a misoperation resulting in an accident.

(3) The influence of workers' arousal level -

The production process with DDFPs is tense, monotonous and repetitive. The process is noisy with a great deal of vibration. Some types of DDFPs keep workers standing throughout their working hours. All this causes workers to feel dull, dreary and tired, and finally low in their arousal level, lax in attention, and tending towards misperformance.

(4) The effect of working habits -

Every firebrick worker has developed his own distinctive working methods through practice. In piecework, to raise efficiency and increase output, most workers prefer cleaning the running hammer and brick-die without stopping the machine, though this is against safety regulations. This practice causes workers to have their hands beneath the hammer and exposed to the danger of getting hurt. Accidents statistics reveal that most hand injuries happen when the workers are cleaning the hammer or the brick-die with the machine running.

(5) Mechanical failures -

Generally speaking, the possible mechanical failures include: abrupt breaking of the spiral screw, which results in the unexpected drop of the hammer possibly hurting workers' hands; failures in the control system of the DDFP, which causes the machine to be out of order; and failures in the brick-push-up. Furthermore, some specially shaped brick-die may prevent workers from operating dexterously.

To illustrate the percentage causes of finger and hand amputation accidents which occurred in the refractory factory of a large-scale steel mill in China from 1959 to 1989, Figure 3 shows that accidents caused by workers' errors (70% of the total) are much more frequent than those caused by mechanical failures (24% of the total).

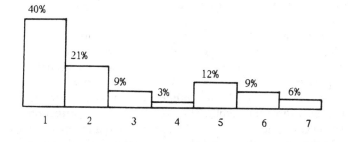

Figure 3. Percentage causes of hand injuries in a refractory factory

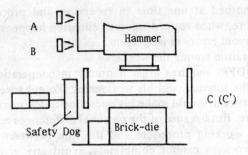

*Figure 4. A safety device for DDFPs*

## An ergonomic design of a safety device

Although the causes of hand injuries are very complicated, some of them can be easily prevented and eliminated. For example, paying more attention to the inspection and daily maintenance of a DDFP may prevent failures both in the control system and the spiral screw. However, the problem lies in the prevention of workers' errors and in the creation of a harmonious man-machine relationship. Therefore, the study should be aimed at the match between man and machine. Detection and feedback of information on workers' errors should be applied taking account of the operators' habits.

Figure 4 shows the design of a safety device. The aim is to design a new man-machine system by combining the device with a DDFP. The device performs the function of monitoring and feeding back to the system the information on the DDFP's running condition and the workers' hand positions. The system can automatically modify the DDFP's working pattern to safeguard the workers.

Four pairs of semiconductor infrared position transducers are fixed at points A, B, C and C'. Transducers at A and B detect the running hammer's position, and C and C' the workers' hands. C and C' are positioned within the area into which workers often put their hands. A is set by the device to be a prewarning point, because it may be dangerous for a worker if he puts his hands beneath the hammer after it has passed this point. B is an arm point from which the hammer is very close to the brick-die, and the worker will be threatened by the hammer if his hands are beneath it. The information picked up at points A, B, C and C' are respectively fed back to the control unit of the device. The control unit carries out a logical analysis of the information to judge yes-or-no for safety for the workers, and controls the machine to prevent accidents.

The new system follows the working pattern programmed by the device as shown in Figure 5.

Suppose a worker puts his hands beneath the hammer when it goes downwards and has just passed point A, the system changes the working state of the DDFP and stops driving the hammer. Then the hammer will continue to go down with a reduced speed due to its motion inertia and gravity, which allows time for the worker to withdraw his hands. If the worker's hands remain beneath the hammer when it keeps moving and passes point B, the system will force the hammer to change its motion direction i.e., goes upwards, and drives the safety dog (chock) to the position under the hammer to stop it, reinforcing safety. In the process described above, once the worker's hands are out of the DDFP, the device will let the machine resume its normal working situation immediately.

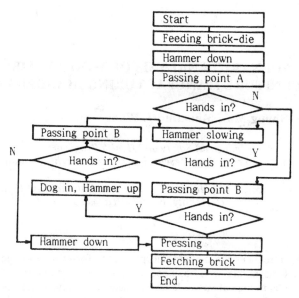

*Figure 5. The working pattern of the new operator-machine system*

The device is designed as an interlocking type. The control unit will automatically make the hammer go upwards and drive the safety dog in whenever the device itself fails to function properly. Such a state will remain unchanged until the failures in the device have been removed.

## Conclusions

(1) The detection and feedback of the information on both the workers' hand positions and the DDFP's working states makes it possible to construct a closed-loop channel for information exchanges between the workers and the machine, which lays the foundation for regulation, harmonization and the ergonomic match for the production system.

(2) The results of the logical analysis of the information in the system reflect the human errors made by the workers in operation. The safety device takes them as guidance to manipulate the DDFP's working states, which provides the machine with some 'intelligence' i.e., to automatically adapt itself to the workers' actions to avoid hand injuries.

(3) The safety device has been put into use in the refractory factory of a large-scale steel mill in China. The operation has proved successful in safeguarding the firebrick workers, and benefiting production.

# 13. SCENARIO ANALYSIS OF WOOD-BAMBOO FURNITURE MANUFACTURING ACCIDENTS

A. Wen-Shung Ma[*], Mao-Jiun J. Wang[**] and Frank S. Chou[*]

[*] Industrial Safety and Health Program
Industrial Technology Research Institute, Hsinchu, Taiwan

[**] Department of Industrial Engineering, National Tsin Hwa University, Hsinchu, Taiwan

## Abstract

This paper introduces the method and procedure for evaluating mechanical injury problems in the wood-bamboo furniture manufacturing industry in Taiwan. Epidemiological surveillance and scenario analysis were performed on 1695 workplace related incidents to identify the general problem areas. Two major hazard patterns were identified: caught in/between, and laceration accidents. Subsequently, two intervention measures were proposed to reduce the potential hazards resulting from unsafe conditions and unsafe acts.

**Keywords:** *scenario analysis; accidents; furniture industry; ergonomics intervention; machine guarding*

## *Introduction*

Wood-bamboo furniture manufacturing is characterized by woodworking machinery operations. Several kinds of machines including saws, drills, planers, presses, and grinding machines, etc. are commonly used. It is not surprising that mechanical injury is the major occupational safety problem in that industry. Accident types like cutting, shearing, crushing, and puncture are due to the operator's body part being exposed to the catch, pinch, shearing or cutting point of the machine resulting in unwanted energy transfer (Marshall, 1982). Therefore, it is obvious that effective machine guarding which can function as a barrier to protect the operator is the most important safety design consideration for woodworking machinery.

From the labour statistics reported in 1988 (Labor Council, 1988), mechanical injuries accounted for 44% of all the reported cases in Taiwan. Further, among the mechanical injury incidents, 75% of the machinery injuries were attributed to two accident types: caught in/between, and cut and lacerations. It is also reported that the accident rate in the wood-bamboo furniture manufacturing industry is two-fold higher than the average accident rate of all other manufacturing industries. This high incidence figure has been most consistent over the past few years with no sign of decreasing. A tremendous amount of cost in terms of workers' compensation, medical costs, and absenteeism is involved. Therefore, a thorough investigation is needed to solve the problem and to develop possible countermeasures. The objective of this study is to carry out a detailed accident investigation and to determine the specific causes of mechanical injuries in the wood-bamboo furniture manufacturing industry and to suggest some prevention measures.

In order to evaluate the problem, a review of historical data is essential. Epidemiological surveillance of occupational injuries from labour insurance claims forms was performed to identify the general problem areas.

## Epidemiological surveillance

For the epidemiological evaluation, more than 2000 reported cases were obtained from the 1987 labour insurance claims forms. In other words, only reported injury data were considered. By filtering out the traffic accident cases, 1695 workplace related accident cases were left for further analysis. Each case contained information pertinent to a specific injury occurrence. After careful evaluation, a coding scheme was developed for representing information about each accident case numerically. The coded data for each case represent a record in the database to be used for subsequent computer statistical analysis. The injuries were coded in accordance with the classification system of the American National Standard Institute (ANSI) Z-16.2 (1969). Each injury was classified in terms of: (1) body part affected, (2) nature of injury. The hazard pattern was coded in accordance with the classification system of the Labor Insurance Agency in Taiwan. Codings for task variables like department, job/task, and machine/equipment were defined through a tally process. Demographic, task, and injury information in each record was thus translated into numerical values following the predefined coding scheme of a set of variables, including: age of victim, victim's sex, victim's department, victim's job task, reported machine/equipment, reported hazard patterns, injured body part, and type of injuries. An epidemiological database was then created.

In order to have a clear understanding of the accident patterns, frequency analysis on the hazard pattern variable was first performed. The analysis was done by SPSS (a statistical analysis package) and the results are shown in Table 1. The two major hazard patterns which accounted for 80% of the incidents are caught in/between, and cut and laceration. And other relatively infrequent accidents are: sprains/strains, slip and fall, step on/strike against, and struck by falling objects. In order to determine whether the hazard patterns occurred randomly or systematically, cross-tabulations between the hazard pattern variable and the task variable as well as the injury variable were calculated. The results are summarized in Table 2. They were all significant (p.<0.001) in Chi-square tests, which implies that some systematic relationships seem to exist between the two variables. The occurrence of certain hazard patterns is simply not due to chance.

*Table 1. Frequency distribution of accident types in wood-bamboo furniture manufacturing*

| Accident type | Frequency | Percentage |
|---|---|---|
| Caught in/between | 749 | 44.2 |
| Cut and laceration | 611 | 36.1 |
| Sprains/ strains | 116 | 6.8 |
| Slip and fall | 92 | 5.4 |
| Fall of person | 61 | 3.6 |
| Step on/strike against | 42 | 2.5 |
| Struck by falling objects | 24 | 1.4 |

Table 2. Cross-tabulation analysis between hazard pattern and other variables

| Variable X variable | $X^2$-value | Significance |
|---|---|---|
| Hazard pattern x task | 2046.99 | *** |
| x body part | 3615.79 | *** |
| x injury type | 6560.88 | *** |
| x agents (machine/equipment) | 6998.52 | *** |

*** $p < 0.0001$

Since this study is mostly interested in determining the causative mechanism of mechanical injuries, accident cases related to the first two hazard patterns (caught in/between, and cut and laceration) were used for further scenario analysis.

## Scenario analysis

A scenario describes the actors (victims), the props (products), the scene (environment), and the action (task) (Drury and Brill, 1983). Words used to describe the scenarios are derived from a statistical analysis of incidents in that scenario. By conducting a series of frequency analyses on accident variables, two scenarios labelled as: caught in/between, and cut and laceration are obtained. These are the generic hazard patterns in the wood-working industry. The results are presented in Tables 3 and 4. In the first scenario (caught in/between), 749 cases were involved (Table 3). The victims were predominantly young males. While performing wood-working operations (90%), the fingers (71%) were caught in/between power transmission equipment (e.g. roller, gears, and chains) (36.8%) or some other types of wood-working machines and materials, and resulted in contusion and crushing injuries (60.5%). For the second accident scenario, cut/contusion, 611 cases were reported (Table 4). Again, the scenario indicates that the victims were mainly young males. While performing woodworking operations (90.7%), their fingers (70.5%) were mostly cut by sawing machines (70.5%) and resulted in lacerations, contusions and amputations (81.0%). The common phenomena observed in the two scenarios are: (1) More than half the victims were young workers (aged from 20 to 40), (2) over 70% of the victims were males, (3) slightly more than 70% of the victims' fingers were injured, and (4) the goal directed activity during the accident was the wood-working operation (90%). The results can help accident investigators foresee the victim population as well as their behaviour before conducting an in-depth investigation. One immediate conclusion from the two accident scenarios is that there is no effective safety guarding to isolate the catch point (or cutting point) of power transmission equipment (or cutting machines) which results in injuries and amputations of many fingers.

*Table 3. Scenario 1 - caught in/between (749 cases)*

| Sex | Age | Task | Agents | Body Part | Injury Type |
|---|---|---|---|---|---|
| male (567) 75.7% | 31-40 (198) 26.4% | woodworking (671) 89.6% | power-transmission equipment (276) 36.8% | finger (532) 71.0% | contusion/ crushing (453) 60.5% |
| female (182) | 21-30 (181) | material handling (40) | wood-material (149) | hand (104) | fracture (139) |
|  | 41-50 (164) | maintenance (38) | punch-press (122) | foot (40) | amputation (134) |
|  | 51-60 (127) |  | sawing-machine (103) | ankle (18) | others (23) |
|  | 15-20 (55) |  | planer (78) | head (16) |  |
|  | > 61 (24) |  | combiner (21) | leg (14) |  |
|  |  |  |  | thigh (13) |  |
|  |  |  |  | forearm (12) |  |

*Table 4. Scenario 2 - cut and laceration (611 cases)*

| Sex | Age | Task | Agents | Body Part | Injury Type |
|---|---|---|---|---|---|
| male (505) 82.6% | 31-40 (192) 31.4% | woodworking (554) 90.7% | sawing m/c (431) 70.5% | finger (448) 73.3% | cut/laceration (211) 34.5% |
| female (106) | 21-30 (150) | decorating (22) | tools (58) | hand (71) | contusion/crushing (184) |
|  | 41-50 (124) | maintenance (13) | planer (51) | forearm (19) | amputation (100) |
|  | 51-60 (85) | inspection (12) | wood materials (24) | wrist (11) | fracture (46) |
|  | 16-20 (37) | material handling (10) | drill (19) | upperarm (7) | others (70) |
|  | 61-67 (23) |  | combiner (19) | leg (9) |  |
|  |  |  | glass (9) | head (6) |  |
|  |  |  |  | others (40) |  |

## Intervention strategies

Two intervention measures are planned to solve the problem resulting from unsafe acts and unsafe conditions. To determine unsafe acts, an ergonomic job analysis will first be performed. Ergonomics analysis consists of defining the operation and discovering what operator behaviours are required. After the analysis, potential hazardous motions will be identified. Suggestions in terms of safe operation, design, and maintenance will be made to avoid the undesired event. Further, a video tape for teaching correct machine operation methods will be produced. The video tape will demonstrate the correct and safe operation method of the major wood-working machines (e.g. saws, planers, drills, etc.) and it will be broadly distributed to the wood-bamboo furniture manufacturing industry as a training material. This project is funded by the Council of Labor Affairs in Taiwan.

For unsafe conditions, as mentioned earlier, the unprotected machine was the centre of the problem. A research project which concentrates on the fact that the wood-working machine is inadequately guarded is also proceeding. Since the purpose of machine guarding is to prevent the operator's body part or clothing from coming into contact with any dangerous moving part of a machine, an effective guarding should not only provide positive protection but also should impose no discomfort or restriction on the operator. The study first surveys the commonly used guarding design for various wood-working machines, and then assesses the feasibility of various safe-guarding designs. The costs associated with each device are also estimated. Finally, some cost-effective guarding design ideas for different types of wood-working machines are recommended. Again the results will be published and propagated to the relevant people in that industry. This study is sponsored by the Industrial Institute in Taiwan.

## Conclusion

This study presents a case study for evaluating mechanical injury problems in wood-working machine operations in Taiwan. Two major hazard patterns were identified: caught in/between, and cut and laceration. Accidents are usually caused by both unsafe conditions and unsafe acts. For prevention purposes, two studies are proceeding to develop a manual of cost-effective safe guarding design and a video tape of demonstrating correct wood-working machine operations to educate the relevant people and to promote safety awareness in the wood-working industry.

## References

American National Standard Institute, 1969. Method of recording basic facts relating to the nature and occurrence of work injuries, ANSI, New York, NY.

Drury, C. G., and Brill, M., 1983. Human factors in consumer product accident investigation, Human Factors, 25(3): 329-342.

Labor Council, 1988. Labor Auditing Annual Report of 1987, Labor Council, Taiwan, ROC.

Marshall, G., 1982. Safety Engineering, Brooks/Cole Engineering Division, Monterey, CA.

# 14. HUMAN ERRORS IN MAINTENANCE

### Shang H. Hsu

*Department of Industrial Engineering and Management
National Chiao Tung University, Hsinchu, Taiwan, 30050*

**Abstract**

Accidents and incidents can result not only from improper operations but also from maintenance errors. It is important that causes of maintenance errors be understood and prevented in the design of a system. This paper summarizes a study which identifies design features contributing to maintenance errors. Fault tree analyses were conducted to identify accidents/incidents due to improper maintenance. Maintenance records associated with these accidents and incidents were then identified and analyzed. Maintenance errors could be classified into three categories: failures to recognize problems; failures to locate faulty components; and deviations from normal maintenance procedures. Field observations and structured interviews were conducted to identify system design features contributing to these three types of errors. Results shed light on areas where future studies on maintenance performance should be concentrated.

**Keywords:** *human error; maintenance; status information; fault diagnosis; accidents; work procedures*

## Introduction

Human error is a major contributor to accidents and incidents. Studies of human errors have focused on analysis of operational errors. Maintenance errors, however, have not drawn as much attention as they should. In fact, maintenance errors are pervasive and costly. Griffon-Fouco and Ghertman (1987) analyzed human error data in nuclear power plants, and found 60% of the errors occurred during periodic tests and maintenance work. Sears (1986) examined airline accident records from 1959 to 1983 and found 12% of the accidents resulted from maintenance and inspection deficiencies. Moreover, many injuries in manufacturing settings occurred due to machine malfunctions resulting from maintenance errors.

Maintenance errors have resulted in many serious accidents and incidents. Examples of disastrous accidents related to maintenance errors have been the Three Mile Island reactor incident and the 1979 aircraft crash at O'Hare Airport in Chicago. The Three Mile Island reactor incident started out with a maintenance error in which the pipe of the alternate water supply was blocked off. Incompleteness of engine installation and negligence of engine inspection resulted in the 1979 DC-10 crash, costing the lives of hundreds of people.

Therefore, it is important that maintenance errors be understood and prevented in the design of a system. The objective of this study was to investigate maintenance errors committed by aircraft technicians and then propose improvements to the design of aircraft systems.

## Method

Aircraft accidents/incidents report data (over a one year period) were gathered. In order to identify causes of these accidents/incidents, a fault tree was constructed for each accident/incident. These fault trees were then validated by aviation safety experts. Among the 250 accidents/incidents, 156 were found to be related to equipment malfunctions.

The causes of these equipment malfunctions were further analyzed to identify maintenance errors. The methodology employed in this portion was a systems approach proposed by Inaba (Fuchs and Inaba, 1981). Since maintenance errors might not have an immediate effect on safety, maintenance records of faulty subsystems and components were gathered - then sorted in terms of the time sequence. From malfunction descriptions and consecutive maintenance work which had been performed on the subsystems/components, 48 events could be linked to maintenance errors. These maintenance errors occurred mostly during the work of checkout, troubleshooting, and removal/installation. They could then be classified into three categories: failures to recognize problems (accounting for 19% of maintenance errors); failures to locate faulty components (60% of maintenance errors); and deviations from prescribed maintenance procedures (21%).

In order to identify causes of these maintenance errors, subsystems (with high and low error rates in each type of maintenance error) were selected for investigation. A questionnaire was developed to enquire about the maintainer's capability, training, design features, and working environment. Maintainers were asked to fill out the questionnaire. Structured interviews with these maintainers were conducted to identify causes contributing to the differences between high and low rates of maintenance error. Although the results of structured interviews provided some insight as to which design variables affect maintenance performance, this information needed to be verified by observing maintainers performing those tasks in the field setting. Maintenance performance on the selected subsystem was then observed and evaluated by subject matter experts, along with human factors specialists.

## Results

Results from interviews and field observations showed most of the failures in recognizing system problems could be attributed to incomplete or improper provisions of system status cues. Incomplete or improper provisions of system status meant the system provided too many or too few cues of system status. Maintainers had too much or too little information to judge whether the system was operational. This problem became worse when maintainers did not possess sufficient knowledge of the system. When maintainers were overloaded with too much information, they failed to utilize this information in interpreting system status and making a judgement. Instead, they discarded this information and employed heuristics. Consequently, they often drew wrong conclusions about the system state and falsely replaced good components. Data showed about 15% of replacements made to electronic subsystems were false replacements, resulting from improper provisions of status indicators.

Failures in locating faulty components resulted from the following design features:
(1) The system was too complicated to be understood. System complexity refers to the number of relevant relationships among components. Due to intricate relationships among components, technicians had difficulty in building a mental model of the system. Consequently, technicians had a hard time keeping track of the progress in the troubleshooting process;

(2) Parts were not easily accessible to test. Technicians usually work under time pressure. Some tests were omitted because test points were not easily accessible. Technicians made decisions based on insufficient test information, resulting in errors;
(3) The troubleshooting procedure was too long. Troubleshooting has been regarded as problem solving behaviour (Rouse, 1978). To locate faulty components, maintainers have to apply a test strategy, evaluate test results, and memorize the results of each test. When the rules and steps of a troubleshooting procedure exceeded the limits of a technician's working memory, the performance level of troubleshooting degraded.

Deviations from normal maintenance procedures could be further classified into two categories: wrong actions were chosen; and failure to act accurately. These errors resulted from the following causes:

(1) The complexity and the length of the maintenance procedure. Since the maintenance procedure exceeded technicians' cognitive capabilities, technicians tended to take short-cuts and consequently an error occurred;
(2) Maintainers' misunderstanding of maintenance instruction. Maintenance instructions were written in a reading level higher than technicians usually have. In addition, the format of maintenance instructions was too complicated and inconsistent;
(3) Maintenance actions demanded a higher proficiency of psychomotor skills and coordination with other maintainers. These maintenance actions were difficult to perform accurately;
(4) Poor working environment (i.e., poor illumination, temperature, etc.) to prevent accurate maintenance actions;
(5) Packing of the system did not allow components to be easily accessible, both visually and physically. Time had to be lengthened, requiring extra effort for technicians to remove and install an inaccessible component. Since these maintenance actions required more physical demand, actions could possibly not be performed as accurately; and components could easily become damaged.

## Discussion

Maintenance work is usually performed under time pressure. The results of this study indicate maintenance errors are due to excessive demands (both cognitive and physical) imposed upon maintainers by the system. Maintenance errors can be prevented by considering human capabilities when designing the human-machine interface, maintenance manuals, and training.

To reduce maintenance errors, several equipment design improvements are recommended:

(1) Modularization and appropriate circuit partitioning. Modularization can reduce the complexity of the system and improve troubleshooting and checkout performance (Wohl, 1981). Also, providing appropriate circuit partitioning and coding will make the circuit-tracking task easier for technicians;

(2) Proper design of status indications. Status indicators should be able to distinguish between different line-replaceable-units. Checkouts and troubleshooting can be enhanced. Status indicators should further be provided to inform maintainers of hazardous components which are not installed properly. Inadvertent actions can be prevented;

(3) Make components, test and checkout points easily accessible. Accessibility is defined as the ease with which the maintainer can reach or see in a component. An accessibility problem usually resulted from improper packaging practice. Repackaging the system to make less reliable components easily accessible can greatly enhance maintainability and yield less damage to the related components. For example, Theisen and Hsu (1982) conducted a study of the effects of physical accessibility on removal and installation. The F-14 was used as a platform system for the study. From the analysis of maintenance data, fire-sensing elements had an excessive maintenance time and error rate. The analysis of the maintenance procedure showed that maintainers had to remove the overwing fairing in order to gain access to the fire sense element. This step was time consuming and error-prone. A design change was proposed to improve accessibility of fire-sensing elements on the F-14. This proposal was introduced in order to provide an adapter for each sensing element to avoid an access requirement external to the engine nacelle. The results of a laboratory study showed the design change can yield a reduction rate of maintenance errors from 35% to 2%;

(4) Provide guards for delicate equipment so damage to equipment can be reduced. In addition, to prevent excessive force being applied to delicate equipment, a warning sign should be provided to maintainers;

(5) Use standardized components to reduce the likelihood of choosing wrong actions and the need to learn different skills and procedures for similar components. Using nonstandard components will impose an excessive mental workload on maintainers in which the maintainers have to remember and select the appropriate maintenance procedure for each component.

The results of this study point out areas where future laboratory studies on maintenance performance should concentrate in the future. One area is the integration of maintenance performance data with system safety measures. At the present time, most of the system reliability prediction techniques take only operational performance into consideration. The role of maintenance performance becomes increasingly important as the system operation is automated. Maintenance performance should then be added to system reliability prediction. To achieve this goal, a maintenance performance reliability database and maintenance system performance model should be developed. The database should allow system designers to predict maintenance error rates from equipment design features. The maintenance system performance model simulates maintenance processes which allows for the integration of maintenance performance with other safety analysis data to predict system safety.

Since most of the maintenance errors have usually occurred in troubleshooting, much effort should be placed on the study of troubleshooting performance. Troubleshooters have complained that the system is too complicated to be understood. One means to solve this problem is to reduce the complexity of the system. To achieve this goal, one has to identify which inherent system characteristics will cause the complexity of the system.

Developing a training aid to facilitate the understanding of the system would be fruitful to maintainers. One should then identify an effective way to represent system knowledge and include that as a training aid.

Another point in enhancing troubleshooting performance is to develop a troubleshooting aid. Before one can develop an effective troubleshooting aid, one should explore the cognitive mechanism underlying troubleshooting. The conditions under which troubleshooters possess different types of knowledge is in contention. How can experts apply their knowledge to troubleshooting? Which troubleshooting strategy imposes the least mental workload upon troubleshooters?

## References

Fuchs, F., & Inaba, K. (1981). Design for the maintainers: Final report. Xyzyx Information Corporation Technical Report.

Griffon-Fouco, M., & Ghertman, F. (1987). Data collection on human factors. In Rasmussen, J., Duncan, K., & Leplat, J. (Eds.) New technology and human error. New York, NY: John Wiley & Sons.

Rouse, W. B. (1978). Human problem solving performance in a fault diagnosis task. IEEE Transactions on Systems, Man, and Cybernetics, SMC-8, pp. 357-361.

Sears, R. L. (1986). A new look at accident contributions and the implications of operational and training procedures. Unpublished report. Boeing Commercial Aircraft Company.

Theisen, C. J., & Hsu, S. H. (1982). An empirical approach to design for human/machine interface in aircraft maintenance. Paper presented at the Annual Convention of the American Psychological Association, Washington, D. C., August 1982.

Wohl, J. G. (1981). System complexity, diagnostic behavior, and repair time: A predictive theory. In Rasmussen, J., & Rouse, W. B. (Eds.) Human detection and diagnosis of system failures. New York, NY: Plenum Press.

# Part III
Posture and Musculoskeletal Disorders

# 15. MUSCULOSKELETAL DISORDERS IN THE WORKPLACE

## Margaret I. Bullock

*University of Queensland*
*Australia*

**Abstract**

While several methods have been used to prevent the occurrence of musculoskeletal disorders, re-design of both the workplace and the work method, according to ergonomics principles, has offered the most effective preventative approach. The importance of encouraging worker participation in outlining solutions to work problems has been recognised, and the need for the acceptance of the introduction of ergonomic design principles by management has been appreciated. Strategies to encourage these new attitudes have had to be developed by ergonomists. This overview paper discusses current thinking in regard to the mechanism of injury to the musculoskeletal system, the assessment of postural load and its effects, the assessment of the injured worker, the relative value of preventative approaches and possible strategies to ensure their implementation.

**Keywords:** *musculoskeletal disorders; job analysis; posture; health care; education in ergonomics; job design*

## Introduction

As the name implies, musculoskeletal disorders comprise a range of problems affecting the muscles, bones or joints of the body. Within the industrial setting, it is the effect of excessive stress which generates the disorder and most often it is within the soft tissues which are unable to combat those stresses that a pathological process develops.

Since the 1960s, the number of patients with musculoskeletal disorders or discomfort has increased and this has implications for the worker, the employer and the ergonomist. For the worker, the pain, discomfort and interruption to their normal living and working pattern can be traumatic. For industry in general, the consequences of work injury and subsequent absenteeism are considerable in terms of productivity and the cost of rising insurance premiums. The effects on both the worker and the industry are of particular significance to the ergonomist, whose contribution to the control of industrially related injuries not only protects the worker, but also maintains an effective level of productivity and ensures the operation of a cost-effective enterprise. Spilling et al. (1986) have already shown that an ergonomically well designed workplace reduces sick leave. They have also performed a cost benefit analysis of the improvement and have shown that the application of ergonomics principles is a good investment.

## Work induced musculoskeletal disorders

Disorders such as sprains, bursitis, tenosynovitis, tendinitis and synovitis may all occur as a result of imposed stresses. Although the back is the most susceptible area to injury, the neck, shoulder and upper limb are also very susceptible to work stresses.

### Factors contributing to musculoskeletal injury

Many factors can contribute to the incidence of musculoskeletal injuries in the workplace and these may emanate from the environment, the workplace design, work postures, work methods, work demands which lead to fatigue, inappropriate incentives, psychological or emotional stress, or the capacity of the worker to cope with the variety of physical and mental stresses imposed.

Either extreme of physical load on the musculoskeletal system can result in a soft tissue disorder. For example, the effects on the back of lifting a load which is too heavy or in a way which, biomechanically, increases the stresses on the intervertebral joints, are well known. On the other hand, the effects of tasks requiring a minimum of muscular effort can be considerable if those tasks are repetitive and the conditions under which the musculoskeletal system is called upon to work are not ideal. Further, muscles not working actively to produce movement can suffer strain if subjected to certain undesirable factors. This can occur when sustaining a particular posture, either when movement takes place elsewhere in the body, or during work requiring monitoring or surveillance. The term 'postural workload', used commonly today, implies both work movements and work postures.

Recent technological changes within industry have been accompanied by a marked decrease in physical activity and the consequential lack of exercise has undoubtedly contributed to a decreasing ability for the musculoskeletal system to withstand imposed stresses.

### Mechanism of injury

Unfortunately, the mechanism by which strain occurs is still unclear. Some researchers have tried to quantify workload and its effect on workers in terms of musculoskeletal systems. Biomechanical studies have also been undertaken, determining that the mechanisms of injury appear to be the overloading of tissues through excessive force, faulty work postures, repetitive movements and inadequate recovery periods.

## Ergonomic assessment

The thorough analysis of a job is a task of considerable magnitude. It includes on the one hand, an assessment of the workplace and the work environment and the likelihood of injury within it, and on the other, a survey of the load imposed on the worker, a comprehensive evaluation of the worker's method of performing the task and the person's capacity to cope with the particular demands of the task. As Luopajarvi (1990) has explained, ergonomic work analysis methods can be divided into two main categories according to the object of the study: methods which are focused on the recording and analysis of the **work** which causes the load in question; and methods which are aimed at assessment of the **effects** of the workload on the person.

Where musculoskeletal injury has already occurred, an evaluation of the functional capacity of the worker is also important, so that appropriate decisions regarding the continuation of or return to work can be made and so that the working environment can be adapted, where appropriate, to match the capacity of the worker.

## Job analysis

Each work situation requires individual analysis so that satisfactory working relationships and conditions may be ensured and so that errors may be anticipated. To identify the need for ergonomic design, the ergonomist can survey the nature of the task, the work surroundings, and the worker-task relationship. The importance of involving workers in this process is widely recognized today. One way of encouraging worker participation and involvement in job analysis is in the use of self-analysis checklists, which enable them to make relevant suggestions as far as their own work and workplace are concerned.

## Assessment of working postures

Several methods of quantitatively assessing work postures which can be used to demonstrate those working positions requiring special attention have been devised. Some methods are based simply on direct visual observation, but other advanced methods, with computerised data recording and processing, have also been developed. The Ovaco Working Posture Analysis System (OWAS) was developed in Finland for the identification and evaluation of unsuitable working postures. Observations of work are recorded at regular intervals using coding digits which refer to previously defined basic posture. These records are then summarized into an overall posture, classified according to the strain it imposes on the musculoskeletal system. Recommendations for the type and timing of action to be taken depend on the degree of strain perceived (Karhu et al., 1977). In the Swedish system, 'ARBAN', details of the workplace are recorded on video tape or film and the posture and load on the body are coded in terms of 'frozen situations' for computerised analysis (Holzmann, 1982). A simpler method, VIRA, developed in Sweden, focuses specifically on movements and postures of the arms and head during short-cycle, seated assembly work with relatively rapid repetitive movement. It involves a type of frequency study in which a video technique registers movements subsequently analysed by micro-computer (Persson and Kilbom, 1983). Another approach developed in Holland provides quantitative data on the curvature of the spine throughout the day. From the graphs recorded, it is possible to calculate the duration of holding one posture, the frequency of taking up each posture and the average time spent in it (Snijders, 1984). Luopajarvi (1990) has described a relatively new approach developed in Finland. This 'Ergoshape' system is constructed within a micro-CAD programme called 'autoCAD' and consists of three parts: anthropometric man models, biomechanical calculations to allow the evaluation of postural stress, and recommendation charts which provide design guidelines for special situations.

## Assessment of the effects of workload on the person

In relation to the development of musculoskeletal injuries in the workplace, it is vital to consider the effects of workload on the person, be they physical, mental or social. Commonly, short-term physical effects of only slightly overloading continuous work are experienced as fatigue, pain or 'discomfort'. In the Nordic countries, a questionnaire has been developed to provide a standardized method for collecting information about workers' complaints and symptoms. It comprises a general questionnaire and two specific questionnaires which focus on the low back and the neck and shoulder region (Kuorinka et al., 1987).

The risk of strain can be increased when new demands are imposed, as when work methods are changed, the person is transferred to a new area, or external factors extend the person beyond their capacity. Psychological and social stresses can produce a sense of agitation which, in turn, can lead to an increase in muscle tension and an increase in static muscle work. Psychological stress can be reflected in poor posture, leading to static load on neck and shoulder muscles which, if sustained, can be a precursor to musculoskeletal disorders.

It is important, therefore, that in the analysis of musculoskeletal load, attention is focused not only on the physical factors but on the whole working environment, including mental and social factors.

## Preventive approaches

Over the years, various methods have been espoused for the prevention of musculoskeletal injuries in the workplace. The importance of both education and workplace design, in particular, continues to be recognized.

### Education

Planning an educational programme entails consideration of the needs of the recipients, the nature of the work and its inherent problems, and the objectives of the education. It is important, also, to be aware of the nature of relevant legislation and union practices, for there is little point in advocating certain practices if laws and industrial agreements prevent their implementation. It is also essential that all participants be made aware of their rights and their responsibilities according to the law.

All the educational programmes must highlight design problems within the workstation. Participants are then able to relate ergonomics principles to their own working environment and to identify potential hazards.

The concept that the worker should be involved in identifying work problems and in suggesting solutions to them is not new and the practice of using a participative approach to ergonomics is growing in many countries throughout the world. Participation in management, quality of work life and employee involvement all define small group activities as part of a new human resource management philosophy (Lifshitz et al., 1989). Education is fundamental to informed participation. Nevertheless, it is still important to offer sound ergonomics advice when it is required. The change in recent years is in the understanding of how to communicate and transform expert knowledge into user/client knowledge and, in turn, convert this knowledge into effective action (Rustad, 1990). Because clients must be able to understand the nature and meaning of the advice given, the 'expert' must have the ability to communicate his/her knowledge effectively.

The implications for musculoskeletal disorders of retaining poor design features in the workplace must be understood, and education in design principles coupled with simple explanation of the biomechanics of injury, should therefore be provided.

### Design

One means of preventing musculoskeletal injury is to apply purely ergonomic measures aimed at reducing the load on the locomotor system, by altering the design of the workplace or the working environment. Above all, an attempt should be made to avoid peak strains and static loads. At the same time, however, an attempt must be made to try to avoid muscular inactivity (Jonsson, 1988). The adequacy of the workplace design can be a major determinant in whether or not a current or persistent pain syndrome develops. The design of workplace and equipment to ensure a correct worker-task relationship so that postures are not extreme or fixed and that unnecessary, jerky or uneven movements may be avoided, must be emphasized. Work process design should allow varied activity by the worker, combining gross and fine movements. The appropriate basis for effective design of work areas and placement of controls is a knowledge of body size and clearance tolerances. People in the workforce must also appreciate the variations which exist, so that they can provide any necessary adjustments to ensure that their own working relationships are optimal.

Unfortunately, if the cost effectiveness of a new design is not appreciated or profits are low, managers are often reluctant to introduce design changes. Both management and working personnel must accept and appreciate the value of redesign, otherwise implementation of new ideas is unlikely to occur. The challenge to the ergonomist is to encourage the recognition of the importance of applying ergonomics principles and to foster an appreciation of the implications for injury of not doing so.

Jonsson (1988) asserts that continuing ergonomic work which relates **solely** to workplace design cannot be expected to have more than purely marginal effects. He suggests that the preventive measures deserving priority for the future are probably the revision of work organization and changes to the mental and social environment. This change in focus and scope of ergonomics practice is a natural consequence of technological development. The macro-ergonomic approach, in which organizational design is a fundamental question, offers an opportunity for putting the human-machine interface, the user-system interface and the organization-machine interface into perspective. Hendrick (1987) has explained the necessity of applying the macro-ergonomic design approach in the development of decisions about the overall organization prior to the effective design of the components.

## Conclusions

Those involved with ergonomics today must acknowledge the changes in technology and work practices. Recognition must be given to the need to alter attitudes, strengthen the participation of workers in problem solving and give due emphasis to the importance of organizational design, in the overall approach to prevention of musculoskeletal injury.

## References

Hendrick H W 1987 Macro-ergonomics: a concept whose time has come. Bulletin of the Human Factors Society 30:2, 1-3.

Holzmann P 1982 Arban - a new method for analysis of ergonomic effort. Applied Ergonomics 13:2, 82-86.

Jonsson B 1988 Facts and hypotheses about musculoskeletal injuries at work. Newsletter of the National Board of Occupational Safety and Health 2/88, 6-7.

Karhu O, Kansi P, Kuorinka I 1977 Correcting work posture in industry. A practical method for analysis (OWAS). Applied Ergonomics 8:4, 199-201.

Kuorinka I, Jonsson B, Kilbom A, et al 1987 Standardised Nordic questionnaires for the analysis of musculoskeletal symptoms. Applied Ergonomics 18:3, 233-237.

Lifshitz Y R, Archer R, Armstrong T J, Smith T D 1989 The effectiveness of an ergonomics program in controlling work related disorders in an automotive plant. Proceedings, 'Marketing Ergonomics' International Conference, Noordvijk.

Luopajarvi I 1990 Ergonomic analysis of workplace and postural load. In: Bullock M I, Ergonomics: the physiotherapist in the workplace. Churchill Livingstone, Edinburgh.

Persson J and Kilbom A 1983 VIRA - a simple videofilm technique for registering and analysing work postures and movements. Report 1983:10. Department of Clinical Work Physiology, National Board of Occupational Safety and Health, Stockholm, Sweden.

Rustad R 1990 Ergonomics - an educational challenge. In: Bullock M I, Ergonomics: the physiotherapist in the workplace. Churchill Livingstone, Edinburgh.

Snijders C J 1984 Faculty of Medicine, Erasmus University, Rotterdam. Personal communication.

Spilling S, Eitrheim J, Aaras A 1986 Cost-benefit analysis of work environment investment at STK's telephone plant. In: Corlett N, Wilson J, Manenica I, The ergonomics of working postures: models, methods and cases. Taylor & Francis, London

# 16. OCCUPATIONAL STRESS AT THE WORKPLACE IN INDUSTRIALLY DEVELOPING COUNTRIES: CASE STUDIES IN CHINA AND THAILAND

Houshang Shahnavaz, Shihan Bao
and Pranee Chavalitsakulchai

*Center for Ergonomics of Developing Countries (CEDC)*
*Department of Human Work Sciences, Lulea University*
*S-951 87 Lulea, Sweden*

**Abstract**

Poor working conditions and the non-existence of an effective work injury prevention programme in many industrially developing countries (IDC), has resulted in a very high rate of work related musculoskeletal disorders (MSD). As a result, most IDCs are paying an unacceptably high price in terms of human suffering, sickness and loss of production due to work related injuries and accidents. In order to understand the causation factors and the relationships between MSD and various work and worker related factors in some IDCs, a comprehensive research project is running at the CEDC. This paper describes some of the findings from case studies carried out in the People's Republic of China and Thailand during 1989. A questionnaire survey was conducted in 11 industries and institutes in the Hubei province of P.R. China, investigating some 1373 industrial workers. The results indicate that "low back" pain is the most prevalent problem among Chinese workers. In second place come "knee" problems followed by "shoulder" and "neck" problems. However, the frequencies of these problems vary between industries and tasks as well as individual operators. In Thailand, some 1000 female workers were studied at 5 different types of industries in three central provinces. The results indicate serious and common problems among the work force. The symptoms of musculoskeletal stresses in these two countries follow a logical pattern in terms of the influence of a few key parameters, such as the prevailing working condition, operator's age, number of years employed, and the type of tasks performed. The findings of this study support the development and execution of a comprehensive programme for improving working conditions and preventing occupational stresses at workplaces in IDCs. Such programmes are best served if they are designed and developed locally by all parties concerned at the workplaces and conducted jointly by management, workers and safety personnel.

**Keywords:** *musculoskeletal disorders; developing countries; absenteeism; back pain*

## *Introduction*

The prevalence of various occupational and work related disorders and diseases is changing with industrial development and the introduction of new technologies. Musculoskeletal disorders (MSD) are known to be one of the main problems in many industrialized countries for the last decades. Several studies indicate that even when ergonomics guidelines, regarding physical aspects of the workplace are followed and the workstations are adjusted to fit individual operators, a variety of posture complaints still arises (Arndt, 1983; Noddeland and Winkel, 1988; Aronsson et al., 1989). The connection between health problems at work and work environment factors, as well as the design aspects of work and the workplace is very complex and is influenced by many different factors.

The scope and importance of MSD as a worldwide problem have been documented in many studies (Magora, 1970; Maeda, 1977; Kelsey et al., 1979; Andersson, 1981; Biering-Soerensen, 1982; Klein et al., 1984; Snook and Webster, 1987). Possible risk factors such as those related to various aspects of the working environment, the workstation, the working method and the operators' characteristics have also been studied by many authors (Chaffin, 1974; Armstrong et al., 1982; Hagberg, 1984; Westgaard and Aaras, 1984; Bammer, 1987; Hildebrant, 1987; Kilbom and Persson, 1987; Gilad and Kirschenbaum, 1988; Wiker et al., 1989).

In order to control and prevent the MSD problem at workplaces, many intervention programmes, which mainly emphasise workstation redesign, work organisation, worker training and education, have also been suggested (Westgaard and Aaras, 1985; Kilbom, 1988; Vincent et al., 1989). Although, some successful programmes have been reported, the incident frequencies of MSD at many workplaces are still very high thus indicating that the problems are far from being controlled. For example in Sweden, where workers, management and the government are very conscious of employees' health and safety, as well as providing good working conditions, the rate of MSD at workplaces is still unacceptably high. During 1983 some 11,918 cases of occupational diseases were reported. Among them 55.7% were attributed to MSD. The situation has, however, further deteriorated since then, creating great concern in Swedish society. This is why the government has appointed a national work environment commission to identify those jobs having the highest risk of health problems such as MSD within various occupations. The commission published the results of its investigation in February 1990 (Arbetsmiljokommission, 1990). Based on the results of this very unique study, a good scientific information base is now available for the systematic improvement of the working situation of those dangerous jobs at Swedish workplaces.

However, most of the studies and activities have been carried out in industrialized countries. This does not mean that the MSD problems in industrially developing countries are less serious or at a lower magnitude.

In order to answer this question, understand the causation factors, investigate the type and extent of MSD at various workplaces and highlight some of the existing problems in IDCs, a comprehensive research project is running at the Center for Ergonomics of Developing Countries (CEDC) on MSD problems in the largest IDC, China, and in Thailand. This paper reports some of the results and findings from this study.

## Material and methods

The methods used were a) a mailed questionnaire in China, and b) a questionnaire-interview in Thailand.

a) In China some 1373 male and female workers were investigated in 11 different industries of the Hubei province with regard to MSD problems. Subjects were selected randomly from workers who were engaged in the main production tasks in the selected industries. The distribution of the subjects in various industries is shown in Table 1.

Table 1. The distribution of subjects in the selected occupational groups

| | MALE | FEMALE | TOTAL | AVERAGE AGE (years) | AVERAGE WORK EXPERIENCE OF THE SAME JOB (years) |
|---|---|---|---|---|---|
| TEXTILE | 8 | 92 | 100 | 34.2 | 10.6 |
| GARMENT | 8 | 216 | 224 | 21.1 | 1.2 |
| PAINT | 32 | 35 | 67 | 30.3 | 6.0 |
| HETALLURGY | 374 | 14 | 388 | 34.3 | 10.0 |
| ELECTRONICS | 59 | 162 | 221 | 29.6 | 7.8 |
| CARRAIGE | 70 | 15 | 85 | 32.8 | 9.1 |
| CONSTRUCTION | 103 | 1 | 104 | 38.6 | 16.1 |
| TRANSPORTATION | 53 | 2 | 55 | 34.1 | 12.1 |
| COMPUTER | 36 | 32 | 68 | 27.9 | 4.6 |
| TEACHERS | 1 | 34 | 35 | 38.4 | 17.0 |
| STUDENTS | 22 | 4 | 26 | 24.8 | 3.5 |
| TOTAL | 766 | 607 | 1373 | 31.2 | 8.3 |

Prior to the study, one of the authors visited the Hubei province to arrange the procedure for the study in cooperation with the Department of Labour Health of the Tongji Medical University, Wuhan (Hubei) and the safety and health personnel of the Hubei province.

The questionnaire used in this study was a modified version of the "Standard Nordic Questionnaire for Analysis of Musculoskeletal Symptoms" (Kuorinka, 1987). Some minor modifications were made, considering the specific situation in China. Basically the questionnaire comprised seven questions for each of the nine parts of the locomotor system (i.e. Neck, Shoulders, Upper Back, Elbows, Low Back, Wrists/Hands, Hips/Thighs, Knees, and Ankles/Feet).

Questionnaires were distributed among the subjects and later collected by the safety and health personnel in the various industries. Subjects completed the questionnaire during the last "break" of their working shift.

The contingency table method was used to analyse the data and the chi-square test was performed for significance testing.

b) In Thailand some 1000 female workers were studied in 5 different industries in three central provinces with regard to MSD problems. Because of the rapidly increasing numbers of female workers in the various types of industries in IDCs, especially in Thailand, the study concentrated only on female employees.

Likewise, in this study, only subjects working directly in production were randomly selected for investigation. However in this study only female subjects were selected with the number of subjects being 200 in each of the five industries.

In this study the same questionnaire was used, but the questions were asked by the investigator and the subjects' responses were recorded. One of the authors conducted the interviews and completed all the 1000 questionnaires in the 5 industries.

## Results

### From China

Table 2 shows the extent of musculoskeletal complaints experienced by workers from various industries. From the results it is evident that the most prevalent problem is low back pain (55.4%), followed by knee troubles (32.4%) for all subjects.

Table 2. Musculoskeletal complaints in different parts of the body and different occupational groups

| INDUSTRY / PART OF THE BODY | TEXTILE | GARMENT | PAINT | METALLILL | ELECTRITR | CARRIACR | CONSTRAR | TRANSNSP | COMPUTER | TEACHER | STUDENT | TOTAL |
|---|---|---|---|---|---|---|---|---|---|---|---|---|
| Total Number of subjects | 100 | 224 | 67 | 388 | 221 | 85 | 104 | 55 | 68 | 35 | 26 | 1375 |
| Neck | 20 | 7 | 20 | 28 | 154 | 18 | 30 | 12 | 17 | 19 | 8 | 333 |
|  | 20.0 | 3.1 | 29.9 | 7.2 | 69.7 | 21.2 | 28.8 | 21.8 | 25.0 | 54.3 | 30.8 | 24.3 |
| Shoulders | 22 | 8 | 27 | 55 | 133 | 20 | 32 | 10 | 10 | 19 | 5 | 341 |
|  | 22.0 | 3.6 | 40.3 | 14.2 | 60.2 | 23.5 | 30.8 | 18.2 | 14.7 | 54.3 | 19.2 | 24.8 |
| Upper back | 8 | 4 | 11 | 32 | 117 | 18 | 25 | 9 | 12 | 18 | 4 | 258 |
|  | 8.0 | 1.8 | 16.4 | 8.2 | 52.9 | 21.2 | 24.0 | 16.4 | 17.6 | 51.4 | 15.4 | 18.8 |
| Elbows | 2 | 3 | 12 | 32 | 74 | 9 | 32 | 2 | 0 | 7 | 1 | 174 |
|  | 2.0 | 1.3 | 17.9 | 8.2 | 33.2 | 10.6 | 30.8 | 3.6 | 0.0 | 20.0 | 3.8 | 12.7 |
| Low back | 75 | 22 | 44 | 213 | 176 | 48 | 92 | 25 | 32 | 24 | 10 | 761 |
|  | 75.0 | 9.8 | 65.7 | 54.9 | 79.6 | 56.5 | 88.5 | 45.5 | 47.1 | 68.6 | 38.5 | 55.4 |
| Wrists/hands | 18 | 7 | 26 | 43 | 113 | 26 | 40 | 2 | 6 | 12 | 2 | 295 |
|  | 18.0 | 3.1 | 38.8 | 11.1 | 51.1 | 30.6 | 38.5 | 3.6 | 8.8 | 34.3 | 7.7 | 21.5 |
| Hips/thighs | 14 | 8 | 6 | 33 | 82 | 7 | 14 | 7 | 5 | 10 | 2 | 188 |
|  | 14.0 | 3.6 | 9.0 | 8.5 | 37.1 | 8.2 | 13.5 | 12.7 | 7.4 | 28.6 | 7.7 | 13.7 |
| Knees | 82 | 17 | 16 | 119 | 88 | 23 | 61 | 5 | 14 | 16 | 4 | 445 |
|  | 82.0 | 7.6 | 23.9 | 30.7 | 39.8 | 27.1 | 58.7 | 9.1 | 20.6 | 45.7 | 15.4 | 32.4 |
| Ankles/feet | 34 | 5 | 9 | 20 | 57 | 13 | 16 | 4 | 0 | 13 | 3 | 174 |
|  | 34.0 | 2.2 | 13.4 | 5.2 | 25.8 | 15.3 | 15.4 | 7.3 | 0.0 | 37.1 | 11.5 | 12.7 |

Note: The upper figure is the number of subjects who had trouble in the corresponding locomotor region in easch industry, and the lower figure is the percentage of workers with complaints.

Using the chi-square test, we found that the frequencies of musculoskeletal problems in each locomotor region are significantly different from industry to industry. Subjects from the construction, textile and electronics industries are at most risk with regard to low back problems (88.5%, 75.0% and 72.1% respectively). Subjects working in the electronics industries have a high incidence of problems in the neck, shoulders and upper back regions (63.6%, 49.6% and 49.6% respectively).

Table 3 presents information on sick leave taken due to MSD in these industries. Complaints due to MSD of different parts of the body in the last 12 months, were generally highest for "low back" pain (46.2%) and thereafter for "knee" pain (26.4%). However, significant variation was found with regard to different occupations. In order to compare the extent of problems in different locomotor regions, Figure 1 shows the results regarding the relationship between frequencies of musculoskeletal complaints and sick leaves. It indicates that low back troubles cause more sick leaves and most operators have had these problems for the last 12 months. Comparing these results in different industries, taking low back and neck trouble as examples, we found that low back troubles are more severe in construction, carriage, textile and electronics industries than in the others.

## From Thailand

The working hours of the female workers in the selected industries were on average more than 48 hours, including overtime. However it was evident that the majority of textile workers usually work continuously in two shifts.

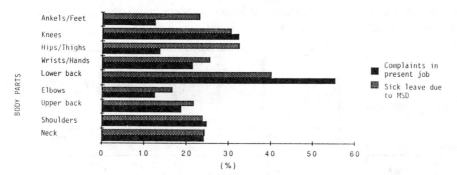

Figure 1. The relationship between frequencies of musculoskeletal complaints and workers' sick leaves

Table 3. Sick leave due to MSD with regard to different parts of the body and industry

| INDUSTRY / PART OF THE BODY | TEXTILE | GARMENT | PAINT | METAL | ELECTRICAL | CARS | CONSTR | TRANSP | COMPUT | TEACHER | STUDENT | TOTAL |
|---|---|---|---|---|---|---|---|---|---|---|---|---|
| Total Number of subjects | 100 | 224 | 67 | 388 | 221 | 85 | 104 | 55 | 68 | 35 | 26 | 1375 |
| Neck | 5 / 5.0 | 1 / 0.4 | 1 / 1.5 | 8 / 2.1 | 50 / 22.6 | 3 / 3.5 | 6 / 5.8 | 2 / 3.6 | 0 / 0.0 | 5 / 14.3 | 0 / 0.0 | 81 / 5.9 |
| Shoulders | 7 / 7.0 | 0 / 0.0 | 3 / 4.5 | 18 / 4.6 | 39 / 17.6 | 5 / 5.9 | 6 / 5.8 | 1 / 1.8 | 0 / 0.0 | 2 / 5.7 | 0 / 0.0 | 81 / 5.9 |
| Upper back | 2 / 2.0 | 0 / 0.0 | 0 / 0.0 | 7 / 1.8 | 36 / 16.3 | 5 / 5.9 | 1 / 1.0 | 2 / 3.6 | 0 / 0.0 | 3 / 8.6 | 0 / 0.0 | 56 / 4.1 |
| Elbows | 1 / 1.0 | 0 / 0.0 | 0 / 0.0 | 7 / 1.8 | 15 / 6.8 | 2 / 2.4 | 4 / 3.8 | 0 / 0.0 | 0 / 0.0 | 0 / 0.0 | 0 / 0.0 | 29 / 2.1 |
| Low back | 40 / 40.0 | 3 / 1.3 | 7 / 10.4 | 67 / 17.3 | 83 / 37.6 | 27 / 31.8 | 62 / 59.6 | 7 / 12.7 | 0 / 0.0 | 7 / 20.0 | 3 / 11.5 | 306 / 22.3 |
| Wrists/hands | 7 / 7.0 | 0 / 0.0 | 6 / 9.0 | 9 / 2.3 | 36 / 16.3 | 6 / 7.1 | 8 / 7.7 | 0 / 0.0 | 0 / 0.0 | 3 / 8.6 | 0 / 0.0 | 75 / 5.5 |
| Hips/thighs | 5 / 5.0 | 2 / 0.9 | 0 / 0.0 | 14 / 3.6 | 31 / 14.0 | 2 / 2.4 | 2 / 1.9 | 4 / 7.3 | 0 / 0.0 | 1 / 2.9 | 0 / 0.0 | 61 / 4.4 |
| Knees | 35 / 35.0 | 3 / 1.3 | 1 / 1.5 | 34 / 8.8 | 35 / 15.8 | 8 / 9.4 | 17 / 16.3 | 0 / 0.0 | 0 / 0.0 | 3 / 8.6 | 0 / 0.0 | 136 / 9.9 |
| Ankles/feet | 7 / 7.0 | 2 / 0.9 | 1 / 1.5 | 5 / 1.3 | 16 / 7.2 | 3 / 3.5 | 5 / 4.8 | 0 / 0.0 | 0 / 0.0 | 1 / 2.9 | 0 / 0.0 | 40 / 2.9 |

Note: The upper figure represents the number of people in the corresponding industry, who had taken sick leave due to the trouble in the corresponding locomotor region. The lover figure represents the percentage of people who had taken sick leave in the corresponding industry. In the last column, the total number of people in the 11 industries, who had taken sick leave (upper figures), and the percentage values are presented.

The musculoskeletal problems were significantly different (p.<0.001) in various industries. However, "low back" pain was the most pronounced problem, indicated by 74.8% of the workers. Thereafter came complaints about "shoulder" pain, by 65.2%, "hip/thigh" pain by 63.4% and pain in the "knees" by 53.9% of the work force.

Table 4. Musculoskeletal complaints of female workers in different parts of the body and different occupational groups

| INDUSTRY / PART OF THE BODY | GARMENT | FERTILIZER | PHARMACY | TEXTILE | CIGARETTE | TOTAL |
|---|---|---|---|---|---|---|
| Total No. of subjects | 200 | 200 | 200 | 200 | 200 | 1000 |
| Neck | 125 / 62.5 | 91 / 45.5 | 70 / 35.0 | 48 / 24.0 | 125 / 62.5 | 459 / 45.9 |
| Shoulder | 133 / 66.5 | 151 / 75.5 | 105 / 52.5 | 145 / 72.5 | 118 / 59.0 | 652 / 65.2 |
| Upper back | 77 / 38.5 | 70 / 35.0 | 67 / 33.5 | 124 / 62.0 | 98 / 49.0 | 436 / 43.6 |
| Elbows | 93 / 46.5 | 95 / 47.5 | 86 / 43.0 | 66 / 33.0 | 48 / 24.0 | 388 / 38.8 |
| Low back | 154 / 77.0 | 144 / 72.0 | 119 / 59.5 | 193 / 96.5 | 138 / 69.0 | 748 / 74.8 |
| Wrists/hand | 93 / 46.5 | 95 / 47.5 | 86 / 43.0 | 66 / 33.0 | 48 / 24.0 | 388 / 38.8 |
| Hips/thighs | 166 / 83.0 | 142 / 71.0 | 97 / 48.5 | 170 / 85.0 | 59 / 29.5 | 634 / 63.4 |
| Knees | 103 / 51.5 | 128 / 64.0 | 50 / 25.0 | 109 / 54.5 | 149 / 74.5 | 539 / 53.9 |
| Ankles/feet | 90 / 45.0 | 115 / 57.5 | 23 / 11.5 | 105 / 52.5 | 74 / 37.0 | 407 / 40.7 |

Note: The upper figure is the number of female workers who had troubles in the corresponding locomotor region in each industry, and the lower figure is the percentage of female workers with complain.

The distribution of pain in different parts of the musculoskeletal system amongst female workers in the five industries is shown in Table 4. Table 5 presents the frequency distribution of sickness leave among workers of different industries because of musculoskeletal problems experienced in various parts of the body. Comparing the present sick leave statistics with the information from last year, it became evident that both the symptoms and the rate of sick leaves due to musculoskeletal problems were higher, especially for workers in the cigarette factory and for textile workers, who also had the highest overtime work. The relationship between the frequencies of musculoskeletal complaints and sick leaves for Thai workers is presented in Figure 2. Again low back pain caused the highest sick leaves.

## Discussion

The results of this study clearly indicate that musculoskeletal problems have a significant impact on the health of the workers in the investigated countries, China and Thailand. However, the extent of the problems and possible injuries varies significantly from occupation to occupation and is different from the type of problems in the industrialized countries. For example, in a report from Australia, Bammer (1987) indicates that up to 79% of VDT operators have common musculoskeletal problems in the neck and shoulder regions. In the present study VDT work in China

*Table 5. Sick leave due to musculoskeletal disorders with regard to part of the body of female workers and industry*

| INDUSTRY / PART OF THE BODY | GARMENT | FERTILIZER | PHARMACY | TEXTILE | CIGARETTE | TOTAL |
|---|---|---|---|---|---|---|
| Total No. of subjects | 200 | 200 | 200 | 200 | 200 | 1000 |
| Neck | 0 / 0 | 0 / 0 | 4 / 2.0 | 1 / 0.5 | 20 / 10.0 | 25 / 2.5 |
| Shoulder | 0 / 0 | 5 / 2.5 | 9 / 4.5 | 0 / 0 | 22 / 11.0 | 36 / 3.6 |
| Upper back | 1 / 0.5 | 4 / 2.0 | 14 / 7.0 | 2 / 1.0 | 20 / 10.0 | 41 / 4.1 |
| Elbows | 2 / 1.0 | 3 / 1.5 | 15 / 7.5 | 0 / 0 | 10 / 5.0 | 30 / 3.0 |
| Low back | 5 / 2.5 | 14 / 7.0 | 25 / 12.5 | 3 / 1.5 | 20 / 10.0 | 67 / 6.7 |
| Wrists/hand | 2 / 1.0 | 3 / 1.5 | 15 / 7.5 | 0 / 0 | 10 / 5.0 | 30 / 3.0 |
| Hips/thighs | 2 / 1.0 | 8 / 4.0 | 11 / 5.5 | 2 / 1 | 5 / 2.5 | 28 / 2.8 |
| Knees | 2 / 1.0 | 6 / 3.0 | 14 / 7.0 | 2 / 1.0 | 28 / 14.0 | 52 / 5.2 |
| Ankles/feet | 1 / 0.5 | 4 / 2.0 | 5 / 2.5 | 1 / 0.5 | 11 / 5.5 | 22 / 2.2 |

Note: The upper figure represents the number of female workers in the corresponding industry, who had taken sick leave due to the trouble in the corresponding lecomotoer region. The lower figure represents the percentage of female workers who had taken sick leave in the corresponding industry. In the last column, the total number of female workers in 5 industries, who had taken sick leave (upper figures), and the percentage values are presented.

was the least risky occupation amongst the twelve occupations. The work content, work organization and working environments in these two countries are however not comparable.

Low back pain appears to be the greatest overall problem in both countries. On average, 74.8% of Thai and 55.4% of Chinese workers had problems with the low back, which caused sick leaves for 22.3% of Chinese, but only 6.7% of Thai workers. This could however, be explained by the differences in social security systems of these two countries. The frequency of the musculoskeletal problems is however dependent upon the exposure time, the type of task performed and the individual characteristics. For example, the incidence of problems was higher in the older group (31-40) of Thai workers with a longer working experience of up to 26 years.

Musculoskeletal disorders are a serious and common problem in most workplaces in IDCs. An urgent ergonomics intervention programme is required for minimising the workers suffering and loss of production as well as improving the quality of working life in Third World countries, a programme which should ensure active participation of workers, occupational health and safety personnel, as well as management representatives. Worker education, providing sufficient, easy, understandable information on musculoskeletal problems with regard to individual specific tasks and how to effectively prevent occupational hazards/injuries should be

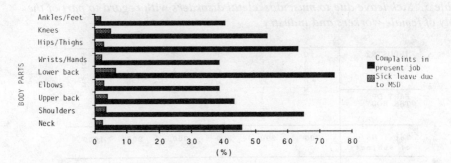

*Figure 2. The relationship between frequencies of musculoskeletal complaints and workers' sick leaves*

given high priority. In parallel ergonomics improvement of the workplace, working environment, as well as work organization, is required while considering workers' needs and preferences.

## References

Andersson G.B.T., 1981, Epidemiologic aspects on low back pain in industry. Spine, 6, 53-60.
Arbetsmiljokommission, 1990, Arbeten utsatta for sarskilda halsorisker. Arbetarskyddsstyrelsen, 171-84 Solna, Sweden.
Armstrong T.J., Foulke J.A. and Goldstein S.A., 1982, Investigation of cumulative trauma disorders in a poultry processing plant. American Industrial Hygiene Association Journal, 43, 103-116.
Arndt R., 1983, Working posture and musculoskeletal problems of video terminal operators. American Industrial Hygiene Association Journal, 44, 437-446.
Aronsson G., Oxenburgh M., Wennberg A. and Winkel J., 1989, Optimizing physical activity in VDU work. An intervention study. Report No. 1989:24, Arbetarskyddsstyrelsen, 171-84 Solna, Sweden.
Bammer G., 1987, VDUs and musculo-skeletal problems at the Australian National University - a case study. In Work with Display Units 86, B. Knave and P.G. Wideback (Eds), Elsevier Science Publishers, 279-287.
Biering-Soerensen F., 1982, Low back trouble in a general population of 30, 40, 50 and 60 year-old men and women. Danish Medical Bulletin, 29, 289-299.
Chaffin D.B., 1974, Human strength capability and low back pain. Journal of Occupational Medicine, 16, 248-254.
Gilad I. and Kirschenbaum A., 1988, Rates of back pain incidence associated with job attitudes and worker characteristics. International Journal of Industrial Ergonomics, 2, 267-277.
Hagberg M., 1984, Occupational musculoskeletal stress and disorders of the neck and shoulder. A review of possible pathophysiology. International Archives of Occupational Environmental Health, 53, 269-278.
Hildebrandt V.H., 1987, A review of epidemiological research on risk factors of low back pain. In Musculoskeletal Disorders at Work, P. Buckle (Ed), Taylor & Francis, 9-16
Kelsey J. L. et al., 1979, The impact of musculoskeletal disorders on the population of the United States. Journal of Bone and Joint Surgery, 51 A, 959-964.

Kilbom A., 1988, Intervention programmes for work-related neck and upper limb disorders, strategies and evaluation. Ergonomics, 31, 735-747.

Kilbom A. and Persson J., 1987, Work technique and its consequences for musculo-skeletal disorder. Ergonomics, 30, 272-279.

Klein B.P., Jensen R.C. and Sanderson L.M., 1984, Assessment of workers' compensation claims for back strains. Journal of Occupational Medicine, 26, 443-448.

Kuorinka I. et al., 1987, Standardized Nordic questionnaire for the analysis of Musculoskeletal Symptoms. Applied Ergonomics, 18, 233-237.

Maeda K., 1977, Occupational cervicobrachial disorder and its causative factors. Journal of Human Ergology, 6, 193-202.

Magora A., 1970, Investigation of the relation between low back pain and occupation. Industrial Medicine and Surgery, 39, 465-471.

Noddeland H. and Winkel J., 1988, Effects of leg activity and ambient barometric pressure on foot swelling and lower-limb skin temperature during 8hr of sitting. European Journal of Applied Physiology, 57, 409-414.

Snook S.H. and Webster B.S., 1987, The cost of disability. Clinical Orthopaedics, 221, 77-84.

Vincent M.St, Tellier C. and Lortie M., 1989, Training in handling. An evaluation study. Ergonomics, 32, 191-210.

Westgaard R.H. and Aaras A., 1984, Postural muscle strain as a causal factor in the development of musculo-skeletal illness. Applied Ergonomics, 15, 162-174.

Westgaard R.H. and Aaras A., 1985, The effect of improved workplace design on the development of the work-related musculoskeletal illness. Applied Ergonomics, 16, 91.

Wiker S.F., Chaffin D.B. and Langolf G.D., 1989, Shoulder posture and localized muscle fatigue and discomfort. Ergonomics, 32, 211-237.

# 17. THE BENEFITS OF ERGONOMICS INTERVENTION WITH PARTICULAR REFERENCE TO PHYSICAL STRAIN

David Stubbs and Peter Buckle

*Ergonomics Research Unit, Robens Institute*
*University of Surrey*
*Guildford GU2 5XH, Surrey, England*

## Abstract

In all respects the prevention or reduction of ergonomic mismatches resulting from physical and cognitive factors would seem to be beneficial to the individual, to organisations and to society as a whole. If this is the case, then what are the benefits and how should they be quantified? This paper attempts to address some of these issues with particular reference to musculoskeletal disorder, and emphasises the need for the evaluation of ergonomic intervention. A series of simple measures is proposed and should be seen as a minimum requirement. As more successful interventions are demonstrated then so the likelihood of longer term studies being undertaken will increase. These will enable a better understanding of the development of musculoskeletal disorders to be established which in turn will improve our ability to intervene more successfully to the benefit of all.

**Keywords:** *musculoskeletal disorders; ergonomics intervention; cost benefit analysis*

## *Introduction*

Prevention or reduction of ergonomic mismatch arising from physical, psychological and social factors would seem to be beneficial to the individual, to industry and to society. Musculoskeletal disorders affect very many people directly, have a major impact on industry and have both a direct and insidious effect on society. At the individual level, musculoskeletal consultations in the UK have increased by 40% over the last five years (Buckle and Stubbs, 1990). For society and industry 46.5 million working days were lost from back pain in 1989 and the cost was estimated at £2,000 million (Lapsley, 1990). Musculoskeletal disorders in Sweden amount to SK16,700 million (Kilbom, 1988) and repetitive strain injuries (RSI) claims in Australia amounted to $A400 million (O'Grady, 1985).

Interaction between the levels of organisation of such systems makes interpretation of such figures more difficult. For example, reductions in staffing levels resulting from injured/sick staff may increase the short-term efficiency (cost of production unit) of an organisation. This may be perceived of benefit to society until it is offset by cost of treatment and care of those injured/sick.

Concerning the aetiology of these disorders, whilst they are considered to be multifactoral, the relative importance of different causative/associated factors is still unknown. Andersson (1982) and Kilbom (1988) present a strikingly similar set of factors for back pain and work related upper limb disorders respectively, (ie. individual, work and psychosocial factors). Kilbom points out that the aetiological fraction (see Hernberg, 1984) of some common neck and shoulder disorders is high

(Hagberg and Wegman, 1987) and that it should be possible to prevent a relatively large proportion of these cases through appropriate ergonomic intervention. Similar arguments have been presented with respect to back pain (Stubbs et al., 1986).

It should be pointed out that the literature relating to the back pain and work related upper limb disorders area is substantial and it is not the intention of this paper to review these studies (see Jayson, 1987 for back pain, and Wallace and Buckle, 1987 for work related upper limb disorders).

## Interventions

The areas most influenced by ergonomics applications are shown in Table 1. Many of these overlap and interact (see Davids and Wall, 1990) and, in addition, some have a major influence on society.

Table 1. Areas influenced by the application of ergonomics

| Organisational | Health and Safety | Production |
|---|---|---|
| Industrial relations | Accidents | Unit costs |
| Management attitudes | Health (signs and disorders) | Labour turnover |
| Worker attitudes | Complaints of fatigue | Morale |
| Motivation | postural stress | Service/product |
| Commitment | unspecified illness | quality |
| Absenteeism | Sickness absence | Productivity |
| Training needs | Wellbeing | Late deliveries |
| | | Staffing levels |

Most ergonomic interventions are neither assessed nor designed with assessment in mind. This is unfortunate, as well controlled and successfully monitored intervention studies can provide models for subsequent implementation of ergonomics principles and, through the generation of improved data sources, better job design. Both help our understanding of how such disorders arise and/or can be reduced or prevented.

Intervention studies take many forms (Kilbom, 1988). These include: studying the same group of subjects before and after the intervention; randomly to assign subjects to either an intervention or a reference (control) group, and to make comparisons between groups. This latter design can often be combined with before and after comparisons in the intervention group. The difficulties encountered in all designs include: the need to control other factors (confounders, modifiers) that are not being manipulated; the time scale required; the problems of variation in staff turnover; the fluctuations in work output and the need to control or at least consider psychosocial variables.

Kilbom (1988) notes that the success or failure of an intervention depends on the effectiveness of the intervention technique ie. whether the intervention does lead to the expected changes and whether this change in work content is sufficient to influence the development of the disorder. She adds that for a full evaluation both steps should be assessed. Time scale is clearly important and many studies report only reductions in musculoskeletal discomfort in short-term evaluations (Ong, 1984). The uncertainty in the relationship between short- and long-term effects of work on the musculoskeletal system requires a cautious approach to the interpretation of such data and also those relating to, for example, sickness absence, accidents and compensation claims. Sickness may be a better measure of workload than of severity of disorder (Stubbs et al., 1986).

In many studies it is unlikely that the ergonomist will be asked to evaluate the intervention but rather to show how ergonomics principles can be applied to a particular industrial problem. If this is accepted, the manager then assumes that his/her problem has been solved. This presents a dilemma for the ergonomist as there is no measure of success or failure (ie. no feedback) and therefore no refinement/enhancement or addition to methods that can be added in future interventions. It is often, therefore, assumed that no feedback means the problem has been resolved, but not that the ergonomic intervention was necessarily the main factor. An additional problem may arise with some organisations who, even where evaluations of intervention are conducted, are unlikely to publish negative or positive effects externally for a number of reasons eg. admitting they had a problem in the first place, or giving away their competitive edge.

## Efficiency and cost benefit applications

Corlett (1988) considers that a total cost benefit assessment of ergonomic job design interventions is feasible although he acknowledges the difficulties. He argues that with the costs and costed benefits, it would be possible to recognise the importance to industry of better work, and to initiate some realistic social impact costings permitting policy decisions on the distribution of the costs between society, organisations and individuals (see also Nichols, 1989). Other reviews that have been conducted into the economic implications of workplace health promotion programmes, including back injury prevention, have identified the difficulty of making such assessments. Warner et al. (1988) concluded that, in general, the claims of health promotion programme cost benefits are based on anecdotal evidence or analyses seriously flawed in terms of assumptions, data or methodology. In relation to back injury programmes they report few health benefits, with the exception of ergonomic interventions. Their review excluded consideration of social costs and benefits that fell outside of the direct economic interests of business!

## What has been measured?

Several ergonomics case studies have shown what can be measured within the context of some of the variables noted earlier. Ong (1984), for example, studied the effects of changes in work space design and work organisation on data entry operators and found both an increase in productivity and a reduction in the symptoms of visual and muscular fatigue and discomfort. Performance as measured by keystrokes per hour increased by 25% and at the same time the error rate dropped from an overall 1.5% to 0.1%.

Clearly therefore, performance measures should be made together with changes in comfort and wellbeing, although where possible more objective measures should be employed (eg. clinically diagnosed conditions). Additionally, short-term and longer term effects have been considered important (Aaras, 1987). This study also highlighted the importance of establishing baseline data prior to an intervention. Such data are rarely routinely available or in the form required and are often difficult to obtain and to interpret retrospectively.

Dressel and Francis (1987) observed improvements in both productivity and employee satisfaction after the implementation of new workstations, and noted the added benefit in economic terms, the cost of providing the workstations was amortised due to space savings and increases in productivity in only 10.8 months. Whilst such a cost benefit argument appears attractive and is best illustrated by Spilling et al. (1986), a cautionary note is sounded by Eklund (1988) in that the greater part of the cost of injuries at work is paid by the community in general and the injured worker, not the organisation.

## The common experience

This raises the question of interventions that are not planned as such but with which many ergonomists are faced ie. where the ergonomist is invited into an organisation as a troubleshooter. Buckle (1990) presents a typical, though frequently unreported scenario, where an ergonomist is called in to solve a problem, in this case upper limb disorders among newspaper personnel. His paper addresses physical factors, psychological factors and work organisational issues, particularly copy flow and deadlines, which he considers of considerable importance in relation to physical and psychological stress. Through ergonomic analysis and user groups feedback and development, improvements in work organisation (notably in copy flow), in training and in ergonomics aspects of the physical work space layout were effected. Similarly, the needs of the sufferers were identified. These resulted in major initiatives being undertaken by the organisation. Whilst Buckle (1990) reports that such an investigation can be advanced along ergonomic lines he acknowledges that it may not be possible, upon completion, to directly attribute interventions to changes in illness rates, as control groups were not available. In addition, with several interventions being introduced simultaneously the identification of the most effective will be impossible.

This and other intervention studies (Mckenzie et al., 1985; Buckle and Stubbs, 1989) stress the need to consider those affected as a separate group from those unaffected. One final point that relates to this and other studies is the use of participatory ergonomics throughout the intervention process (see Shipley, 1990).

## Feasible measures and priorities

Measures of success of ergonomic interventions commence with whether an ergonomist has been asked to intervene, whether the intervention was accepted and user participation included and whether the intervention was implemented. A positive response to each of these questions would seem to imply some perceived need and benefit to the instigator of the intervention. Further measures might include whether the intervention has been evaluated through subjective assessment (eg. discomfort and pain). If possible this should be supported by pre-intervention responses and a control group. More objective measures could also be considered and reported, including short-term physiological variables and long-term health. Both are more complex and require longer term measures both before and after an intervention. The lack of baseline data on musculoskeletal disorders can be compensated for by using either general population health trends or data from other industrial populations, although difficulties exist with this approach (Buckle and Stubbs, 1990). As many musculoskeletal disorders are age and sex related, care must be taken in comparing rates before and after interventions and between data sets. The extent to which reductions in accidents or compensation claims arising from injuries are true reflections of ergonomic intervention through either workplace re-design or organisational changes is unclear. If, however, these are reported they should include before and after measures over a sufficiently long time scale.

There is also a need to use criteria separately for healthy/unhealthy workers and ensure that the prior and post intervention health status is available for each employee. Moving from health and safety issues to the issues of production and efficiency, the examples cited in this paper of improvements in performance (eg. computer data entry), are perhaps the easiest to quantify and report. It is, however, more difficult to assess improvements when considering for example, the quality of a service or of a product. Reductions in labour turnover should be reported, as should changes in staffing levels, although both are measures that may be influenced by many factors that are outside the control of either the ergonomist and perhaps even the organisation. Control data should be sought from national trends or from other comparable organisations/institutions if at all possible. The attitudes of workers to their jobs and the degree of satisfaction may also be assessed. Similarly, management response to the intervention should be monitored.

The economic justification for ergonomics has been considered by Simpson (1988) and every ergonomist involved in intervention and assessments of interventions should attempt to cost the effects of any improvements in health and safety and performance/productivity. Reductions in staffing levels may increase the productivity of an organisation which in turn may be of benefit to society arising from the increased taxes paid to governments. This may however be offset by increased unemployment and its associated costs to society and individuals.

## Conclusion

Clearly the benefits of ergonomic intervention with respect to musculoskeletal disorders have to be considered in relation to the individual, the organisation and society. Currently there is no model that considers all of these dimensions either separately or in an interactive way. In order to develop an understanding of how each can be influenced by ergonomic interventions more data are required and more reporting is needed. Simple measures of success, as outlined, should be considered a minimum requirement. As more successful interventions are demonstrated then so the likelihood of longer term studies being undertaken will increase. These will enable a better understanding of the development of musculoskeletal disorders to be established which in turn will improve our ability to intervene more successfully in the future, hopefully to the benefit of all.

## References

Aaras, A. 1987, Postural load and the development of musculoskeletal illness. PhD Thesis STK/Institute of Work Physiology, Oslo.

Andersson, G.B.J. 1982, Occupational aspects of low back pain. Proceedings of the International Symposium on Low Back Pain and Industrial and Social Disablement. Ed. M. Nelson. Back Pain Association, 45-51.

Buckle, P.W. 1990, Upper limb disorders among newspaper personnel. Presented at Occupational Disorders of the Upper Extremities, March 29-30, University of Michigan. Programme Director T.J. Armstrong.

Buckle, P.W. and Stubbs, D.A. 1989, The contribution of ergonomics to the rehabilitation of back pain patients. Journal of the Society of Occupational Medicine, 39, 2, 56-60.

Buckle, P.W. and Stubbs, D.A. 1990, Epidemiological aspects of musculoskeletal disorders of the shoulder and upper limbs. Contemporary Ergonomics. Ed. E.J. Lovesey. (London: Taylor & Francis), 75-78.

Corlett, E.N. 1988, Cost benefit analysis of ergonomic and work design changes. International Reviews of Ergonomics: 2. Ed. D.J. Oborne. (London: Taylor & Francis), 85-104.

Davids, K. and Wall, T.D. 1990, Advanced manufacturing technology and shopfloor work organisation. Irish Journal of Psychology (in press).

Dressel, D.L. and Francis, J. 1987, Office productivity: contribution of the workstation. Behaviour and Information Technology, 6, 3, 279-284.

Eklund, J. 1988, Hindrances to ergonomic improvements in industry. Proceedings of IEA Congress. Eds. A.S. Adams, R.R. Hall, B.J. McPhee and M.S. Oxenburgh. (London: Taylor & Francis), 181-183.

Hagberg, M. and Wegman, D.H. 1987, Prevalence rates and odds ratios of shoulder-neck diseases in different occupational groups. British Journal of Industrial Medicine, 44, 9, 602-610.

Hernberg, S. 1984, Work related diseases - some problems in study design. Scandinavian Journal of Work, Environment and Health, 10, 367-372.

Jayson, M.I.V. 1987, The lumbar spine and back pain. Churchill Livingstone, Edinburgh.

Kilbom, A. 1988, Intervention programmes for work related neck and upper limb disorders: strategies and evaluation. Ergonomics, 31, 5, 735-747.

Lapsley, 1990, The economics of back pain. Talkback, National Back Pain Association, 16.1.3.

McKenzie, F., Storment, J., VanHoote, P. and Armstrong, T. 1985, A program for control of repetitive trauma disorders associated with hand tool operations in a telecommunication manufacturing factory. American Industrial Hygiene Association Journal, 46, 11, 674-678.

Nichols, T. 1989, The business cycle and industrial injuries in British manufacturing over a quarter of a century. The Sociological Review, 37, 3, 538-550.

O'Grady, C. 1985, Clockwork jobs plague high tech office workers. The Computer Magazine, 12 September.

Ong, C.N. 1984, VDT workplace design and physical fatigue. A case study in Singapore. Ergonomics and Health in Modern Offices. Ed. E. Grandjean. (London: Taylor & Francis), 484-494.

Shipley, P. 1990, Participation, ideology and methodology in ergonomics practice. Ergonomics of Working Posture. Eds. E.N. Corlett, J. Wilson and I. Manenica. (London: Taylor & Francis), 819-834.

Simpson, G.C. 1988, The economic justification for ergonomics. International Journal of Industrial Ergonomics, 2, 157-163.

Spilling, S., Eitrheim, J. and Aaras, A. 1986, Cost benefit analysis of work environment investment at STKs telephone plant at Kongsvinger. Ergonomics of Working Posture. Eds. E.N. Corlett, J. Wilson and I. Manenica. (London: Taylor & Francis), 380-397.

Stubbs, D.A., Buckle, P.W., Baty, D., Fernandes, A., Hudson, M.P., Rivers, P.M., Worringham, C.J. and Barlow, C. 1986, Back pain in nurses: summary and recommendations. University of Surrey Press, England.

Wallace, M. and Buckle, P.W. 1987, Ergonomic aspects of upper limb disorders. International Reviews of Ergonomics: 1. Ed. D.J. Oborne. (London: Taylor & Francis), 173-200.

Warner, K. E., Wickizer, T.M., Wolfe, R.A., Schildroth, J.E. and Samuelson, M.H. 1988, Economic implications of workplace health promotion programs. Review of the literature. Journal of Occupational Medicine, 30, 2, 106-112.

# 18. WHAT IS AN ACCEPTABLE LOAD ON THE NECK AND SHOULDER REGIONS DURING PROLONGED WORKING PERIODS?

Arne Aaras

*Alcatel STK A/S*
*P.O. Box 60, Okern N-0508 Oslo 5*
*Norway*

**Abstract**

This paper is a summary of several papers published in different journals and conference proceedings. The contents deal with the incidence of load related musculoskeletal illness of groups of female workers exposed to different workloads. Based on the results of this epidemiological study, a prediction of acceptable load on the neck and shoulder is discussed.

**Keywords:** *musculoskeletal disorders; posture; static workload; electromyography; absenteeism*

## Introduction

A suggested threshold level for acceptable load on the musculoskeletal system of the neck and shoulder regions should consider both intensity and duration of such a load in order to assess the work pattern. In addition, the duration of very low load levels (micropauses) could also be assumed to be important for recovery of muscle function. The influence of spontaneous pauses on reducing absenteeism has been discussed by Grandjean (1980). Thus, in order to determine a threshold level for acceptable load on the neck and shoulder all the above parameters influencing the load need to be considered.

However, acceptable load should furthermore be related to health criteria such as acceptable limits for development of musculoskeletal illness among a group of workers. Factors influencing the incidence of such illness, such as working hours per day, total time of employment, age and gender within the group, must all be taken into account when establishing acceptable load levels.

Defining an acceptable load according to health criteria such as the incidence of musculoskeletal sick leave must be based on the prerequisite that there is a relationship between the load due to the work and the development of musculoskeletal illness (Aaras, 1987b). In order to explore a workload/musculoskeletal injury relationship, comparable groups of female workers with different development of musculoskeletal sick leave were studied. Furthermore, to assess the influence of prolonged workload periods on the incidence of musculoskeletal illness, the above groups were compared with a control group of female workers without continuous workload.

## Methods

### Measurements of load on the musculoskeletal system

The load on the neck and shoulder muscles was quantified by electromyography (EMG) from the descending part of m. trapezius. Workload on local body structures was also assessed by recording body movements and posture at the workplace. The EMG and the postural angle methods are described and the methodological limitations discussed by Ericson et al. (1978), Jonsson (1982), Aaras et al. (1987a), Aaras et al. (1988a, 1988b & 1988c). For the measurements of both EMG and postural angles, we used the Physiometer (Premed A/S, Oslo).

### Estimation of risk of injury

It was decided to use musculoskeletal sick leave for quantification of injury, since such statistics give reliable information concerning both the development of musculoskeletal sick leave and the variation of such sick leave as a function of time. The procedure is described by Aaras (1987b). Musculoskeletal sick leave is also well-defined in terms of consequences for the company. In particular, it is possible to indicate the financial cost due to such sick leave (Spilling et al., 1986).

### Statistics

The statistical methods are described in detail by Aaras (1987b) and Aaras et al. (1988c).

## Results

### Sick leave statistics

The study covered a long epidemiological investigation of comparable groups of female workers who carried out different electromechanical assembly tasks (8B system, cableforming, Minimat system, 10C system and 11B system) (Westgaard et al., 1984).

Workers at the 8B system, Minimat system and cableforming had a statistically significant higher incidence of musculoskeletal sick leave compared to workers at the 10C and 11B systems with time of employment between 0-2 years (about 25% vs 0%) and 2-5 years (about 55% vs 20%) (p.<0.01) (Aaras et al., 1987a). For the workers employed for more than 5 years, a statistically significant difference was found between workers at the 8B system and workers at the 10C/11B systems (about 65% vs 30%) (p.<0.05).

The influence of the workload on the musculoskeletal system was assessed to be located mostly in the shoulder and neck regions. The development of musculoskeletal illness, resulting in sick leave, was for these body areas for workers at the 8B system and 10C/11B systems, 0-2 years (18% vs 0%), 2-5 years (45% vs 20%) and more than 5 years (50% vs 28%) respectively (Aaras et al., 1987a). Workers at the 10C and 11B systems are considered to be a homogeneous group as the EMG recordings showed nearly identical trapezius load for the workers at the two systems and the development of musculoskeletal sick leave was very similar. The workers at the 8B system had a significantly higher incidence of musculoskeletal sick leave than those workers employed at the 10C/11B systems, considering the period of employment between two and five years (p.<0.05) (Aaras et al., 1987a).

The development of musculoskeletal sick leave for workers at the 10C/11B systems was higher than the incidence of such illness, regardless of body location, in a control group. In the latter group, only one of 35 workers had had musculoskeletal sick leave when working less than 5 years. For those employed more than 10 years, 30 to 40% of the group had been on musculoskeletal sick leave (Westgaard et al., 1984). This percentage is similar to the incidence of such sick leave for workers at the 10C/11B systems who were employed more than 5 years. The control group consisted of females with a great variety of administrative and clerical work, indicating that the musculoskeletal load from the work was not a continuous one. Thus, the load on the musculoskeletal systems of the workers at the 10C/11B work tasks seem to approximate an acceptable load level, reflected by a near similar incidence of musculoskeletal sick leave for workers at the 10C/11B systems and the workers in the control group.

## Clinical symptoms and signs of musculoskeletal illness

When suggesting an acceptable workload on the musculoskeletal system, information about health effects or complaints other than musculoskeletal sick leave is also important. Such parameters are intensity and duration of pain from the musculoskeletal system. Assessed by such parameters, the effect of the load on the musculoskeletal system for workers at the 10C/11B systems seems unacceptable. Every fourth worker at these systems experienced pain at an intensity level necessitating pauses during working hours over the last year before the interviews in 1984. Every fifth worker reported that the pain did not disappear during pauses and the pain made it difficult to carry out work. The duration of such periods with pain was more than 30 days in total over the last year before the interviews in 1984 (Aaras, 1990a). These results were further supported by the fact that every third worker reported continuous tenderness or pain after cessation of isometric and endurance muscle tests which were performed during clinical examination at the time of the interviews (Aaras, 1990a).

The work groups employed at the different electromechanical work systems were comparable with regard to age, gender, time of employment and working hours per day (Aaras, 1987b). However, psychosocial problems and activities outside work may have influenced the development of musculoskeletal illness for the individual and at the group level. However, a coarse quantification of some important psychosocial factors, spare time activities and living habits of the workers, showed no statistically significant difference with regard to these factors between the groups in 1984 (Aaras, 1984, unpublished). Other situational variables, such as the foremen and the factory layout, were also unchanged during the period of investigation (1967-1983). Therefore, the confounding factors discussed above probably had a similar influence on each work group in terms of development of musculoskeletal illness, making the groups equivalent when comparing the relevant variable; load on the musculoskeletal system.

## The workload

Postural load clearly influenced the development of musculoskeletal sick leave. The median value across the group of static trapezius load was similar for workers employed at the 8B, cableforming and the Minimat systems, 4.3 to 5.5% MVC, regardless of sitting or standing work positions (Aaras et al., 1987b). For the 10C and 11B systems, the workers showed a much lower median group value of static muscle load for the predominantly standing work position, 1 to 1.3% MVC, compared with those workers employed at the 8B, cableforming and Minimat systems. More than 75% and 50% respectively of the work at the 11B and 10C systems was performed in a standing position (Aaras et al., 1987b), which created less load on the shoulder compared with a seated posture. The median group value of static trapezius load, for sitting position on the 10C and 11B, was 3.8% MVC and 4.1% MVC respectively (Aaras et al., 1987a). The median and peak loads on the m. trapezius were low for all systems and did not normally exceed 11 and 22% MVC, respectively.

The mechanical workload on the shoulder was also assessed by measuring flexion/extension and abduction/adduction in the gleno-humeral joint as well as head flexion. The abduction/adduction of the upper arm in the gleno-humeral joint were small for all systems (Aaras et al., 1988b). Thus, when considering the shoulder loading in terms of the postural angles in the shoulder joint, the difference between the 8B system and the 10C/11B systems was mainly in the values of flexion and extension.

Workers at the 10C/11B systems had slightly less upper arm flexion, compared with those who worked at the 8B system for the predominant work position. It was also a clearly different pattern of movement of the upper arm in the sagittal plane for most workers of the 10C/11B systems vs. the 8B system. The range of the median flexion (i.e. values of the angles for 50% of the recording time) was for 10C ($-14°$ to $8°$), 11B ($-11°$ to $15°$), and 8B ($7°$ to $44°$) respectively (Aaras et al., 1988b). Thus, for the 8B system, workers showed only flexion in the shoulder regardless of the working position, while most workers at the 10C/11B systems performed their work with both flexion and extension of the upper arm. That means that workers at the 10C/11B systems reduced the period of activity of flexors and extensors of the shoulder joint by distributing the workload between the two muscle groups in standing. In addition, the hand tool used by workers at the 10C/11B systems had a weight of 0.5 kg less than the one used by workers at the 8B system.

By studying the pattern of movement of the upper arm in the sagittal plane, the operators at the 10C/11B systems had a more dynamic movement pattern compared to the workers at the 8B system (Aaras et al., 1988c).

The head flexion influenced the trapezius load significantly less than arm position. The workers at the 10C and 11B systems had greater head flexion compared with the 8B system. The measurements showed a variation of the median flexion of $39°$ to $58°$ for 10C, $15°$ to $48°$ for 11B and $9°$ to $31°$ for 8B respectively (Aaras et al., 1988b).

**The shoulder moment and the musculoskeletal illness**

The shoulder moment seems clearly to influence the incidence of musculoskeletal sick leave. Table 1 shows the recorded postural angles of the upper arm in the gleno-humeral joint for workers at the 8B, 10C and 11B systems. In addition, a group of female workers performing electromechanical assembly work at the DF system is included. The median value of the flexion angles in the elbow joint was estimated for different systems from video recordings (see Table 2). The recordings were taken from the side (perpendicular to the sagittal plane of the operators) and straight from behind the workers (in the sagittal plane).

The importance of the weight of the hand tool regarding the load on the shoulder and neck regions is documented by biomechanical calculations both for static and dynamic movements by Ashton-Miller (1986). According to his studies, the shoulder moment increases dramatically by increasing the weight of the hand tool. Furthermore, Sigholm et al. (1984) have documented increasing activity in the shoulder muscles (Deltoid, Infraspinatus, Supraspinatus and Upper Trapezius) by increasing the hand load both for flexion and abduction movements of the upper arm in the shoulder joint. Table 3 shows the weight of the hand tool for the different work systems.

Table 1. Postural angles of the upper arm in the gleno-humeral joint

| Work system | No. of subjects | Flexion/Extension median | range | Abduction/Adduction median | range |
|---|---|---|---|---|---|
| 8B* sitt | 6 | 19° | 7°-44° | 8° | 5°-14° |
| 8B* stand | 6 | 17° | 8°-43° | 5° | -3°-12° |
| 10C* sitt | 3 | 8° | 0°-28° | 7° | 0°- 8° |
| 10C* stand | 3 | -11° | -14°- 8° | 7° | 2°-19° |
| 11B* sitt | 2 | 11° | 8°-14° | -1° | -4°- 2° |
| 11B* stand | 4 | 13° | -11°-15° | -4° | -8°- 2° |
| DF sitt | 6 | 39° | 11°-58° | 60° | 23°-77° |

\*   Aarås, Westgaard and Stranden, 1988

Table 2. Flexion angle in the elbow joint

| Work system | Median flexion |
|---|---|
| 8B sitt | 80° |
| 8B stand | 75° |
| 10C sitt | 70° |
| 10C stand | 90° |
| 11B sitt | 70° |
| 11B stand | 70° |
| DF sitt | 90° |

Table 3. Weight of the hand tool

| Work system | Weight (N) |
|---|---|
| 8B | 8,5 |
| 10C | 3,5 |
| 11B | 3,5 |
| DF | 0,2 |

A rough calculation of the static shoulder moments (Nm) was performed from the recorded and estimated **median** postural angles of the arm, in the shoulder and elbow joints for different work systems. Thus, the values of the shoulder moments are the values for 50% of the recording time and are shown on the vertical axis in Figure 1. The time in the sitting and standing position as a percentage of total work time is indicated on the horizontal axis. The calculations were based only on the group median value of the flexion/extension angles in the shoulder joint except for another telephone exchange system (the DF system), where abduction was included in the shoulder moment calculations. Lacking data on body dimensions for each subject in the group, we used anthropometric information from Humanscale (Diffrient, 1987), regarding the values for centre of gravity locations and body segment weights for medium build female workers. A body weight of 60 kg was used in all calculations.

The employees at the 8B and DF systems had to perform work with a higher shoulder moment (5Nm), compared with workers at the 10C and 11B systems (7Nm). Workers at the DF system performed electromechanical assembly work in different plant from the rest of the workers. However, the groups were comparable in terms of age, working hours per day and time of employment. Furthermore, in all groups the psychosocial factors seemed not to be an important contributor to the development of musculoskeletal sick leave. Serious psychosocial problems were reported by less than a quarter in the groups (Aaras, 1990a; Westgaard et al., 1982). The incidence of musculoskeletal sick leave was very similar for workers at the 8B system and operators at the DF system, regarding period of employment less than 5 years (Westgaard et al., 1986). For workers employed for more than 5 years, the incidence of musculoskeletal sick leave was higher for workers at the 8B system compared with operators at the DF work task. About 65 percent of total workers at risk had been on musculoskeletal sick leave at the 8B system. The corresponding value for the DF workers was about 45 percent.

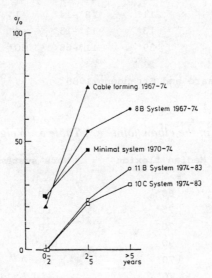

Figure 1. The static shoulder moment is calculated on the basis of the group median value of postural angles of the upper arm in the gleno-humeral joint and flexion in the elbow joint for different work systems on the Y-axis. Time required to maintain these moments as a percent of total work time is indicated for sitting and standing on the X-axis

The influence of the weight of the hand tool on the shoulder moment is clearly demonstrated by comparing workers of the DF and the 8B systems. Table 1 shows that both the median group flexion and abduction of the upper arm in the gleno-humeral joint were greater for workers at the DF system compared to the same angles for workers at the 8B system. However, the latter group had a hand tool of about 850 g compared with workers at the DF task, where the hand tool had a weight of only about 20 g. The shoulder moment for workers at the two systems was similar.

## Low trapezius activity and shoulder moment

In order to explore further the work pattern of the different work tasks, an analysis of EMG activity recorded from m. trapezius was undertaken to establish any evidence of low muscle activity. It can be argued on a physiological basis that pauses or short breaks in the muscle load pattern are necessary for recovery of the muscle function.

In the predominant work position, the total duration of seconds per minutes of low level periods for workers at the 8B system as a group median value was: 0.4 (% MVC), 1.6 (% MVC), and 10.1 (% MVC) which is less than for workers at the 10C system: 0.9 (% MVC), 3.9 (% MVC) and 34.8 (% MVC) or for 11B workers: 3.0 (% MVC), 7.6 (% MVC) and 27.1 (% MVC) (Aaras et al., 1988c; Aaras, 1990b).

Low postural angles between $5°$ flexion and $5°$ extension in the shoulder joint are associated with low moment about the shoulder joint (Sigholm et al., 1984). For the predominant working position the duration of activities in seconds per minute for flexion/extension less than $5°$ was as a group median value: 0.5 (8B) vs. 13.7 (10C) and 5.6 (11B) (Aaras et al., 1988c). These results reflect the high static loading for workers at the 8B system which produces fewer breaks in the high activity periods compared to workers of the 10C/11B systems.

In addition, real breaks in the load pattern in terms of spontaneous pauses varied individually for workers at the 10C and 11B systems and amounted to 30 to 60 minutes for each worker, for an eight hour work day (Aaras, 1987b).

## Important stress factors outside plantwork

A discussion of acceptable muscle load must also consider other factors which influence the load on the musculoskeletal system and the incidence of musculoskeletal illness (Kvarnstrom, 1983). These include the workload outside the plant, psycho-social problems, spare time activities and living habits (Ogar, 1977; Kogstad et al., 1978). In the present study, workload at home and psychological problems were the only two factors considered as potential contributors to the overall load on the musculoskeletal system. Much work at home and indications of serious psychological problems were reported by 20% of the workers. But a combination of these factors was experienced only by 5% of the workers. Factors considered to strengthen the musculoskeletal system, such as sport and other recreational and leisure activities, were practised only occasionally by less than 35% of the workers.

## *Discussion*

This investigation strongly supports a postural load/musculoskeletal injury relationship, making it possible to relate the workload to the incidence of musculoskeletal illness. A relationship between the incidence of musculoskeletal sick leave and static trapezius load has been confirmed as a median value for groups of female workers (Aaras et al., 1987a). Furthermore, a postural load/musculoskeletal injury relationship was found as a group median value for the number and the duration of periods with trapezius activity below 1% and 2% MVC and the development of musculoskeletal illness (Aaras et al., 1988c). Low trapezius load level i.e. below 5% MVC is assumed to be important in reducing the development of musculoskeletal illness (Aaras, 1990b).

The shoulder moment in terms of the magnitude of flexion/extension of the upper arm in the shoulder joint, the number and duration of very low postural angles (an arm position between $+5°$ (flexion) and $-5°$ (extension), the distribution of muscle load between subgroups of muscles such as flexors and extensors and the dynamic pattern of work are factors assumed to influence the incidence of musculoskeletal illness (Aaras et al., 1988c).

The work pattern in terms of the amount of dynamic or static work, may be an important load variable, influencing the development of musculoskeletal illness. Dynamic work pattern increases work capacity compared with static work (Hagberg, 1981). Furthermore, Kilbom et al. (1986) documented that a more dynamic work pattern was negatively related to neck and shoulder symptoms.

According to the load related parameters discussed above, the loading of the substructures of the shoulder joint was therefore less for workers at the 10C/11B systems than operators at the 8B system. There was also clear evidence of less musculoskeletal sick leave for workers at the 10C/11B systems compared with those employed at the 8B system. Other factors influencing the musculoskeletal sick leave had similar influence on each group. The load on the upper part of the body for workers at the 10C/11B systems seems to approximate an acceptable threshold level, reflected by nearly the same development of musculoskeletal illness as the control group. The incidence of musculoskeletal sick leave for these two groups was less than 20% and 30% for workers employed between 2 and 5 years and more than 5 years respectively. However, every fifth worker from the two groups reported serious pain both in terms of intensity and duration, indicating that the load on the neck and shoulder regions may still be higher than acceptable. However, the values of the load parameters for workers at the 10C/11B systems given below, seem to approximate a rough indication of an acceptable trapezius load, upper arm position and rest periods. The **static** trapezius load was as a group median value about 1 to 2% MVC for most of the work day. The total duration of trapezius load was below 5% MVC for about 40% of the total work time. The trapezius load as a group median was below 1% MVC (micropauses) for less than 5% for the major part of the working day.

Shoulder loading assessed by measuring the median flexion/extension of the upper arm in the gleno-humeral joint (i.e the value of the angle for 50% of the recording time) was usually less than $15°$, while the abduction/adduction was mostly less than $10°$ (Aaras et al., 1988b). The duration of low shoulder moment, defined as flexion/extension of the upper arm in the gleno-humeral joint between $+5°$ and $-5°$ (as a median group value), was mostly less than 20% of the work time (Aaras et al., 1988c).

The results from a rough calculation of the static shoulder moments in this study indicate that these moments should be low (Nm for females) for the predominant part of the work day (Figure 1). Static shoulder moments above 5Nm for the major part of the work day seem unacceptable for female workers, reflected by the incidence of musculoskeletal sick leave at the 8B and DF systems. The great difference in shoulder moment for workers at the 10C and 11B systems in standing was unexpected. However, the workers at the 11B system had a much more dynamic work posture due to the fact that they had to perform work at the top of the 100 cm high work piece, then the next moment kneel down to work at the extreme lower parts (Aaras et al., 1988b). This variety of work positions for operators at the 11B system may contribute to the lower incidence of musculoskeletal sick leave, making it almost similar to recorded incidence for workers at the 10C system. Considering female workers with a more static work posture of the lower part of the body such as for operators at the 10C system, the static shoulder moment was very low (Nm as a group median value), for most of the workday. These workers had a nearly acceptable incidence of musculoskeletal sick leave. The importance of keeping the weight of the hand tool at a minimum can be deduced from the results in Tables 1 and 3. Two groups of workers (DF and 8B) had a similar incidence of musculoskeletal sick leave for time of employment less than 5 years. The operators in the DF group worked with high flexion/abduction angles of the upper arm in the gleno-humeral joint, but the hand tool had a very small weight (0.2N). The opposite was found for workers at the 8B system. The postural angles of the upper arm in the gleno-humeral joint were smaller for workers at this system, but they had to perform work with a much heavier hand tool (8.5N). The results indicate that the two variables, the magnitude of the postural angles of the upper arm in the gleno-humeral joint and the weight of the hand tool, influenced the magnitude of the shoulder moment and the incidence of musculoskeletal sick leave.

Real breaks in the load pattern in terms of spontaneous pauses varied for individual workers and amounted to 30 to 60 minutes for each worker in an eight hour working day (Aaras, 1987b).

Workers at the 10C system had considerable median flexion of the head (between $40^{\circ}$ to $60^{\circ}$) which seems to have a small influence on the load of the neck/shoulder and the development of musculoskeletal illness in the same regions.

The aetiology of load related musculoskeletal illness is considered to be multifactorial (Kvarnstrom, 1983). Less than a quarter of the workers reported other factors which might be considered to influence the incidence of musculoskeletal illness, such as workload outside the plant or serious psychological problems.

## Conclusion

For groups of female workers, a quantitative relationship has been established between the postural load, in terms of median value across the group, of static trapezius load, trapezius load below 5% MVC and shoulder moment, and the development of musculoskeletal sick leave related to time of employment. This postural load/musculoskeletal injury relationship makes it possible to relate the discussion of an acceptable postural load to what is acceptable incidence of musculoskeletal illness among a group of workers. The results from this study suggest that the load on the upper part of the body for two groups of workers seems to approximate an acceptable threshold load level, reflected by nearly the same development of musculoskeletal illness as the control group without continuous work load. However, in these groups many workers suffered serious pain and clinical symptoms and signs of musculoskeletal illness were found during interviews and clinical examination. This result may indicate that the postural load was higher than acceptable. Thus, the postural load/workload measured and estimated for the two groups of workers (10C/11B) underlines the importance of reducing the static muscle load to a munimum. On the other hand, increasing the amount of dynamic muscle work may be beneficial for the musculoskeletal system.

## References

Aaras, A., 1984, Psycho-social factors, sparetime activities and living habits of workers at STK's telephone plant in Kongsvinger, (Unpublished).

Aaras, A. and Westgaard, R.H., 1987a, Further studies of postural load and musculo-skeletal injuries of workers at an electromechanical assembly plant. Applied Ergonomics, 18, 3: 211-219.

Aaras, A., 1987b, Postural load and the development of musculo-skeletal illness. Scandinavian Journal of Rehabilitation Medicine, Supplement No 18.

Aaras, A. and Stranden, E., 1988a, Measurement of postural angles during work. Ergonomics, 31, 6: 935-944.

Aaras, A., Westgaard, R.H. and Stranden, E., 1988b, Postural angles as an indicator of postural load and muscular injury in occupational work situations. Ergonomics, 31, 6: 915-933.

Aaras, A., Westgaard, R.H., Stranden, E. and Larsen, S., 1988c, Postural load and from the incidence of musculo-skeletal illness. In: Proceedings from the NIOSH workshop "Promoting Health & Productivity in the Computerized Office" (London: Taylor & Francis Ltd.), (In press).

Aaras, A., 1990a, Load related musculo-skeletal illness: Is ergonomic workplace design a sufficient remedy? In: Work Design in Practice, edited by C.M. Haslegrave, J.R. Wilson, E.N. Corlett and I. Manenica, (London: Taylor and Francis Ltd.).

Aaras, A., 1990b, Acceptable muscle load on the neck and shoulder regions assessed in relation to the incidence of musculo-skeletal sick leave. International Journal of Human-Computer Interaction, 2, 1: 29-39.

Ashton-Miller, J.A., 1986, Department of Mechanical Engineering and Applied Mechanics, University of Michigan, U.S.A. Personal communication.

Diffrient, N., Tilley, A.R. and Bardagjy, J.C., 1987, Humanscale 1/2/3. The MIT Press, Massachusetts Institute of Technology, Cambridge, Massachusetts.

Ericson, B.E. and Hagberg, M., 1978, EMG signal level versus external force: a methodological study on computer aided analysis. In: Biomechanics VI-A, edited by E. Asmussen and K. Jorgensen, (Baltimore: University Park Press), pp 251-255.

Grandjean, E., 1980, Fitting the task to the man. (London: Taylor & Francis Ltd.).

Hagberg, M., 1981, Electromyographic signs of shoulder muscular fatigue in two elevated arm positions. American Journal of Physical Medicine, 60, 111-121.

Jonsson, B., 1982, Measurment and evaluation of local muscular strain in the shoulder during constrained work. Journal of Human Ergology, 11, 73-88.

Kilbom, A., Persson, J. and Jonsson, B. 1986, Disorders of cervicobrachial region among female workers in the electronics industry. International Journal of Industrial Ergonomics, 1, 37-47.

Kogstand, O. and Hartvig, P., 1978, The influence of psychiatric and social factors in disablement benefit of rheumatological illness. The Journal of the Norwegian Medical Association, 98, 943-945 (In Norwegian).

Kvarnstrom, S., 1983, Occurrence of musculo-skeletal disorders in a manufacturing industry, with special attention to occupational shoulder disorders. Scandinavian Journal of Rehabilitation Medicine, Supplement No 8.

Ogar, B., 1987, Patients treated by general practitioners in Norway. Universitetsforlaget Oslo-Bergen-Tromso (In Norwegian).

Sigholm, G., Herberts, P., Almstrom, C. and Kadefors, R., 1984, Electromyographic analysis of shoulder muscle load. Journal of Orthopaedic Research, 1, 379-386.

Spilling, S., Aaras, A. and Eitrheim, J., 1986, Cost-benefit analysis of work enviroment investment at STK's telephone plant at Kongsvinger. In: The Ergonomics of Working Postures, edited by N. Corlett, J. Wilson and I. Manenica, (London: Taylor & Francis Ltd.).

Westgaard, R.H., Jansen, T.H. and Aaras, A., 1982, Development of musculo-skeletal illness among females working with static muscle load. Institute of Work Physiology (In Norwegian).

Westgaard, R.H. and Aaras, A., 1984, Postural muscle strain as a causal factor in the development of musculo-skeletal illness. Applied Ergonomics, 15, 3: 162-174.

Westgaard, R.H., Woersted, M., Jansen, T. and Aaras, A., 1986, Muscle load and illness associated with constrained body postures. In: The Ergonomics of Working Postures, edited by N. Corlett, J. Wilson and I. Manenica, (London: Taylor & Francis Ltd.).

# 19. PERCEIVED WORKLOAD AND MUSCULOSKELETAL STRAIN AMONG HEALTH CARE WORKERS: A FOUR-YEAR FOLLOW-UP STUDY

Tuulikki Luopajarvi and Clas-Hakan Nygard

*Institute of Occupational Health*
*Helsinki, Finland*

**Abstract**

The aims of the study were to examine the perceived workload and musculoskeletal strain of 981 middle-aged women in eight health care occupations and to follow the development of the problems during four years. Data were collected with a postal questionnaire. The feeling of responsibility for people and use of information were reported as load factors by 80 - 90% of all subjects. In addition, many physical, work equipment and environment hazards, poor working hours and problems in the organizational and social environment were reported. The perceived workload increased in all occupations during the four years. Only 4% of the women did not report any musculoskeletal strain symptoms in either enquiry, whereas 4% reported much strain in both examinations. Neck and shoulder pain was the most common strain symptom in all occupational groups. During the four years, the strain symptoms had become less severe or were cured in 5% of the subjects. The incidence of new strain symptoms or old ones worsening was 18%. The perceived dynamic workload and poor work postures correlated significantly ($p.<0.001$) with musculoskeletal strain, and perceived responsibility with neck pain among bath attendants. Ergonomics programmes should cover the organization and methods of work, as well as the physical environment. Quality of information should be improved in order to decrease the responsibility load in health care work.

**Keywords:** *physical workload; musculoskeletal disorders; health care; posture; nursing; stress; back pain*

## Introduction

Today, at least in Finland, the continual shortage of staff and the growing number of elderly and disabled workers have become big problems in the health care service. The content of work and working conditions have therefore become important.

Health care involves a number of different jobs with specialized tasks. The work differs fundamentally from industrial and most other types of work, but in addition, the occupations and their content also differ from each other. In health care work back disorders caused by lifting and manual handling of patients have been the generally known health problems, because they cause most sick leaves and require most care.

Workload has different effects on different people (Hildebrandt, 1987). When the quality and quantity of the load are right, the load is beneficial. Overload of short duration may cause strain, fatigue and other similar temporary symptoms. If the situation continues for a long time, diseases or even structural changes may occur.

Today, local or holistic underload is also encountered in many jobs. The demands of the work are so low that employees need only a limited amount of their capacity, which in time may decrease. This can be especially harmful for older workers, who already suffer from arthrosis or other skeletal problems (Andersson, 1984).

In order to achieve ergonomic improvements and to prevent overload, the exposures and hazardous factors should be known as well as possible. In addition to the ergonomic defects which can be observed, and for which recommendations are available, the perception of workload is extremely important for one's well-being and motivation at work.

The aims of this study were:
1. to examine the perceived workload among women in eight health care occupations
2. to examine the extent to which there were differences in the preceived musculoskeletal strain among women
3. to describe the development of musculoskeletal disorders during four years
4. to analyze the role of reported work hazards for planning preventive ergonomics measures.

## Subjects and methods

The study of health care workers' perceived workload and musculoskeletal disorders is part of a multidisciplinary four-year follow-up project aimed at determining the retirement ages in municipal occupations (Ilmarinen, 1987). Data were collected with a postal questionnaire, which included both standardized and open questions. The first enquiry was carried out between December 1980 and March 1981, and the follow-up inquiry during February to August 1985.

*Table 1. Age and occupations of the subjects in the follow-up study*

| Occupational groups in 1981 | 1985 N | % | Age, years X | range |
|---|---|---|---|---|
| Orderly, n=240 | 160 | 67 | 54 | 48-60 |
| Bath attendant, n=155 | 85 | 55 | 53 | 49-59 |
| Reg. nurse, n=327 | 224 | 68 | 53 | 49-60 |
| Assist. nurse, n=280 | 194 | 69 | 53 | 59-59 |
| Psych. nurse, n=97 | 58 | 60 | 53 | 49-57 |
| Childr. nurse, n=88 | 61 | 69 | 53 | 49-57 |
| Sister, n=171 | 130 | 76 | 54 | 49-62 |
| Matron, n=83 | 69 | 83 | 53 | 49-59 |
| Total 1441 | 981 | 68 | 53 | 48-62 |

The subjects were middle-aged female health care workers, who in 1981 had continued in their present occupation for the previous five years in hospitals and municipal health care centres in Finland. In 1981 their mean age was 50 (range 44 - 58) years. The follow-up analysis in 1985 comprised those who still continued in the same occupation (Table 1). From the 460 drop outs in 1985, 323 had retired, 49 had changed occupation, 13 had died and 75 did not answer the second enquiry. In 1985, the subjects were divided into three subgroups: physical work (n=245), mental work (n=199) and mixed work (n=537).

## Results

The perceived workload was surveyed by 58 questions which covered the following aspects: work environment hazards, strain and heaviness of work, content of work tasks, job satisfaction, and social relations and management.

Factor analysis was used in the analysis of the data to combine separate questions. The reliability was calculated by Cronbach's alpha or correlation coefficients.

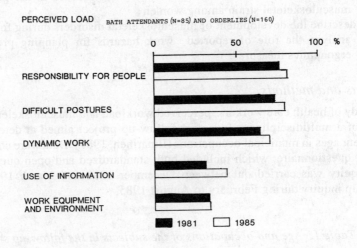

Figure 1. The perceived workload in the physical work group

**Perceived workload**

The most common perceived workload factor in all groups was the feeling of responsibility either for the patients or the staff. In addition, the orderlies and bath attendants reported many physical hazards, and about one third of them felt that the use of information, work equipment and the environment were strain factors in their work. The perceived information load had increased about 10% during the four years (Figure 1).

Every sixth nurse in the category of various nurses stated that the use of information, heavy dynamic work, and poor work postures were prominent load factors. The amount of perceived physical work had increased more than the other load factors. In both enquiries, about 40% of the nurses also complained about the poor working time (Figure 2).

The sisters and matrons felt responsibility mainly for the staff. In addition, they reported much handling of information and, unlike the others, they complained about sitting, as well as problems in their organizational position and social environment (Figure 3).

The perceived workload had increased in all groups during the four years.

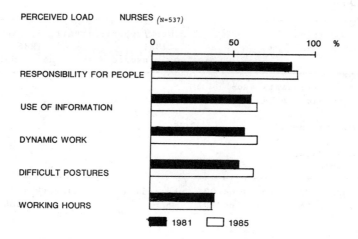

Figure 2. The perceived workload in the mixed workload group

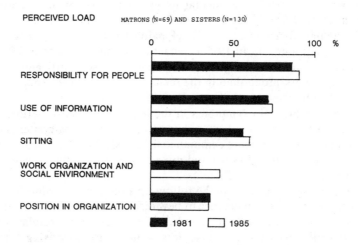

Figure 3. The perceived workload in the mental workload group

## Musculoskeletal strain

The musculoskeletal strain was investigated by 10 questions. In order to increase the validity, the following factors were formed from the data: the neck/upper extremities, the back and lower extremities. The values of the factors were as follows: 0 = no strain symptoms, 1 - 5 = some strain symptoms, 6 - 10 = numerous strain symptoms.

Only 4% of the women did not report any musculoskeletal strain symptoms in either enquiry, whereas 4% reported a considerable amount in both examinations.

Table 2. The % of subjects reporting strain symptoms in the neck, shoulders, arms and hands

| Occupational group | Neck/upper extremity symptoms, % | | | |
|---|---|---|---|---|
| | 1981 | | 1985 | |
| | no | numerous | no | numerous |
| Bath attendants and orderlies, n=245 | 13 | 13 | 8 | 25 |
| Nurses, n=537 | 26 | 7 | 12 | 13 |
| Sisters and matrons, n=199 | 32 | 3 | 22 | 5 |

Chi-square 22.45 df= 1 p<.001, physical and mental work 1981
Chi-square 38.27 df= 1 p<.001,          -"-                 -"-         1985

In 1981, the prevalence of constant neck pain varied from 3% (sisters) to 19% (bath attendants). In 1985, constant neck pain was reported by 5% (sisters) to 35% (bath attendants). Strain in the neck and arms was significantly more common among the physical work group than the mental group. Nine out of ten orderlies and bath attendants suffered from these complaints to some extent (Table 2).

In 1981 the prevalence of constant back pain varied from 2% (sisters and matrons) to 19% (assistant nurses). In 1985 constant back pain varied from 7% (matrons) to 24% (bath attendants). Although the back strain was also significantly more common in the physical work group than in the mental group, the differences were not quite so big and they increased less during the follow-up (Table 3).

The prevalence of strain symptoms in the legs varied from 3% (mental group) to 15% (orderlies) in 1981, and from 6% (matrons) to 16% (bath attendants) in 1985. Also strain in the legs was more common in the physical work group than in the mental group. However, in the physical and mixed work groups the amount of strain had remained almost unchanged (Table 4).

All in all, in 5% of the cases in the follow-up study, the disorders had become less severe or were cured. The incidence of new disorders or old ones worsened was 18%.

The perceived dynamic workload and poor work postures correlated significantly (p.<0.001) with musculoskeletal disorders, which is not surprising. However, 60% of those who had a great deal of neck and upper extremity strain, had also reported much responsibility for people in their work. The correlation was significant (p.<0.01) among bath attendants. There was also a less effective correlation (p.<0.05) between back strain and responsibility for people among nurses and musculoskeletal disorders and work environment (p.<0.05) among all subjects.

## Conclusions

Mental strain appears to be the major perceived hazard in health care work, even though physical load factors do play an important role in inducing musculoskeletal disorders.

Perceived musculoskeletal strain symptoms were significantly more common in the physical work group than in the mental work group. The number of symptom-free sisters and matrons, however, decreased more than others during the four years, which might be due to increasing age.

Table 3. The % of subjects reporting strain symptoms in the back

| Occupational group | Back symptoms, % | | | |
|---|---|---|---|---|
| | 1981 | | 1985 | |
| | no | numerous | no | numerous |
| Bath attendants and orderlies, n=245 | 38 | 9 | 33 | 15 |
| Nurses, n=537 | 34 | 9 | 30 | 12 |
| Sisters and matrons, n=199 | 58 | 2 | 45 | 7 |

Chi-square 12.19 df=1 p<.001, in 1981
Chi-square 8.45 df=1 p<.01, in 1985

Table 4. The % of subjects reporting strain symptoms in the lower extremities

| Occupational group | Under extremity symptoms, % | | | |
|---|---|---|---|---|
| | 1981 | | 1985 | |
| | no | numerous | no | numerous |
| Bath attendants and orderlies, n=245 | 39 | 15 | 37 | 16 |
| Nurses, n=537 | 47 | 7 | 47 | 9 |
| Sisters and matrons, n=199 | 67 | 3 | 63 | 6 |

Chi-square 23.67 df = 1 p<.001, physical and mental work 1981
Chi-square 17.90 df = 1 p<.001,      -"-           -"-      1985

Back pain has been known as an important disorder, but in this study neck and shoulder pain was the most common disorder among all the studied workers, except orderlies, who complained more about symptoms in the legs.

As the amount of perceived load and disorders had increased in all the occupational groups, the age of 50 years seems to be an important borderline, which should be paid special attention when health promotion and prevention programmes are planned and implemented.

Preventive ergonomic measures should be comprehensive, covering the organization and methods of work, as well as the physical environment. Attention should be paid not only to handling of patients, but also to the use and quality of information to improve decision making in health care work and decrease the responsibility load.

## References

Andersson J.A.D. 1984. Arthrosis and its relation to work. Scandinavian Journal of Work, Environment and Health 10, 429-433.

Hildebrandt V. 1987. A review of epidemiological research on risk factors of low back pain. In: Musculoskeletal Disorders at Work, edited by P. Buckle. Taylor & Francis, London, pp.9-16.

Ilmarinen J. (ed). 1987. Health, work capacity, and work conditions in municipal occupations. Institute of Occupational Health, Helsinki.

Takala E.-P. and Kukkonen R., 1987. The handling of patients on geriatric wards. A challenge for on-the-job training. Applied Ergonomics, 18.1, 17-22.

# 20. WORKING CONDITIONS AND OCCUPATIONAL INFLUENCES ON LOW BACK PAIN AMONG JAPANESE TRUCK TRANSPORTATION WORKERS

### Akinori Hisashige[*] and Shigeki Koda[**]

[*]*Department of Public Health*
*Kochi Medical School, Japan*

[**]*Department of Hygiene*
*Okayama University, School of Medicine, Japan*

**Abstract**

To identify and estimate recent working conditions and job content in truck transportation as well as the prevalence of low back pain and its occupational influences, a questionnaire survey and time study were carried out. Subjects were 144 truck drivers and 31 terminal workers. Results of rate ratio and multivariate analyses indicated that specific working conditions and environments such as heavy physical work, frequent unnatural postures, lifting heavy materials and vibration were the main risk factors for low back pain. However, it should be noted that the main risk factors indicated in this study consisted not only of these ergonomic problems but also basic working conditions such as working time, shift work and a shortage of rest breaks.

**Keywords:** *back pain; truck drivers; accidents; physical workload; heart rate; posture; manual materials handling; job analysis*

## *Introduciton*

The amount of domestic transportation of goods in Japan reached 6 billion tons. While train transportation is decreasing, truck transportation is increasing enormously with its proportion exceeding 80% of domestic transportation (Yamanobe et al., 1988). Moreover, the system and content for truck transportation have drastically changed, both quantitatively and qualitatively since the 1970s (Yamanobe et al., 1988; Hisashige et al., 1989). The structure of Japanese industry has changed from heavy industries such as the steel, metal and petrochemical industries to the electrical and precision machinery industries. Consequently, highly valuable, lighter, and smaller products have dramatically increased. Moreover, according to changes in social structure and national living standards, needs among individuals and households rather than industries have grown (Hisashige et al., 1989). To cope with these changes, a computerized network for quick transportation has been established all over the country, and most goods are delivered in a day except for extremely long-distance deliveries.

This situation is suspected of having a large impact on working conditions and job content in the truck transportation industry. The number of truck drivers in Japan, including both long-haul and local pickup and delivery drivers, is estimated to be more than 800,000 (Yamanobe et al., 1988; Hisashige et al., 1989). Truck transportation has been pointed out as being a hazardous and stressful occupation and drivers have a higher rate of job-related illnesses than do workers in other fields (LaDou, 1988). However, there is no information or research on recent working conditions, occupational health care or their state of health, because 90% of the 400,000 truck transportation firms belong to small enterprises (Yamanobe et al., 1988; Hisashige et al., 1989). The main purpose of this study was to identify and estimate the recent working conditions and job content in truck transportation. Moreover, the prevalence of low back pain, which is thought to lead to occupational disease among workers, and its occupational influences were estimated.

## Subjects and methods

A questionnaire survey was carried out among 194 workers at a truck transportation firm in the Kochi prefecture. The questionnaire consisted of general demographic characteristics, working conditions and environments, job content and low back pain symptoms. The valid responses (and response rate) were 175 (90.0%). Subjects consisted of 144 truck drivers delivering and collecting goods (Figure 1), and 31 truck terminal workers, as a control group, supporting truck drivers in handling heavy materials and managing clerical work at a truck terminal. The mean age and duration of job experience among truck drivers were 45.5 (SD=7.9) and 16.0 years (SD=7.8), respectively. Those among terminal workers were 40.0 (10.1) and 16.0 (8.4).

Figure 1. Loading and unloading goods at a truck terminal

To estimate work content and loads, a time study was carried out among 22 local pickup and delivery truck drivers. Work content, the number of goods and the weight to be handled were observed during work. Working postures were recorded every minute during a day of work. Simultaneously, three subjects were monitored by a Holter ECG (Fukuda SM28) which was fastened to a belt worn around the waist. The heart rate recordings were analyzed with a microcomputer (Fukuda SCM280 and NEC 9801RX21). They were estimated for every period of 10 R-R intervals and their mean value was calculated. Moreover, daily records of truck transportation kept by a firm were analyzed for 17 long-haul truck drivers to estimate work content, the weight to be handled and sleeping time.

Table 1. Wages and working hours of the truck drivers

|  | Truck drivers (N=144) | Terminal workers (N=31) |
|---|---|---|
| Wage | % | % |
| daily and monthly | 97.9 | 90.3 |
| others | 2.1 | 9.7 |
| Piece work | 86.9 | 9.7*** |
|  | mean (sd) | |
| Working time (hrs/day) | 13.1 (2.3) | 9.9 (1.5)+++ |
| Overtime work (hrs/month) | 106 (58) | 65 (23) +++ |

\*\*\* $p<0.001$ (age-adjusted by Mantel-Henszel's method)
+++ $p<0.001$ (by t test)

Table 2. Working loads and environments

| Items | Truck drivers (N=144) | Terminal workers (N=31) |
|---|---|---|
| Working conditions | % | % |
| overload of work | 48.6 | 16.1** |
| long working hours | 81.3 | 38.7*** |
| shortage of recess time | 61.1 | 16.1*** |
| difficulties in taking recess | 75.7 | 29.0*** |
| Handling heavy materials | | |
| over 20 kg | 81.3 | 19.3*** |
| Working posture | | |
| squatting | 25.7 | 9.7* |
| bending | 72.2 | 6.5*** |
| deep bending | 34.7 | 3.2** |
| standing | 43.8 | 25.8* |
| twisting | 41.7 | 3.2*** |
| Working environments | | |
| outdoor work | 76.4 | 35.5*** |
| footing instability | 23.6 | 0 ++ |
| vibration | 66.0 | 0 +++ |
| narrow working space | 25.7 | 9.7* |
| Job frustration | | |
| concentrated attention | 86.1 | 48.4*** |

\* $p<0.05$, \*\* $p<0.01$, \*\*\* $p<0.001$ (age-adjusted by Mantel-Henszel's method)
++ $p<0.01$, +++ $p<0.001$ (by Fisher's direct method, two tailed)

Prevalence rates for low back pain were compared between truck drivers and truck terminal workers. The statistical significance was estimated by the Mantel-Henszel age-adjusted method or Fisher's direct method. In analyzing occupational influences on low back pain, rate ratios for its prevalence rate, which is similar to the concept of relative risk, and their confidence intervals (95%) were computed (Miettinen, 1985). The prevalence rate of the ordinal or interval category, which was considered to produce workload at the lowest level within each item, was used as a standard (1.0). Finally, to examine simultaneously the effect of multiple risk factors for low back pain and confirm univariate rate ratio analysis, discriminant analysis was performed (SPSSx, version 2). Prevalence of low back pain was used as a dependent variable while occupational factors where rate ratios were relatively high were used as independent variables. However, since many ordinary variables were included among explanatory variables, dummy variables were used in this analysis (Kleinbaum et al., 1985).

## Results

### Working conditions and environments in truck transportation

As shown in Table 1, the main wage type for truck drivers and terminal workers was daily and monthly payment. However, a system of piece work was mainly applied to truck drivers, simultaneously. Truck drivers worked over 13 hours on average per day under an irregular shift system. Moreover, mean overtime work for truck drivers was 106 hours per month which was approximately twice as high as that for terminal workers ($p.<0.001$). These working conditions obviously violate the Labor Standard Law and its Enforcement Ordinance. In addition, Table 2 shows that rates for working conditions such as overload of work, shortage of rest break time and so on were higher for truck drivers than for terminal workers ($p.<0.01$ or $p.<0.001$).

Table 2 also shows working loads and environments for truck transportation workers. The rate of handling heavy materials over 20 kg was higher among truck drivers than among terminal workers ($p.<0.001$). The frequency of unnatural working posture, such as bending, twisting, squatting and so on, was also higher among truck drivers than among terminal workers ($p.<0.05 - p.<0.001$). The rates of environmental issues pointed out, such as instability of footing, vibration, narrow working space and outdoor work, were higher for truck drivers than for terminal workers ($p.<0.05 - p.<0.001$) (Table 2). The rate of concentrated attention among job frustrations during work was higher among truck drivers than among terminal workers ($p.<0.001$).

### Characteristics of truck transportation

Table 3 shows the characteristics of truck transportation. Under a one-man transportation system, driving, loading and unloading goods were performed by a single truck driver. The average actual driving time per day and driving distance per month were 7.8 hours (SD=2.4) and 5,600 kms (SD=2.4) respectively. The average amount of goods to be handled per day was 9.0 t (SD=2.7). These figures for long-haul drivers were statistically significantly higher than those for local drivers ($p.<0.001$).

Table 3. Characteristics of truck transportation

| Truck transportation | mean | (sd) |
|---|---|---|
| Driving time (hr/day) | 7.8 | (2.4) |
| Driving distance ($10^3$ km/month) | 5.6 | (2.4) |
| Amount of goods handled (t/day) | 9.0 | (2.7) |

Among truck drivers (N = 144)

Table 4. Total amount of weight for loading and delivering goods in the three-day round trip among long-haul truck drivers

| Course | Total weight handled average (max-min) |
|---|---|
| Going forth | t |
| before departure | 9.2 (12.1-6.5) |
| during delivery | 3.1 (10.2- 0 ) |
| after arrival | 6.1 (10.1- 0 ) |
| Going back | |
| before departure | 9.3 (11.5-5.2) |
| during delivery | 3.9 (10.2- 0 ) |
| after arrival | 5.4 (11.4- 0 ) |

Among 17 line-haul truck drivers

Table 5. Time analysis of picking up and delivering goods during a day of work among local truck drivers

| Time | Total weight handled average (max-min) |
|---|---|
| In the morning | kg |
| before departure | 2339 (4940- 975) |
| delivering, collecting | 3273 (7340 1105) |
| after arrival | 705 (2690- 0) |
| In the afternoon | |
| before departure | 655 (3285- 0) |
| delivering, collecting | 2973 (5875- 860) |
| after arrival | 1331 (3920- 255) |

Among 22 local pickup and delivery truck workers

As shown in Table 4, the average total weight of loading and delivering goods during a three-day round trip amounted to 37.1 t. Truck drivers picked up and delivered the goods manually, usually by themselves without any mechanical aids. Only at a truck terminal, did they sometimes load or unload with the aid of a fork-lift truck or a wheelbarrow depending on the bulk and weight of the goods. The median of time for sleeping or taking a nap during this three-day trip was only 5 hours with a wide range from 2 to 12. Moreover, they usually rested in their trucks rather than in sleeping facilities.

On the other hand, the results of a time study on handling goods during a day of work showed that total weight handled amounted to 11,276 kg which exceeded that obtained by self-reports (Table 5). Figure 2 shows working postures during loading and unloading at a terminal. The proportion of unnatural working postures such as bending or squatting reached more than 30% of total time, excluding the work after arrival in the morning.

The average heart rate during work is shown in Figure 3. It varied between 80 and 150 beats per minute. Its change mainly depended on the physical load, such as handling goods. Those for handling goods, driving and walking, and resting were 120-150, 100-110 and 80-90 respectively.

**Prevalence of low back pain and occupational influences on it among truck transportation workers**

Table 6 shows the comparison of prevalence rates for low back pain between truck drivers and terminal workers. Prevalence rates for low back pain (present, last month and life-time) were statistically significantly higher among truck drivers than among

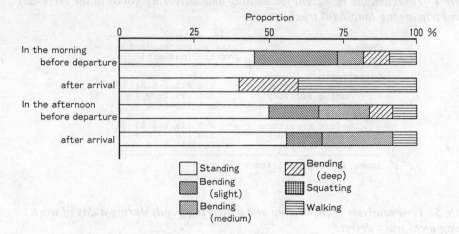

Figure 2. Working postures during loading and unloading among 22 local truck drivers

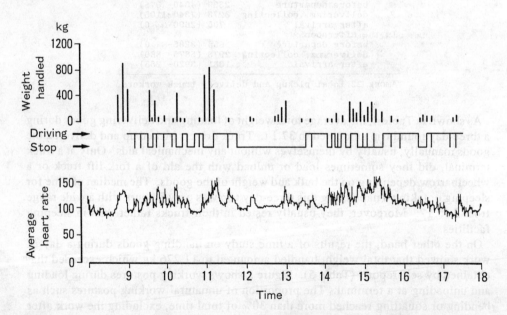

Figure 3. An example of time analysis of truck transportation with heart rate recording

terminal workers (p.<0.001). Proportions for the pattern of developing low back pain gradually among 126 truck driver cases and 12 terminal worker cases were 50.8 and 41.7%, respectively. There was no statistically significant difference observed. However, the proportion of low back pain which resulted in absence or rest break during work was higher among truck drivers than among terminal workers (p.<0.05). Moreover, only 7.9% of cases among truck drivers had improved in comparison with 33.3% among terminal workers. The greater part of the rest had experienced either aggravation, relapsing symptoms or no particular change.

Table 6. Prevalence of low back pain

|  | Truck drivers (N=144) % | Terminal workers (N=31) % |
|---|---|---|
| Low back pain present | 51.4 | 9.7*** |
| last month | 70.8 | 22.6*** |
| always | 16.0 | 0 |
| sometimes | 54.8 | 22.6 |
| life-time | 87.5 | 38.7*** |

*** $p<0.001$ (age-adjusted by Mantel-Henszel's method)

Table 7. Rate ratios for occupational factors influencing low back pain

| Items | Rate ratio (95% CI) |
|---|---|
| Weight of goods handled | 5.7 (2.6-12.1) |
| Difficulties in taking a recess | 3.6 (2.0-6.3) |
| Duration of heavy work experience | 1.9 (1.4-2.7) |
| Working time | 1.8 (1.3-2.5) |
| Number of days for night work | 1.7 (1.3-2.2) |
| Bending posture | 1.5 (1.2-1.9) |
| Vibration | 1.5 (1.1-1.8) |
| Outdoor work | 1.5 (1.1-1.8) |
| Shortage of recess time | 1.4 (1.1-1.8) |
| Twisting posture | 1.4 (1.1-1.8) |

High ranking 10 items ($p<0.05$)
Rate ratios are crude without adjustment

High ranking rate ratios of the prevalence rate for low back pain is shown in Table 7. They consisted of weight of goods handled (RR=5.7), difficulties in taking a break (3.6), duration of heavy work experience (1.9) and so on. On the contrary, factors recently receiving attention, such as the method of lifting (RR=1.2), physical exercise (0.9), obesity (1.3) or drinking (1.0) and smoking (0.7) habits did not show any statistically significant excess of risk for low back pain.

As shown in Table 8, although there are several exceptions, discriminant analysis confirmed these results of rate ratio analysis. The main contributing factors for discriminating low back pain were not only ergonomics factors, such as handling heavy materials and unnatural working postures, but also basic working conditions, such as rest breaks, piece work, shift work, working time and so on. Moreover, working environments were also suggested to be important factors.

## Discussion

Truck transportation is a major industry all over the world. It is also a major source of concern for health and safety (LaDou, 1988). Truck transportation workers are exposed to various hazards, not only physical but also psychosocial stresses (Hisashige et al., 1989; LaDou, 1988). Truck drivers are shown to have a higher rate of job-related injuries and illnesses than do workers in other fields (LaDou, 1988).

In this study, irregular shift work, long working hours and overtime work were observed among truck drivers (Tables 1 and 2). Although these characteristics were pointed out as an inherent problem for transportation work (Kogi et al., 1980), there has been no remarkable improvement in these working conditions until now. Since truck transportation firms are highly competitive in Japan, they usually try to survive by containing wages and increasing workload. Moreover, piece work accelerates this overwork. These poor working conditions violate the Labor Standard Law and its Enforcement Ordinance. However, supervision of the work standards inspection office has seldom reached these problems.

Table 8. Results from the discriminant analysis

| Items | Discriminant function coefficient |
|---|---|
| Difficulties in taking a recess | 0.5522 |
| Piece work | 0.4576 |
| Number of days for night work | 0.4289 |
| Working time | 0.4247 |
| Duration of heavy job experience | 0.4230 |
| Irregular shift system | 0.4222 |
| Bending posture | 0.3994 |
| Vibration | 0.3852 |
| Twisting posture | 0.3681 |
| Outdoor work | 0.3639 |
| Shortage of recess time | 0.3535 |
| Weight of goods handled | 0.3372 |
| Squatting posture | 0.2793 |
| Narrow working space | 0.2793 |
| Concentrated attention | 0.2580 |
| Overall correct rate (31 all items) | 74.29 % |
| Wilk's lambda | 0.721 |
| Chi-square (df-31) | 51.36 |
| Significance | 0.012 |

High ranking 15 items

As regards workload, handling heavy materials and unnatural working postures were indicated in this study as characteristics of workload and job content for truck transportation (Tables 2-5, Figures 2 and 3). These items are well known as risk factors for low back pain (Yu et al., 1984; Kelsey et al., 1988). Moreover, average heart rates during work reached a high level of 150 beats/min (Figure 3), indicating fairly intensive physical load. On the contrary, even in heavy work such as the construction and steel industries, average heart rates have recently decreased to under 120 beats/min due to mechanization, saving manual power and shortening working time (Yamaji, 1982).

On the other hand, the vibration from motor vehicles is a well known risk factor for health hazards including spine degeneration (LaDou, 1988). In this study, several working environments as well as vibration were pointed out as specific problems for truck transportation. These results were consistent with former reports (LaDou, 1988; Kogi et al., 1980; Hedberg, 1985). In addition, mental load such as concentrated attention due to truck driving was observed (Table 2).

To meet a deadline for delivery, cutting rest break or sleeping time as well as extending working time were observed (Tables 2 and 3). However, rest breaks and sleep play an important role not only in recovery from strain and fatigue but also in driving performance (Forbes et al., 1953; Heimstra et al., 1966). Their shortage would reinforce the effect of poor working conditions and physical or mental stresses.

One of the leading occupational diseases among truck drivers is low back pain (Hisashige et al., 1989; LaDou, 1988; Hedberg, 1985). In this study, the prevalence rate of low back pain was two or five times higher among truck drivers than among terminal workers (Table 6). Moreover, results of rate ratio analysis and multivariate analysis indicate that specific working conditions and environments for truck transportation as mentioned above influenced the incidence of low back pain among truck drivers. Risk factors such as heavy physical work, frequent unnatural postures, lifting heavy materials and vibration were consistent with former findings (Yu et al., 1984; Kelsey et al., 1988). However, it should be noted that the main risk factors indicated in this study consisted not only of these ergonomics problems but also basic working conditions such as working time, shift work and a shortage of break time. Therefore, in examining the measures for preventing low back pain, it is essential to evaluate the occupational influences systematically and comprehensively.

Moreover, the results must be cautiously estimated because the design of this study was cross-sectional. They may be vulnerable to information bias as well as selection bias (Rothman, 1986). As to selection bias, results of this study would underestimate relative risks of occupational factors. The largest workload category did not necessarily show the highest prevalence rate among several items when examining a dose-response relationship even when this relationship existed. On the other hand, as to information bias, there are no data for examining it. It is also very difficult to rule out the possibility that those cases with low back pain estimated workload more severely than did healthy workers. Therefore, risk factors suggested in this study must be examined after a more powerful study design (e.g. a cohort study) focusing on examining causal relationship.

## References

Forbes TW et al, 1953, Sleep deprivation effects on components of driver behavior, Highway Research Record, 28:21.
Hedberg GE, 1985, Physical strain in Swedish lorry drivers engaged in the distribution on goods, Journal of Human Ergology, 14:33-40.
Heimstra NW et al, 1966, The effects of stress fatigue on performance, Ergonomics, 13:209.
Hisashige A et al, 1989, Issues on job contents and occupational health care among truck transportation workers, Occupational Health Journal, 12:92-99.
Kelsey J & Hochberg MC, 1988, Epidemiology of chronic musculoskeletal disorders, American Review of Public Health, 9:379-401.
Kleinbaum DG & Kupper LL, 1985, Applied regression analysis and other multivariable methods, Duxbury Press, Boston.
Kogi K & Nozawa H, 1980, Jidosha unten rodo (Work of steering wheels), Rodo Kagaku Kenkyusho, Tokyo.
LaDou J, 1988, The health of truck drivers, Zenz C ed, Occupational Medicine, Year Book Med Pub, Chicago.
Miettinen OS, 1985, Theoretical epidemiology, John Wiley & Son, NY.
Rothman KJ, 1986, Modern epidemiology, Little Brown, Boston.
Yamaji K, 1982, Sinzou to Supotu (Heart and sports), Kyoritu, Tokyo, (in Japanese).
Yamanobe Y & Kono S, 1988, Rikuun gyokai (Surface transportation industry), Kyoikusha, Tokyo.
Yu T et al, 1984, Low-back pain in industry, Journal of Occupational Medicine, 26:517-524.

# 21. A FIELD SURVEY ON PRACTICAL LOAD CARRYING LIMITS

## Albert Soued

*C.N.A.M.T.S. Social Security Board*
*Occupational Hazards and Diseases Prevention*
*Paris, France*

### Abstract

This survey has been conducted with the aim of providing employers and employees with guidelines for reducing manual material handling by the introduction of mechanical and automatic equipment and by screening limits beyond which it is not worthwhile maintaining manual operations.

**Keywords:** *accidents; manual materials handling; mechanical materials handling*

## Some figures

In 1988 hazard compensation covered more than 95,000 accidents linked to back diseases (hernia, lumbago, pain etc). Half of these accidents are related to manual material handling (MMH) in the working place and 15% to manual load carrying (MLC). The accidents due to MMH and to MLC represent 30% of all occupational accidents, costing nearly $2 billion/year and 6 million lost working days.

More than 1.5 million workers out of 13.5 million in our compensation system are involved in some way in MMH or MLC, in the industry, trade and transport field. Half this population usually handle or carry weights and we estimate there are between 50,000 and 100,000 difficult and straining jobs which cause the bulk of accidents and diseases. But what are the boundaries between efficient and non-efficient manual handling?

Our intention was not to establish scientific and theoretical tables and diagrams, which are already numerous but difficult to use on the spot; our purpose was to gather enough information from the field about what is considered a difficult manual handling job by both workers and their supervisors and management, and then fix limits beyond which the hazard risks begin.

## The limits research

In 1986, we conducted a statistical survey to determine which activities were most affected by back diseases in MMH and MLC. We selected 7% of the branches of industry representing more than 60% of the accidents and 67% of the lost working days.

In 1988 and 1989, a field survey has been undertaken among these selected branches (metal, wood, plastic, ceramic products etc) concentrating mainly on transport, haulage and the trade business. A questionnaire and an interview draft were distributed to obtain information from 126 different companies all over the country and to screen more than 500 working posts. In each case workers, direct supervisors, company doctors and hazards prevention technicians have been interviewed and helped to fill out the questionnaire by our regional agents. Finding the difficult workposts was easy, as in most cases the management was already looking into mechanization projects; in the other cases workers involved have been injured or had sick leave due to back pain, during the last three years. Some jobs could not be performed by one person for a complete shift: in general difficult manual handling is of a two hour duration after which there is a break period or the end of a half shift period. Difficult jobs are also combined with other less straining operations.

## Resistance to mechanization

Replacing a manual handling job by a mechanized one meets with many difficulties. First, technical dificulties:
 - job analysis is not easy as one cannot determine with precision the consequences on other posts, on production, on quality and on delays,
 - standard equipment does not usually fit to each case and costly studies are necessary to adapt the available equipment or to build more suitable equipment,
 - mechanical equipment is not as flexible and adaptable as a manual handling operation.

Secondly, there are financial difficulties:
 - mechanization costs from $10,000 to more than $200,000 per post and the employer is not willing to invest if the return on capital lasts more than two to three years, especially when one is not absolutely sure that the newly designed equipment will be as reliable as manual handling,
 - when series are short, when the market is fluctuating, when production is diversified, manual handling is more flexible than a machine.

Resistance might also be social as difficulties might emerge when suitable jobs are not available for the idle replaced worker. On the other hand MMH and MLC are monotonous, exhausting and unrewarding.

## Rule of thumb

Management is looking for easy "rule of thumb" measurements which would allow them to determine quickly whether a MMH or a MLC is beyond safe limits or not. Field surveys and job screening have permitted us to determine practical rules which might be easily implemented. We have taken into account only three parameters:
 - daily output or yield (y in ton/day) which includes the frequency
 - weight of the handled load (w in kilogram)
 - distance (if any) of load carrying (d in metre).

These parameters are determined as average daily figures and might fluctuate within the +/-50% range.

Other usual criteria have not been taken into account because they are difficult to implement and because they have already been included in the frequency factor. We have ruled out physical and physiological factors, the nature, the dimensions and the levels of the load, the run difficulties (floor, shoes, gradient, stairs etc) and the environmental factors such as noise, dust, heat etc.

We have noticed two specific simple situations: MLC and MMH, the first one involves mainly male workers, the second one does not include carrying, or only for a short distance (less than 2 metres).

1. "MLC": $w * d * y < 800$
   $w < 30$ kg    in 75% cases with a maximum 50 kg
   $d < 20$ m    in 90% cases with a maximum 30 m
   $y < 8$ t/day    in 98% cases with a maximum 10 t/day

2. "MMH":    $y < 14 - w/2.5$ for a male with a 30 kg limit
   $y < 8 - w/2.5$ for a female with a 15 kg limit.

## Recommendations

To avoid or to reduce MMH or MLC accidents or back diseases, one has to implement on a national level preventive measures in different directions:
- employers' information about MMH and MLC hazards and cost, about safe limits and the necessity of job screening
- financial incentives to invest or to accelerate investment in mechanical equipment in order to improve manual handling work conditions or to mechanize manual jobs.
- emphasizing EEC and national weight limitations of 30 kg for men and 15 kg for women
- standards and incentives to produce safely designed, well-packed and easily-handled products
- MMH and MLC kinetic training promotion with lifting and carrying techniques.

# 22. SKILL, EXCESS EFFORT, AND STRAIN

## Michael Patkin[*] and John Gormley[**]

[*]The Whyalla Hospital
South Australia

[**]South Australian College of Advanced Education

**Abstract**

Musculoskeletal strain and injury at work remains an unsolved and important problem, despite intense ergonomics and medical studies. One reason is failure to consider faulty techniques, especially excessive muscular effort or ineffective muscular effort due to poor timing. This study examines hand and arm pains in office workers and back pain from manual handling. It proposes several mechanisms for these causes. They are co-contraction or inappropriately timed and graded co-contraction, excessively forceful grip, and poor choice of body posture. In this paper, lifting at work is discussed in terms of a process rather than a relatively uncomplicated skill.

**Keywords:** *musculoskeletal disorders; motor skills; back pain; manual materials handling; posture*

## *Introduction*

Upper limb pain in office workers and back pain from lifting are the two commonest types of musculoskeletal strain at work today. Existing medical and ergonomics approaches have failed to solve these problems. We propose that such pains and injuries sometimes arise from unskilful methods of work, not considered in traditional models of ergonomics or medicine. Lack of skill in such situations may cause the use of excess force, poor timing, or poor preparation for movement, and can take several forms:

1. Unnecessary force applied to the task, for example pressing too hard with a pen, or hitting too hard with a hammer or onto a keyboard.
2. Excessively forceful grip, for example holding a handle more tightly than is needed.
3. Co-contraction, or excessive tension in two groups of muscles which oppose one another. A simple example is a tense but empty hand. (Sometimes "co-contraction" is applied to the normal simultaneous contraction of synergist muscles, but this is not the meaning used here.)
4. Abrupt application of force instead of smooth increase during a movement, for example hitting with a hammer in a jerky way or lifting without taking advantage of the momentum of the body, or an unexpected load from a slip or fall.
5. Poor choice of body posture, for example position of the body in relation to loads, failure to adjust seating correctly, and working with wrists severely abducted or dorsiflexed.
6. Poor choice of equipment or material, so that more force is required for performance of the task.

Often it is easy to solve these problems by simple demonstration and explanation.

## Handwriting as an example

Pains in the hand and arm have been associated with handwriting for hundreds of years, and many theories of causes and methods of cure have been proposed (Sheehy and Marsden, 1982). These include "focal dystonia" (spastic grip and jerky movements), relearning to write using shoulder movements, and psychoanalysis. Published descriptions and photographs (Sarkari et al., 1976) suggest that widely different phenomena have been studied, different from the so-called RSI epidemic seen in Australia during the 1980's.

These studies considered 133 subjects with symptoms attributed to repetitive work, so-called RSI (repetitive strain injury) or OOS (occupational overuse syndrome). Ten of these attributed hand pains to their work. They confirmed earlier observations (Patkin, 1984) that "trying too hard" was the major factor responsible. The excess effort had separate components of grip, pressing, and co-contraction, and could be demonstrated by palpation of muscle, indentation of writing paper resting on cloth, and electromyography using simple equipment.

Associated factors included use of ball-point pens with a poor or unpredictable flow of ink, multiple copies of documents, poor teaching of children at school, and tense habits during other activities. These observations are complemented by diagrams in old textbooks of handwriting which show much more relaxed hand-grips. Pressing too hard with traditional types of pen is unnecessary for legible writing, and causes damage to the point. Eight of the ten patients in the series achieved quick and complete recovery by attention to these factors.

In computer terminal operators an extra factor is traction on the brachial plexus caused by holding the neck forward instead of upright. This causes tingling or discomfort in the forearm due to irritation of the nerves supplying skin in the same area as underlying tense muscles. Brachial Plexus Tension Tests (BPTT) (Elvey, 1979) are similar to tests for sciatic nerve irritation, though their validity has been disputed. A poked forward position of the head may be due to habit, poor legibility of the screen, or reading-glasses with the usual focal length of 30 cm instead of twice that distance to suit office work. Here, skill is a good strategy for choice of posture rather than muscle tension.

## Skill and timing

During a task, force can be applied too quickly or too slowly. When pushing a child's swing, for example, it takes little force to gradually increase the amplitude of the swing and its combined kinetic and potential energy, provided that force is not applied too quickly, or out of phase with the movement of the swing.

Case example: A clerk working poorly was transferred to chipping scale inside a blast furnace undergoing overhaul. He was clumsy using a chipping hammer as well as angry at his change of job.

Within a few weeks he developed forearm pain. On examining his style of hammering, he had short jerky strokes, accelerating and stopping the movements of the hammer too quickly.

A similar example is chopping wood, where the tense novice is advised to "let the axe do the work". Using a sledge hammer to drive heavy nails to secure railway lines, the skilled worker will use the recoil or bounce of the hammer to reduce the amount of muscle work needed to raise the hammer again for the next stroke. Good timing can be defined as "the rate of application of force which maximizes the rate at which energy is transferred from one part of a system to another", and is another way of describing resonance, or coupling.

Poor timing transfers energy to the structures of the body rather than the external object. Strains and tears of muscles and tendons occur when their tensile strength is exceeded. This depends on the size of the load, its direction, the mechanical advantage or disadvantage of the applied force, and the rate at which it is applied, since biological tissues have a relatively slow rate of stretch and creep apart from the contractility of muscle. Such strain may cause mild ache or serious tearing injury.

## Other mechanisms of strain and injury

Two cases of lateral epicondylitis occurred in cleaners. They used a heavy floor-polishing machine and developed their injury pulling the polisher up a step while keeping the elbow bent. Their supervisor however kept the elbows straight, using her body weight to counterbalance the weight of the machine. This allowed the extensor muscles of the forearm to stay relaxed. By contrast, flexion of the elbows was accompanied by strong contraction of the forearm extensor muscles just below their attachment to the lateral epicondyle of each elbow.

Cleaners using floor polishers unskilfully have a second risk factor. The rotating head of the polisher is kept slightly tilted so that it moves first to one side and then to the other. If the tilt is altered too quickly at the end of each transverse movement, the operator has to "fight" the polisher and grip it more fiercely, also straining the origin of the forearm extensor muscles as in the common circumstances of "tennis elbow".

## Lifting weights

Injury may arise not only from errors of technique but also from poor preparation, planning, practice, or experience. We propose that there are important differences between the skills of competitive weightlifting and the process of lifting loads in the workplace. These include physical conditioning and mental training, and the control of variables.

Expert weightlifters use similar patterns of movement, as shown by observation, and analysis of videotapes and EMG recordings. However they also show individual characteristics, comparable to differences in handwritten signatures, speech, or choice of words. It becomes possible to consider a particular motion "signature" revealed in kinetic energy, or a "grammar" of movement. Common factors in the patterns of movement of weightlifters include:

1. Preparatory movements (power movements in sport are often preceded by movements in the opposite direction, or backswing).
2. The sequence in which different groups of muscles become active.
3. Reproducible intra- and inter-weightlifter force-time patterns but with sufficient discrimination to identify an individual lifter. (These are shown with force platforms.)

In contrast to weightlifters, Gormley, Sedgwick and Smith (1989) found that warehouse workers and "ordinary" people used techniques and strategies in lifting that varied to meet changing situational demands and as the task proceeded. Subjects did not operate in the invariant manner of a highly skilled weightlifter or machine. Also they did not perform lifting actions in the single plane of motion often assumed in biomechanical models and normative standards. These observations support the notion that lifting is a complex set of behaviours regulated according to the changing circumstances of the environment, loads, performer status and the actual performance of the task as well as by the original intent underlying the action. Lifting requires the operator in the process to code, order, and time the actions according to the syntax and lexical requirements of the lift. The grammar of lifting is acquired through experience and practice in a similar manner to language acquisition. Attempts to teach lifting as though it is merely a mechanical/biomechanical problem have not been successful. Yet such an approach to competitive weightlifting has been found to be highly successful. Why?

## Lifting within limited constraints

The highly constrained conditions of competitive lifts performed with particular styles or techniques allows the development of invariance in the strategies and actions used. Competitive weightlifters learn the codes to produce the actions required to overcome the kinetic and topological constraints through progressive practice and repetition. When a sufficient level of mastery has been achieved the 'lifting action sequence' can be performed at will with only minor modification being needed as the loading demands increase. The action sequence is hierarchically organized and executed with the kinematic and kinetic features becoming increasingly invariant the higher the efficiency of the lifter. Enoka (1988) showed in a comparison of skilled and less skilled Olympic weightlifters that success of the more skilful performers was due to their ability to both generate a sufficient magnitude of joint power and to organise the phases of power production and absorption into an appropriate temporal sequence rather than through a mere quantitative scaling of power production.

## Lifting: perception of task demands

Lifting as it occurs in the garden, on the building site, in the hospital ward and the office is not as constrained as in the sport of weightlifting. The person must respond to the particular circumstances according to their perceptions of and experience in meeting the task demands. Often the situation can be quite novel. In the lifting required of warehouse workers, the operator has to deal with a complex dynamic set of relationships. The lift actions selected involve assessment of the environmental and task demands, action planning and estimation of risks attached to particular decisions made under temporal, spatial and kinetic limitations associated with success/failure probability predictions.

## Choice of lifting strategy

As reported by Gormley, Sedgwick and Smith (1989), the actions selected as appropriate by warehouse workers are often less than optimal in terms of mechanical effectiveness, safety and control. Yet this study found that the task goal was achieved in the minimum time, with loads moved over minimum distances at low levels of perceived effort. The strategy often chosen was to minimize effort rather than maximize safety and control. The importance of experience and knowledge in effectively using such a strategy is supported by the results of Patterson et al. (1987). The study investigated the effects of load knowledge on the stresses at the lower back during lifting. Loads were lifted under conditions of verbal or visual knowledge of load magnitude or no load knowledge. Results showed experienced lifters had lower stress levels at L4/L5 and utilized two technique strategies that were dependent upon the load knowledge conditions, whereas the non-lifters used the same strategy for all lifts. Maximum moment values were significantly higher for the inexperienced lifters under all conditions. Experienced lifters were able to distribute the load more effectively between body segments.

## Optimal solutions to lifting outcomes

Lifting, except under the most constrained circumstances is a process in which the solution outcome can be achieved in a variety of equally successful ways. Optimal solutions vary from task to task and from individual to individual as well as between and within performances of a particular lifting task. Individuals should be recognised as having divergent intra- and inter-individual abilities, knowledge and capacity to undertake the perceptual, cognitive and physical challenges inherent in the lifting process. Teaching of so-called 'safe' or 'correct' lifting methods fails to recognize the indeterminacy of the problems confronting those who have to produce the actions.

## Coping with chaos in lifting

Space does not allow the role of co-contraction in skilled actions to be considered in detail here, but strain and injury are obviously more likely when appropriate synergist muscle groups fail to be recruited at a rate to match the demands of the task. Examples of this are hamstring injury in sprinters, and errors in estimating the weight of an object.

Epidemiological evidence (Magora, 1973; Andersson, 1981) suggests that workers exposed to sudden unexpected loads are particularly vulnerable to low-back problems. Troup et al. (1981) suggested that slipping accidents were often due to sudden jerking or twisting actions. Ergonomic adjustment of the environment and the loads goes only part of the way towards reducing the risk of such events occurring. Intra- and inter-person variability cannot be avoided, and makes the prediction of such events notoriously difficult and unreliable. Since variability is part of the lifting process it follows that education and training programmes should focus on teaching the individual to cope with variation and that adaptability of response rather than single modes of correct technique is the desirable outcome. Such approaches will still not make it possible to predict with high probability estimates when accidents will occur and who the victims may be.

## Acquisition and exercise of skilled activity

The three-stage model of skill acquisition developed by Glencross (1977) has been adapted to hand movements in surgery and micro-surgery (Patkin, 1988), with three stages described as coding, modelling, and timing (the last corresponding to the hierarchical level of skill of Glencross). Errors during skilled activities can be considered as slips or mistakes, according to Reason's (1987) Generic Error-Modelling System. This approach can be combined with Rasmussen's (1980) distinction between skill-based, rule-based and knowledge-based levels of performance to describe error types in a three by five matrix. Tuning allows smoother and more accurate actions, placed in the rule-based performance category.

## Conclusions

Data on pain and skill related to activities as diverse as handwriting and lifting weights provide good evidence that some strains and injuries result from poor skills. Given their widespread nature and high cost in the workplace, it is important to apply existing knowledge to reduce these problems by teaching improved motor skills and to identify the sources of process-based errors in skilled performance at work.

## References

Andersson, G.B. (1981) Epidemiologic aspects of low back pain in industry. Spine, 6, 53-60.

Elvey, R.L. (1979) Brachial plexus tension tests for the pathoanatomical origin of arm pain. In Aspects of Manipulative Therapy, Lincoln Institute of Health Sciences, Melbourne, Australia, 105-110.

Enoka, R.M. (1988) Load and skill-related changes in segmental contributions to a weightlifting movement. Medicine and Science in Sports and Exercise, 20(2), 178-187.

Glencross, D.J. (1977) The control of skilled movements. Psychological Bulletin, 84, 14-29.

Gormley, J.T., Sedgwick, A.W. and Smith, D.S. (1989) Effective, safe and controlled lifting. A mechanistic or process approach to training? Paper, Annual Conference of the Australian Council of Rehabilitation Medicine, Sydney, March 1989.

Magora, A. (1973) Investigation of the relation between low back pain and occupation. 4. Physical requirements: bending, rotation, reaching and sudden maximal effort. Scandinavian Journal of Rehabilitation Medicine, 5, 191-196.

Patkin, M. (1984) "Trying too hard". An aspect of overuse injury. Proceedings, 21st Annual Conference, Ergonomics Society of Australia and New Zealand, 289-301.

Patkin, M. (1988) Hand and arm pain in office workers. Modern Medicine of Australia, 31(10), 66-76.

Patterson, P., Congleton, J., Koppa, R. and Huchingson, R.D., (1987) The effects of load knowledge on stresses at the lower back during lifting. Ergonomics, 30(3), 539-549.

Rasmussen, J. (1980) What can be learned from human error reports? In J. Duncan, M. Gruneberg and D. Wallis (editors), Changes in Working Life, Wiley, London.

Reason, J. (1987) Generic error modelling system (GEMS): A cognitive framework for locating common human error forms. In J. Rasmussen, K. Duncan and J. Leplat (editors), New Technology and Human Error, John Wiley, London.

Sarkari, N.B.S., Mahendru, R.K., Singh, S.S. and Rishi, R.P. (1976) An epidemiological and neuropsychiatric study of writers cramp. Journal of the Association of Physicians of India, 24, 587-591.

Sheehy, M.P. and Marsden C.D. (1982) Writers' Cramp - a focal dystonia. Brain, 105, 461-480.

Troup, J.D., Martin, J.W. and Lloyd, E.F. (1981) Back pain in industry: A prospective survey. Spine, 6, 61-69.

# 23. CONSEQUENCES OF VARIABILITY IN POSTURES ADOPTED FOR HANDLING TASKS

### Christine M. Haslegrave

*Institute for Occupational Ergonomics*
*University of Nottingham, U.K.*

**Abstract**

The hazards associated with manual handling tasks are often assessed by using biomechanical models to evaluate the loads imposed on the body. However, these are difficult to use at the workplace design or planning stage, since there is very little information available on actual working postures adopted by operators. The present study has analysed the postures adopted while exerting maximal forces in standing and kneeling tasks, using a group of experienced miners as test subjects. Considerable variation in posture was observed, which was found to have large consequences for the biomechanical loadings on the spine and joints. In addition, various postural strategies were identified which the subjects used to maximise force exertion. The results suggest that workplace assessments need to take account of the effects of individual variability in behaviour as well as in the population size characteristics.

**Keywords:** *posture; biomechanics; manual materials handling; muscular strength*

## Introduction

Manual handling tasks are strenuous and have high injury rates, since they put considerable biomechanical and physiological stresses on the body. Industrial engineers and safety inspectors are becoming more aware of these problems, and legislation is being introduced in Europe which will force employers to evaluate their own workplaces where manual handling tasks are performed, taking account of the physical effort, characteristics of the load and ergonomics factors in the workplace (HSC, 1988).

The hazards associated with these tasks and workplaces may be assessed by using biomechanical models which evaluate the loads imposed on the body in terms of compressive forces on the spine and torques at the major joints. These have been developed to a high level of sophistication, and are also being used to predict maximum force capability in specified tasks for individuals and for population percentiles (Chaffin and Andersson, 1984). In either application, the models are difficult to use at the workplace design or planning stage since it is difficult to specify the postures which will be adopted by operators in industrial situations, especially when they are constrained by the task layout or by working in confined spaces. If the operators cannot be observed directly, there is at present almost no information in the literature to describe actual working postures. If the jobs are to be accurately evaluated, a better understanding is needed of how operators approach handling tasks.

## Variability of strength and posture in force exertion

In order to improve the knowledge of working postures, a study was undertaken to analyse postures adopted while exerting maximal isometric forces in standing and kneeling tasks (Haslegrave et al., 1988). This involved a series of experiments using groups of subjects both experienced and inexperienced in manual handling. Most of the results presented here come from tests on a group of six experienced miners (all male), who ranged in age from 23-46 years, in weight from 56-101 kg and in stature from 1653mm-1815mm.

The subjects were presented with a series of experimental task layouts and each subject was asked to exert his maximal force single-handed on a testpiece which was positioned at shoulder, waist or floor level. The subjects were otherwise left completely free to choose their own postures in each test condition. The influence of the testpiece location on posture was therefore related to each individual's body dimensions and was the same for each subject. The testpiece was a cylindrical rod of 25mm diameter, and was oriented vertically for push/pull exertions and horizontally for lift/press exertions.

The results of the experiments showed clearly the complexity of the effects of task layout. The postures varied with direction of exertion and with location of the testpiece, just as strength varies with these factors. The reasons lie largely in the biomechanics of the task, since a skilled subject adopts a posture which will permit the maximum advantage to be gained from both muscle strength and use of body weight. The magnitude of the muscular force which can be exerted depends on the length and mechanical advantage of the muscles which are involved, and so ultimately on the joint angles and whole body posture. It also depends on the support or additional bracing which can be obtained by contact with the ground and other external interfaces.

Thus posture in fact determines the maximal strength which can be exerted in a given task. The primary determinants of strength capability in a whole body exertion (probably involving many muscle groups) are the muscles brought into action, the workplace described by the task factors, the posture adopted for the task, and the anthropometric characteristics of the individual.

The experiments showed further that there was considerable individual variability in the postures adopted by the subjects. This can be seen in a few simple measures of posture, such as foot locations and trunk flexion, when subjects were pushing and pulling at shoulder height, as illustrated in Figure 1. However, although the postures varied, they did not represent a continuous range of foot positions or degrees of twist of the body - in most test conditions there were two or possibly three distinct types of postures evident among the group of subjects, as has earlier been reported by Haslegrave and Corlett (1988). Although the postures differed, the subjects were often exerting similar degrees of strength, and the variability seemed rather to reflect the use of different strategies for maximising force. The extent of the individual variability suggests that the choice is not always clear so that in many situations there may not be one "best" posture for exerting a maximal force.

## Postural strategies for force exertion

Postural strategies for maximising force exertion which were identified during the experiments included various methods of orienting the arm and shoulder girdle and of stabilising the skeletal framework in order to increase the ability to transmit high forces across joints which were not actively involved in generating the muscular force.

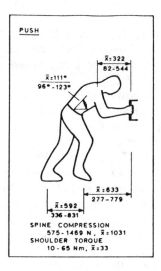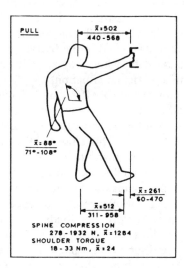

*Figure 1. Range displayed in some posture measures when pushing and pulling at shoulder height (dimensions in mm)*

Subjects also made considerable use of bracing obtained from the free arm either against the floor or against parts of their own bodies (knee, thigh or chest), particularly in kneeling tasks. It is suggested that such bracing may provide greater stability for the shoulder girdle and trunk.

The study also showed that there were differences in the behaviours of experienced and inexperienced operators when exerting maximal strength. Direct comparisons between the two groups of subjects was not possible, so that the differences could not be attributed to specific aspects of the workplace, but the impression from the tests was that some inexperienced subjects took less care in positioning themselves and did not take account of the danger of slipping or of the testpiece itself giving way. This has also been noted in lifting tasks by Patterson et al. (1987) who observed that even experienced workers took less care with light loads.

## Biomechanical effects

The variation in posture had important consequences for the biomechanical loadings on the spine and joints. The resulting variations in spinal compression at the level of the L3/L4 disc and in shoulder joint torque are shown in Figure 1. In some task situations, the biomechanical loads varied by as much as 1:10 in the group of six subjects. Of course, the maximal strengths of the subjects and forces being exerted also varied but to a smaller degree (with forces of 154-453N when pushing and 151-300N when pulling).

The subjects were free to choose their own postures and it was observed that adjustments to posture were complex - they did not respond to a change in task height or direction of exertion by simply changing foot positions or by leaning forward. The posture of the whole body was likely to be changed when any aspect of the task layout altered. The situation was therefore rather more complicated than suggested by previous laboratory experiments which have studied foot placement during force exertion (Ayoub and McDaniel, 1974; Chaffin et al., 1983). In addition, comparison of the tests in the present study with similar tests by Warwick et al. (1980) in which foot position was constrained suggested that maximal force may be significantly increased by relatively small adjustments to posture (Haslegrave, 1990). If this is so, maximal strengths quoted in the literature may well underestimate the true strength of industrial workers when they are allowed to use their skill to increase the force they exert in a particular situation.

Thus, the biomechanical effects of differences in posture are not easy to predict and it is not at all clear how subjects judge their own capabilities and tolerance levels in exerting maximal forces, other than through experience. Chaffin (1969) showed that his subjects appeared to have limited their spinal compression forces in lifting tasks to a "fairly constant magnitude", but Grieve (1987) noted that 10-15 percent of subjects were prepared to exceed normally accepted safety criteria for loads on the lumbar spine during maximal voluntary exertions in the sagittal plane. In the present study also, 9 percent of subjects exceeded the NIOSH (1981) recommended limit of 3400N for spine compression loads when they had a free choice of posture, and the maximum compression imposed by a subject was 6410N. However, both spine compression and shoulder torque were generally kept well within "acceptable" levels, and analyses of the biomechanical measures showed that task factors of direction of exertion and height of testpiece in fact had very little effect on many of measures. This suggests that subjects were conscious of biomechanical loads whatever the sensory mechanisms involved in their judgements. However, there are as yet no generally agreed criteria for assessing the safety or acceptability of joint torques, nor of shear and torsional forces in the spine which could also be very important.

## Evaluation of tasks involving force exertion

Variability in body size, weight and strength are already considered in the design of workplaces, and there would seem to be a corresponding need to take account of the variety in postures adopted by different operators undertaking the same tasks. This would permit better evaluation of jobs, taking into account the effects of individual variability in behaviour as well as in the population size characteristics. This will require observation and recording of tasks in real situations to define posture databases to match the anthropometric and strength databases which already exist.

However, this is only one aspect of the problem. Proper assessment of the hazards of tasks involves evaluation of the effects of the postures using biomechanical models. Current strength prediction models simulate the articulation and musculature of the human body, and take account of many of the factors governing strength capability (muscle strength, body weight and gravity lever arm, the effects of foot and other interfaces). However, the assumptions used in the models to optimise posture are based solely on the strength of muscle groups spanning the major body joints, using established torque/angle relationships. Analysis of the postures adopted by the subjects in the present study has indicated that other factors are equally important, and that these optimisation criteria are not sufficient to define the postures adopted in real work situations (Haslegrave, 1990).

The most important factor which needs to be incorporated in biomechanical models is the requirement for joint stability in bracing the skeletal framework. In order to exert torque about a joint, muscles need to act against a rigid base, and this is as important at individual joints in the body as it is at the foot/ground interface, as was originally recognized by Dempster (1958). This appears to be the limiting factor on maximal force exertion in many situations, and further research is needed to understand the mechanisms involved in order to represent these in the models.

This will also involve study of the reasons for choosing postures, so that prediction methods can be developed which will identify not only the optimum posture or postures, but also indicate some of the less advantageous and/or more hazardous postures which may be adopted by unskilled operators or trainees. This is important to give a better estimate of injury risk, and would in addition help in developing improved methods of training operators for handling tasks.

## References

Ayoub, M M, and J W McDaniel, 1974. Effects of operator stance on pushing and pulling tasks. AIIE Transactions, 6, 185-195.

Chaffin, D B, 1969. A computerized biomechanical model - development of and use in studying gross body actions. Journal of Biomechanics, 2, 429-441.

Chaffin, D B, and G Andersson, 1984. Occupational Biomechanics, New York: Wiley, pp 221-227.

Chaffin, D B, R O Andres and A Garg, 1983. Volitional postures during maximal push/pull exertions in the sagittal plane. Human Factors, 25, 5, 541-550.

Dempster, W T, 1958. Analysis of two-handed pulls using free body diagrams. Journal of Applied Physiology, 13, 469-480.

Grieve, D W, 1987. Demands on the back during maximal exertion. Clinical Biomechanics, 2, 34-42.

Haslegrave, C M, 1990. Postural Constraints on Force Exertion. PhD Thesis, University of Nottingham.

Haslegrave, C M, and E N Corlett, 1988. Factors determining the posture adopted for forceful manual tasks. In: Proceedings of the 10th Congress of the International Ergonomics Association, Sydney, 1-5 August 1988.

Haslegrave, C M, M Tracy and E N Corlett, 1988. Influence of working posture on strength capability. In: Contemporary Ergonomics 1988, (Ed: E D Megaw), Taylor and Francis: London.

HSC (Health and Safety Commission), 1988. Handling loads at work - proposals for regulations and guidance. Consultative Document, Health and Safety Executive, UK.

NIOSH (National Institute for Occupational Safety and Health), 1981. Work practices guide for manual lifting. NIOSH Technical Report, DHHS (NIOSH) Publication No 81-122, US Department of Health and Human Services.

Patterson, P, J Congleton, R Koppa and R D Huchingson, 1987. The effects of load knowledge on stresses at the lower back during lifting. Ergonomics, 30, 3, 539-549.

Warwick, D, G Novak, A Schultz and M Berkson, 1980. Maximum voluntary strengths of male adults in some lifting, pushing and pulling activities. Ergonomics, 23, 1, 49-54.

# 24. SEATED POSTURE AND WORKSTATION CONFIGURATION

## Marvin J. Dainoff and James Balliett

*Center for Ergonomics Research*
*Department of Psychology, Miami University*
*Oxford, Ohio, USA*

**Abstract**

A multidisciplinary integrated approach to the problem of seated work posture is clearly required. It is argued that the theoretical framework of ecological psychology will provide such an approach. An outline of the approach is provided, with specific application to seated posture, and a supporting programme of research is described.

**Keywords:** *posture; seating; chair design; discomfort; static workload*

## Introduction

The static load from prolonged seated fixed work postures can have serious effects on worker health and productivity (Hettinger, 1985; Grieco, 1986). Controlling this problem will clearly require an integrated approach involving improved workstation design, reasonable rest breaks, job redesign, and employee education. On the other hand, this kind of integrated approach will, in many organizations, require a fundamental change in management philosophy. It is unlikely that these changes will come about merely because ergonomists and other occupational health experts say that it is a good idea. Management will no doubt need carefully documented details of how such programmes can be implemented. A particularly difficult problem relates to the inevitable trade-off questions: that is, it would be best to redesign both the task and the workstation, but if we can't do both, which one should receive priority?

What are the scientific bases for answering such global, integrative questions? A good place to begin is with Corlett's (1989) recent and comprehensive overview of methods and approaches to research on seated posture. An overall seating model is proposed which links functions (task, sitter, seat); effects (e.g., discomfort, pressure, pain); and a large variety of measures (physiological, behavioural, geometrical). Corlett concludes that, despite progress, we do not yet have an effective working seat.

Corlett's model has the value of integrating a large and diverse body of literature, while focusing on practical concerns in designing and selecting work chairs. Our approach, although aimed at exactly the same goals, is complementary. We propose an integrated theoretical framework from which we derive a long-term systematic programme of research, utilizing whichever methodological techniques are most useful in testing specific derived hypotheses. We believe that only a focused, systematic attack on the complex problems of the integrated workplace will achieve the end results which Corlett desires. In this paper, we provide a brief overview of this framework, indicate a research strategy derived from this framework, and present some preliminary results.

## Theoretical framework

Our conceptual approach to the specific problems of seated work posture is based on the broader framework of ecological psychology elucidated by Gibson (1979). The brief description which follows here is based on the discussion in Mark et al. (in press) and Dainoff and Mark (1989).

### Ecological ergonomics

The Gibsonian framework, out of which has developed a large body of basic research, focuses on understanding the coupling of perception and action. The performance of commonplace actions such as reaching, climbing, and sitting, depend critically on the individual's (or **actor's**) ability to perceive properties of the environment related to such actions. This framework can be described as **ecological** in the sense that it rests on a complimentarity between attributes of the individual described in terms of the environment and attributes of the environment defined in terms of the individual. Insofar as the primary focus of ergonomics should be on the **fit** between people's goal-directed actions and their environments, **ecological ergonomics** is a natural and ultimately practical development.

The research agenda for ecological ergonomics would be to provide information about human action capabilities, and the concomitant environmental properties required to support performance of specific actions. If we regard the task of the designer to be, broadly stated, the building of environmental objects whose attributes take into account the capabilities and needs of prospective users, then the results of ergonomics research ought to feed directly into support of the designer's objectives. In a sense, the designer's final product embodies a theory, either explicit or implicit, of the relationship between the user's action capabilities and environmental support of those capabilities. Ideally, the environmental ergonomist's efforts to describe/measure/understand that same relationship could refine the designer's theory and ultimately lead to an improved fit. How might this ideal state be brought about?

### Constraints on action

A starting place for ecological analysis is to describe the **constraints** on those actions which might possibly be carried out by an actor. Some constraints, called **effectivities**, are attributes of the individual actor. Effectivities can be classified broadly in three ways: geometric, kinetic, and task. Geometric constraints result from the relationship of the actor's body dimensions (anthropometry) to the size/scale of relevant aspects of the environment. For example, an actor whose popliteal height is 35 cm is constrained from sitting comfortably in a nonadjustable chair with a 45 cm seatpan height. Kinetic variables (mass, force, friction, elasticity) also constrain potential actions. A very weak or obese individual may be unable to get up from a deep soft armchair.

Geometric and kinetic constraints, which are intrinsic to the physical properties of the actor, establish boundaries on a large range of potential actions. Task constraints, being functional/psychological, determine which of these actions are actually performed. Task constraints entail goal-directed intentionality; the seated postures (actions) required for high-speed data entry will likely differ from those involved in recreational television watching.

### Environmental possibilities for action: affordances

Effectivities describe the range of goal-directed activities an individual can perform within a given environment; they specify the action capabilities with respect to the environment. The complement of an effectivity is an **affordance**. Affordances refer to the properties of the environment (object/system) taken with reference to an individual's action capabilities and goals.

Thus, a 45 cm high seatpan does not afford comfortable sitting to a person with a 35 cm popliteal height effectivity (assuming that "comfortable" entails the feet resting on the floor), but would, in fact, afford comfortable sitting to a person with a 45 cm popliteal height. For a chair designer, producing a chair which affords "comfortable sitting" for persons of different action capabilities (popliteal heights between 35 and 45 cm) requires an adjustability range between 35 and 45 cm.

Finally, a central point in ecological analysis is that the actor directly perceives the affordances of the environment, picking up information as to whether the structured layout of the environment will allow selected actions to be performed. Further, the extent to which such structure is, in fact, picked up, depends on **exploratory behaviour** by the actor. The willingness and ability to engage in such behaviour is, of course, an effectivity of a particular actor.

### A case in point: the forward-sloping chair

The use of ecological analysis may provide a new perspective from which to view some of the controversy surrounding the use of the forward-sloping chair. According to the literature, whereas unsupported upright sitting on a horizontal seat tends to result in a (potentially stressful) lordotic posture in the lumbar spine, the use of a forward-sloping seatpan restores the (less stressful) lordotic posture (Zacharkow, 1988; Mandal, 1981). However, forward-sloping chairs may require additional muscular effort in the lower legs to maintain stability at the cost of relaxation (Corlett, 1989; Branton, 1969); and have been found by some workers to be less preferred than conventional chairs (Drury and Francher, 1985; Bendix et al., 1988).

From an ecological perspective, the functional advantage of the forward-sloping chair may be to afford lower back support in work situations such as high-speed data entry, or manuscript typing from paper copy. A conventional ergonomic chair with vertical backrest and lumbar support may be provided; however, the relatively small typeface typical of paper copy requires a shorter viewing distance than that appropriate for VDU screen displays. Hence, the actor is constrained to bend forward, the lumbar support is minimally effective, if at all, and a kyphotic lumbar spine ensues (Mark et al., in press; Dainoff and Mark, 1989).

The difficulty may arise in that the actor may not perceive the affordance of the forward-sloping seatpan for lumbar support. Such chairs have not, until recently, been a common part of most people's worlds. If asked to try sitting in a forward slope, kinetic constraints (body centre of mass moving forward) may result in a disagreeable perception of sliding out of the chair. In this case, the designer (experimenter) must provide a higher order affordance (instruction and training) resulting in a cognitive effectivity (willingness to explore so as to discover the affordances of the chair; understanding the need to use lower leg muscles to find a balance point). We have, in our laboratory, successfully conducted such training with 15-20 min videotaped presentations (Dainoff and Mark, 1987).

## Degrees of freedom problem

Should the combination of effectivities (task constraints, intention) and affordance structure of the chair allow, the actor may shift from a forward to a backward leaning posture. This working posture, with both seatpan and backrest angles sloping rearward, will also provide support to the lower back and seems to be naturally preferred by users (Grandjean et al., 1983). However, as discussed above, the affordances of the task structure (e.g., text size of visual display) should not constrain the head to move forward. In our work, a screen-intensive editing task affords use of the backward-sloping posture insofar as the typical VDU text can be read without leaning forward (Dainoff and Mark, 1987).

To carry out this manoeuvre, the actor must make three separate adjustments of chair surfaces. Seatpan and backrest angles must be rotated from forward to rear, and seatpan height must be lowered to compensate for the raised forward portion of the seatpan.

There are two ergonomics problems here. The first is the classic "human factors" concern of mapping control mechanisms (input) to desired actions (output). The design of chair adjustment mechanisms which are easy to use, and obviously and logically related to intended action, is an area which most furniture manufacturers could greatly improve.

A second, deeper concern can be considered a special case of the general **degrees of freedom** problem conceptualized by the kinesiologist N. Bernstein (1967). While the theory of the general case is too involved to discuss here (see Turvey et al., 1982; Kugler and Turvey, 1987), the approach may be useful in thinking about the operator's control problem.

Simply stated, the usual response of ergonomically oriented designers towards dealing with a variable environment (variability in size/shape of users; variability in task/operator demands) is to provide more flexibility through additional possibilities for adjustment (degrees of freedom). To the three degrees of freedom already described (seatpan and backrest angle, seatpan height) can be added backrest height adjustability, tension control, and armrest height adjustability; parallel sets of height and angle adjustments may be applied to keyboard and VDU support surfaces. While it may seem uncharitable for an ergonomist to complain about too much flexibility, it must be recognized that each independent degree of freedom added to a system has a (cognitive) cost associated with the processing required for its control. At some point, the cognitive overload will be such that the user will simply ignore the adjustment. There is some evidence that this is now occurring with users of advanced ergonomic chairs (cf. Oman, 1988).

Bernstein's solution to the degrees of freedom was to specify **equations of constraint** so as to allow reduction of degrees of freedom to a manageable number. One can imagine the problem of driving an automobile with four independently controllable wheels; the problem is solved by constraining the degrees of freedom (through axles and differential gears) to a single direction. How to accomplish this in the case of workstation design is more problematic.

Two approaches to constraining degrees of freedom which have been seen in the marketplace involve **linking** of seatpan and backrest angles, and introduction of so-called "passive" ergonomics in which the seatpan and backrest surfaces respond dynamically to movements of the body. We have argued elsewhere that the latter solution lacks the requisite stability for many types of work demands, whereas the former constrains users to postures they would not naturally take up if given independent control (Dainoff, 1989).

A systematic ecological approach to dealing with the degrees of freedom problem in seated work posture forms the basis of our overall research strategy. Indeed, we argue that this problem is hardly limited to the arena of seated work posture, but underlies a large domain of current issues in human-computer interaction, and control/management of complex systems.

## Research strategy

### Analytic framework

We regard the analysis of seated working posture as a special case of postural **orientation** (Riccio and Stoffregen, 1988). The actor/operator engages in a controlled interaction with the surfaces of support (seatpan, backrest, floor, work surface) with the goal of maintaining dynamic equilibrium. The fine structure of this interaction is determined by four classes of constraints inherent in the ecological analysis, affordances and effectivities, of the work situation:

1. System Goals
2. Workstation Characteristics
3. Operator Characteristics
4. Chair Characteristics.

The nature of the task (e.g., data entry vs. editing) may require very different postural orientations depending on the viewing distance of the copy/screen, the physical dimensions of the operator, and the nature/range of workstation adjustments. Within these constraints, it is assumed that an operator will minimize muscular efforts in order to maintain equilibrium.

Our approach has borrowed certain theoretical formalisms from dynamic systems theory (e.g., Abraham and Shaw, 1982). Given a chair with adjustable seatpan and backrest surfaces, and assuming height to be constant, we can, to a first approximation, describe specific postural configurations in terms of a particular combination of seatpan and backrest angles. A three-dimensional plot of seatpan against backrest angle on the X and Y axes, and dwell time per configuration on the Z axis defines a **configuration space**. At a given point in time, we can define a region within the configuration space where posture is stable, as evidenced by high levels of dwell time. This region represents an optimal zone or **attractor** in which there is a (temporary) state of biomechanical balance, and the energy expended in maintaining posture is minimized (Kugler and Turvey, 1987).

Since any fixed posture will generate fatigue, all postural attractors are necessarily temporary. Nevertheless, we believe that there is a limited number of attractors in the configuration space for seated posture, each of which can be specified by the set of four constraints described above (Mark et al., in press).

The conception of a limited number of stable attractors as optimal postures has several direct ergonomics implications. On the one hand, the designer can concentrate on designing affordances which will facilitate stable posture in the most likely regions of optimality, while simultaneously allowing for ease of movement between attractor regions. On the other hand, user instruction can now be couched in terms of encouraging exploration of the configuration space so as to find the individual's unique set of attractors.

In this context, examining the problem of perception will become a long-term but crucial concern. The primary perceptual system to be involved is likely to be **haptic**, a combination of active touch and kinaesthesis. The haptic system will have two roles; providing information which signals a buildup of fatigue (ischemia?) sufficient to require a shift in position, and guiding the self-regulation of body and chair surfaces to achieve a new stable attractor.

## Initial investigations

The goals of the current research programme are to examine the interactions between changes in work posture, adjustability of workstations, musculoskeletal load (assessed both objectively and subjectively) and the time course of work performance.

Our research strategy involves two basic steps. Step one is essentially calibration. Operators will perform computer-based tasks in fixed seated work postures. Some of these postures will involve relatively low biomechanical loads, assessed through biomechanical modelling. Others will involve loads which are clearly greater than most standards and recommendations will allow. We will use a combination of measures: subjective discomfort, subjective effort and work performance variability. The differential time course of these variables will be examined as a function of postural load. We will use this data to establish a set of extreme limits of postural adjustments, which can serve, in turn, as a basis for step two of our approach.

Step two will involve a more complex design in which operators will initially be placed in high load/uncomfortable postures, but then be allowed to adjust themselves into more comfortable orientations. We will investigate the time course of this self-regulation process using the same measures established in step 1.

### Method

Subjects alternated between two working postures in counterbalanced order while performing a two finger tracking task using the keyboard and display screen of an IBM Personal Computer. In Posture 1, the subject was placed into a 90 degree working posture with forearms and thighs horizontal, upper arms and lower legs vertical, and trunk firmly against the lumber support of the backrest. In Posture 2, the upper arm was flexed 30 degrees and the upper arm-forearm angle maintained at 115 degrees. Posture 2 was clearly outside the range of acceptable posture as defined by the recent U.S. Standard (ANSI/HFS 100-1988). To achieve these postures, an advanced ergonomic chair (Fixtures Furniture) and a motorized adjustable bi-level worksurface (Haworth) were adjusted for the subject by the experimenter, and postures verified through video analysis.

The nominal duration of the experimental session was two hours, with a 15 minute pause between two working posture periods. Subjects worked at the tracking task in five minute blocks. At the conclusions of each five minute block, they gave ratings of subjective effort, and subjective discomfort, localized by body part. Each set of rating was on a 10 point scale, and the effort ratings had been previously calibrated for each subject using weights. Subjects were asked to maintain postures throughout the experiment, but were told to immediately stop work if subjective discomfort reached a value of 4.

### Preliminary results

Preliminary data from four subjects were examined; two from each of the two counterbalanced order of presentation. Of the two subjects who worked under the less stressful posture first, Subject 1 completed both sessions, but her rated effort in the second, more stressful posture, was clearly higher. Subject 2 completed 70% of the less stressful session, but only 30% of the more stressful session. Of the two subjects who worked under the more stressful posture first, Subject 3 completed only 20% of the more stressful session, but 80% of the less stressful but later session. The final subject completed only 50% of either session.

Examination of total discomfort scores at the time of stopping work in either session indicates that the primary difference appeared to be between first and second session.

These very preliminary data indicate that, for three of the four subjects, the effect of the postural variable was being felt. It also seems to be the case that order effect was most likely interacting with the posture effect; indicating the possibility of asymmetric carry-over. This would require that either the rest break be increased, or an independent groups design be utilized.

## Discussion

The purpose of step one of this initial study is to find a simple laboratory procedure in which differential responses to postures with clearly different biomechanical loads can be established. Preliminary results indicate we might well be successful in that effort. The next step would be to examine the impact (both objective and subjective) of an active search for an optimal region in the same task environment. This research paradigm could be used to systematically explore the impacts of all four classes of constraints (system, workstation, operator, and chair) on the dynamic orientational activity of the operator. It should, in principle, be possible to triangulate on a set of conditions which might be simultaneously optimal for operator comfort and productivity. Such conditions would be reflected by high levels of performance, low levels of subjective discomfort/effort, relatively long dwell times between movements within the configuration space, and low levels of biomechanical load at the resulting attractor sites.

## References

Abraham, R. H., and Shaw, C. D. (1982). Dynamics - the geometry of behavior: Part One: Periodic Behavior. Santa Cruz, California: Aerial Press.

ANSI/HFS 100-1988: American National Standard for Human Factors Engineering of Visual Display Terminal Workstations. Santa Monica, California: Human Factors Society.

Bendix, A., Jensen, C. V., and Bendix, T. (1988). Posture, acceptability, and energy consumption on a tiltable and knee support chair. Clinical Biomechanics, 3, 66-73.

Bernstein, N. (1967). The coordination and regulation of movements. London: Pergamon Press.

Branton, P. (1969). Behaviour, body mechanics and discomfort. In E. Grandjean (Ed.), Sitting Posture. London: Taylor and Francis Ltd.

Corlett, E. N. (1989). Aspects of the evaluation of industrial seating. Ergonomics, 32, 257-270.

Dainoff, M. J. (1989). Three propositions about ergonomic seating which seem correct are actually doubtful. International Conference on Science of Seating, Waseda University, Tokyo, November 1989.

Dainoff, M. J., and Mark, L. S. (1987). Task and the adjustment of ergonomic furniture. In B. Knave and P-G. Wideback (Eds.), Work with Display Units. Amsterdam: North-Holland.

Dainoff, M. J., and Mark, L. S. (1989). Analysis of seated postures as a basis for ergonomic design. In M. J. Smith and G. Salvendy (Eds.), Work with Computers: Organizational, Management, Stress, and Health Aspects. Amsterdam: Elsevier Science Publishers.

Drury, C. G., and Francher, M. (1985). Evaluation of a forward-sloping chair. Applied Ergonomics, 16, 41-47.

Gibson, J. J. (1979). The ecological approach to visual perception. Boston: Houghton-Mifflin.

Grandjean, E., Hunting, W., and Pidermann, M. (1983). VDT workstation design: Preferred settings and their effects. Human Factors, 25, 161-175.

Grieco, A. (1986). Sitting posture: An old problem and a new one. Ergonomics, 29(3), 345-362.

Hettinger, T. (1985). Statistics on diseases in the Federal Republic of Germany with particular reference to diseases of the skeletal system. Ergonomics, 28, 17-20.

Kugler, P. N., and Turvey, M. T. (1987). Information, natural law, and the self-assembly of rhythmic movement. Hillsdale, New Jersey: Erlbaum Associates.

Mandal, A. C. (1981). The seated man (homo sedens). The seated work position. Theory and practice. Applied Ergonomics, 12, 19-26.

Mark, L. S., Dainoff, M. J., Moritz, R., and Vogele, D. (in press). An ecological framework for ergonomic research and design. In R. R. Hoffman and D. A. Palermo (Eds.), Cognition and the Symbolic Processes: Volume III. Hillsdale, New Jersey: Erlbaum Associates.

Oman, P. W. (1988). College survey: Are VDT operator complaints real? The Office, 8, 61-64.

Riccio, G. E., and Stoffregen, T. (1988). Affordances as constraints on the control of stance. Human Movement Science, 7, 265-300.

Turvey, M. T., Fitch, H. L., and Tuller, N. (1982). The Bernstein perspective: I. The problem of degrees of freedom and context conditioned variability. In J.A.S. Kelso (Ed.), Human Movement Behavior; An Overview. Hillsdale, New Jersey: Erlbaum Associates.

Zacharkow, D. (1988). Posture: Sitting, Standing, Chair Design and Exercise. Springfield, Ill.: Charles C. Thomas.

# 25. THE EFFECT OF SIX DIFFERENT KINDS OF GLOVES ON GRIP STRENGTH

## Mao-Jiun J. Wang

*Department of Industrial Engineering*
*National Tsing Hua University*
*Hsin Chu, Taiwan, 300 ROC*

### Abstract

While gloves are often indispensable from the standpoint of job safety, wearing them impairs grip strength. An experiment was conducted with 52 subjects wearing six types of gloves often used in industry. They were rubber, cotton, leather, surgical, open-finger and asbestos gloves. Bare hand grip strength, measured with a hand dynamometer, was used as the criterion for comparison. The mean grip strengths for four of the six gloved conditions were significantly less than mean barehand grip strength. The leather gloves had the largest reduction in grip strength by 18%. The extent of reduction in grip strength was correlated significantly with glove thickness and finger discomfort rating. This effect was consistent across subjects although their absolute strengths varied greatly. A possible explanation for this strength reduction and its implications in workplace design are discussed.

**Keywords:** *grip strength; gloves; clothing materials; discomfort*

## Introduction

Sargent first reported the use of hand grip as a test for determining strength in 1897 (Sargent, 1897). Since that time, grip strength has been correlated with hand dominance, overall physical fitness, and normal growth. It has also been used to determine the seriousness of upper extremity injuries and the success of rehabilitation programmes in restoring grip strength. In 1964, Toews concluded the grip strength difference between the major and minor hand for three classes of tested group was 5.8 percent with the major hand predominating (Toews, 1964). Lunde, Brewer and Garcia (1972) performed a grip strength study on 57 college women using body weight (W) and height (H) as independent variables to predict strength for both major and minor hands. The equation for major hand grip strength was $G_{mj} = 6 + 0.58H + 0.19W$ and minor hand was $G_{mi} = -9 + 0.83H + 0.12W$. The $R^2$ values were low at 0.32 and 0.27 respectively showing that less than one third of the variation could be explained by the model. For age factor, Schmidt and Toews (1970) reported that grip strength is directly proportional to age up to 32 years, and inversely proportional from then on. Also, it is well recognized that men have greater grip strength than women and retain this superiority throughout adult life (NIOSH, 1981).

While grip strength had been the main focus in the above mentioned studies, bare hands are not often used in day to day work in industry. Many industrial tasks involve people wearing gloves. The primary function of gloves is to protect hands from the environment. However, wearing gloves impairs certain human capabilities such as fine finger movements, touch and grip strength. In this regard, Griffin (1944) showed that the manipulability of hands and fingers is reduced with gloves. Stump (1953) indicated that it takes longer to adjust a knob when wearing gloves than without gloves. Further, Bradley (1969) evaluated the effect of gloves on the manipulation of 5 types of controls and concluded that the control operation time is affected by type of gloves worn. As for grip strength, Hertzberg (1955) found a consistent reduction of grip strength among flying personnel when flying gloves were worn. He showed that there was a 20% reduction in grip strength with standard heavy flying gloves (wool liner inside a leather shell). Recently, Riley et al. (1985) assessed single, double and no glove conditions for four different force tasks. They concluded that the single glove condition was superior to the conditions of no glove and double gloves. Subsequently, Cochran et al. (1986) and Wang et al. (1987) compared different kinds of gloves applied in industry, and both reported that a significant reduction in grip strength was found. One limitation on these results is the rather small number of subjects (7 and 10 respectively) used in both studies. These studies indicated that gloves do have a significant effect on grip strength as well as on finger dexterity. This must be considered in equipment and job design if people will be wearing gloves on the job.

The objective of this study was twofold: (1) to compare the effects of different kinds of gloves on grip strength, (2) to evaluate whether glove thickness has an effect on grip strength reduction. Six kinds of gloves that are commonly used in industry were chosen. They were leather, asbestos, rubber, cotton, open-finger, and surgical gloves. The material characteristics and industrial applications of the six different gloves are described in Table 1.

Table 1. *The material characteristics and industrial applications of six different types of gloves*

| Type of Gloves | Material | Industrial Application |
|---|---|---|
| Asbestos | Asbestos | For thermal environment |
| Leather | Leather | For welding and thermal environment |
| Rubber | Rubber | For chemical environment |
| Cotton | Cotton | For packaging and other handling |
| Open-finger | Synthetic leather | For fine assembly |
| Surgical | Synthetic rubber | For medical environment or electronics assembly |

## Method

Fifty two male college students volunteered for the experiment. They were all in good health, had no known hand disability, and were well-motivated to exert maximum effort during each test. For each subject, seven grip strength conditions (bare hand, asbestos, leather, rubber, cotton, surgical, and open-finger gloves) provided a total of 364 (7x52) conditions. Due to the availability of gloves, only two sizes of each kind were provided. The better fitting one was worn to perform the test. For each of these conditions the dynamometer span was set at 5 cm (2 in.). The subject was asked to exert his/her maximum force and to sustain it for three seconds using the standard strength testing methodology (Caldwell et al., 1974). To reduce the order effect of testing, three replications were carried out for each hand and for each glove, with a minimum of two minutes rest between successive replications. The maximum observed grip strength of the three replications was taken for each condition. Also, at the end of each gloved condition, the subject was asked to evaluate, using a 7-point scale, the finger discomfort when applying force. This procedure was repeated for all 364 conditions in random order.

Analysis of variance, Duncan's multiple range test and correlation analyses were performed to evaluate the results.

## Results

Maximum voluntary grip strength is affected by the anthropometry of the individual. Owing to a large between-subject variability, the percentage reduction in grip strength with gloves as compared to bare hand grip strength was treated as the primary dependent variable. A one-way analysis of variance was performed on the grip strength data. As shown in Table 2, the effect of gloves was significant at $p. < 0.001$. Table 3 gives the mean percentage reduction in grip strength across 52 subjects for each type of gloves.

*Table 2. One-way ANOVA result*

| Source | D.F. | SS | MS | F | F-Prob |
|---|---|---|---|---|---|
| Between groups | 6 | 2529.31 | 421.55 | 6.02 | 0.0000 |
| Within groups | 357 | 25015.21 | 70.07 | | |
| Total | 363 | 27544.52 | | | |

In order to evaluate the effect of gloves further, Duncan's multiple range test was performed on the means. The results in Table 3 indicate that the mean grip strengths of leather, asbestos, rubber and cotton gloves were significantly less ($p. < 0.05$) than bare hand grip strength. As compared to the bare hand condition, leather gloves reduced grip strength by 18% while asbestos, rubber and cotton gloves reduced it by 17.1%, 12.1%, and 10.4%, respectively. No significant difference was found among leather, asbestos, rubber and cotton gloves. Open-finger gloves and surgical gloves which had a reduction of 6.7% and 3.7% respectively showed no significant difference on mean grip strength with bare hand grip strength. For asbestos, rubber and cotton gloves, the level of reduction was less than the result of the previous study (Wang et al., 1987). It is primarily due to the differences in glove thickness. Since the gloves thickness data were not measured in the earlier study, no further quantitative comparison can be made.

Table 4 shows mean finger discomfort rating, and thickness data for six different types of gloves. Three correlations were performed among the three variables (glove thickness, mean discomfort rating, and percent grip strength reduction). It is interesting to note that they were all significantly intercorrelated with each other. Firstly, the significant correlation between glove thickness and percent grip strength reduction ($r = 0.884$, $p. < 0.01$) indicates that the thicker the gloves, the larger the reduction in grip strength. Secondly, the significant negative correlation between glove thickness and mean discomfort rating ($r = -0.917$, $p. < 0.01$) demonstrates that the thicker the gloves, the greater degree of finger discomfort experienced when exerting maximum hand force. Lastly, the significant negative correlation between mean discomfort rating and percent grip strength reduction ($r = -0.698$, $p. < 0.05$) reveals that the smaller the degree of finger comfort, the higher the level of grip strength reduction. It seems that there is a causal relationship among them which is discussed next.

Table 3. Results of Duncan's multiple range test

|  | Type of Gloves | | | | | | |
|---|---|---|---|---|---|---|---|
|  | Leather | Asbestos | Rubber | Cotton | Open Finger | Surgical | Bare Hand |
| Mean (Kg) | 15.87 | 16.04 | 17.01 | 17.34 | 18.05 | 18.63 | 19.35 |
| Percent reduction (%) | 18.0 | 17.1 | 12.1 | 10.4 | 6.7 | 3.7 | -- |
| Duncan's test result | ---------------------------------- | | | | | | |
|  |  |  |  | -------------------------- | | |  |
|  |  |  |  |  | -------------------- | |  |
|  |  |  |  |  |  | -------------------- |  |

Table 4. Mean finger discomfort ratings and glove thickness

|  | Type of Gloves | | | | | |
|---|---|---|---|---|---|---|
|  | Leather | Asbestos | Rubber | Cotton | Open-Finger | Surgical |
| Mean discomfort rating | 2.75 | 2.67 | 3.23 | 4.42 | 4.48 | 4.40 |
| Glove thickness (mm) | 1.25 | 1.75 | 0.75 | 1.15 | 0.55 | 0.33 |

## Discussion and conclusion

At the outset, the observed results indicate that:
a) There is a reduction in grip strength when wearing gloves.
b) The extent of reduction depends on the thickness of gloves.

In this experiment the distance between the dynamometer's heel and the trigger was controlled at 5 cm (2 in.) which is approximately the optimal span setting (Hertzberg, 1955; Wang, 1982), and the subjects were asked to exert their maximum force. With gloves, the angle subtended by the middle phalanges on the trigger changes, which would cause a reduction in grip strength. Further feedback on the extent of gripping

is given by interdigital nerve endings. When fingers press against the heel of the hand and against each other, discomfort is felt. With gloves this feedback is generated at an earlier stage of the grip due to glove thickness. Hence, the subject reaches a maximum at a lower grip strength. Therefore, with gloves:
1. There is an apparent increase in grip span, that is, the distance between the dynamometer's heel, and the trigger.
2. There is earlier pressing of the fingers against each other.

The first will increase grip strength up to a point, and then decrease it (Hertzberg, 1955), and the second will reduce the grip strength. In this experiment, it appears that interdigital pressure is the primary determinant of the reduced grip strength when gloves are worn. The level of feedback of interdigital pressure which can be indicated by the degree of finger discomfort rating was directly related to the thickness of the gloves.

Because only two sizes of gloves were used for all subjects, part of the between-subject variations might be caused by the improper fit of gloves. However, such a thing is not uncommon in industry, where two or three glove sizes are stocked for the full range of hand sizes in a department.

In conclusion, the findings of this study suggest that the thickness of the gloves was the major factor for promoting grip strength reduction. Among the gloves used, leather gloves had the largest reduction which was about 18%. It was consistent across subjects despite wide variations in grip strength. Products or equipment that are likely to be handled by people wearing gloves should be designed with this strength loss in mind, and the required forces should be reduced accordingly. Further research is needed to determine the exact relationships among material characteristics of gloves, glove geometry, and the reduction in strength with gloves.

## References

Bradley, J. V., 1969, Effect of gloves on control operation time. Human Factors, 11(1), 13-20.

Caldwell, L. S., Chaffin, D. B., Dukes-Dobos, F. N., Kroemer, K.H. E., Laubach, L. L., Snook, S. H. and Wasserman, D. E., 1974, A proposed standard procedure for static muscle strength testing. American Industrial Hygiene Association Journal, 35, 201-205.

Cochran, D. J., Albin, T. J., Bishu, R. R. and Riley, M. W., 1986, An analysis of grasp force degradation with commercially available gloves. Proceedings of the Human Factors Society - 30th Annual Meeting, pp.852-855.

Griffin, D. R., 1944, Manual dexterity of men wearing gloves and mittens. Fatigue Lab., Harvard University, Report No. 22.

Hertzberg, T., 1955, Some contribution of applied physical anthropometry to human engineering. Annals of the New York Academy of Science, 63, 621-623.

Lunde, B. K., Brewer, W. D. and Garcia, P. A., 1972, Grip strength of college women. Archives of Physical Medicine & Rehabilitation, 53, 491-493.

NIOSH., 1981, Work Practice Guide for Manual Lifting, DHHS (NIOSH) Publication No. 81-122, Cincinnati.

Riley, M. W., Cochran, D. J. and Schanbacher, C. A., 1985, Force difference due to gloves. Ergonomics, 28, 441-447.

Sargent, D. A., 1897, Strength test and the strong men of Harvard. American Physiological Education Review, 11, 108.

Schmidt, R. T. and Toews, J. V., 1970, Grip strength as measured by the Jamar dynamometer. Archives of Physical Medicine & Rehabilitation, 51, 321-327.

Stump, N. E., 1953, Manipulability of rotary controls as a function of knob diameter and control orientation. WADC in WCRD 53-12 Dayton, Ohio, Wright Patterson AFB.

Toews, J. V., 1964, A grip strength study among steel workers. Archives of Physical Medicine & Rehabilitation, 45, 413-417.

Wang, M. J., 1982, A study of grip strength from static efforts and anthropometric measurements. Unpublished Master thesis, University of Nebraska-Lincoln.

Wang, M. J., Bishu, R. R. and Rodgers, S. H., 1987, Grip strength changes when wearing three types of gloves. Interface 87 Proceedings, Human Factors Society, Santa Monica, California, pp.349-355.

# 26. A METHOD TO MEASURE THE FORCES EXERTED BY THE FINGERS WHEN WRITING WITH A BALL-POINT PEN

Yoshio Ishida

Department of Mechanical Engineering
Tokyo Metropolitan Technical College
Tokyo, Japan

**Abstract**

The position of the fingers on the penholder varies with subjects. Subjects sometimes change the position of their fingers when writing and do not always hold a pen with the fingertips. This paper reports a method to measure the force of the fingers holding a ball-point pen while writing characters.

**Keywords:** *posture; biomechanics; writing*

## Introduction

In Japan, the word processor is widely used in business and in homes. But a man without a word processor must write characters by hand. A character should be legible since it is a means of communication (Oka, 1980; Sasanuma, 1979).

A sentence in Japanese consists of kanji, a Chinese character, and hiragana. One of the factors for judging legibility of Chinese characters handwritten with a ball-point pen is the balance of the strokes (Ishida, 1982; Ishino, 1981; Sasanuma, 1979; Tada, 1977; Ueguchi, 1986). Most of the strokes of a Chinese character are horizontal and vertical (Kiuchi, 1984). In order to draw a straight horizontal or vertical line, it is important to study how the forces of three fingers, that is, the forefinger, the middle finger and the thumb, should be balanced. Furthermore, it is said that good balance among the finger forces is necessary to reduce hand fatigue in writing (Tomiya, 1977).

To study how the forces of the three fingers should be balanced, it is necessary to develop a method of measurement. But the following problems exist (Ishida, 1982): 1) the position of the fingers holding a penholder varies with subjects, 2) subjects sometimes change the positions of the fingers holding a penholder while writing, 3) subjects do not always hold a penholder with the fingertips.

Pens which have been developed in Japan to measure the finger forces have small pressure converters on the penholder (Ohsawa, 1985) or they have cantilevers, each of which has a strain gauge, on the penholder (Iwata, Moriwaki and Miyake, 1986). But these pens cannot cope with the above problems, because the positions of the fingers holding the penholder must be fixed. A method for attaching a pressure sensor to a finger has been also proposed. But, it is thought that the finger may feel uncomfortable.

The purpose of this paper is to introduce a pen which can measure the finger forces while writing characters without experiencing the problems mentioned above.

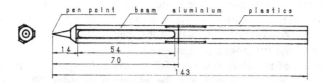

*Figure 1. The external appearance of the pen*

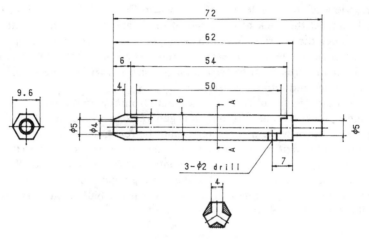

*Figure 2. The cross section of the aluminium holder*

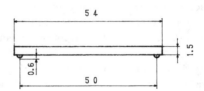

*Figure 3. The side view of the simple beam*

## The construction of the pen

The principle of dynamics that the amount of load to a simple beam is equal to the sum of the magnitudes of reactions generated in both ends is applied to this pen. Basically, it must be possible to measure the forces of the three fingers separately by attaching a simple beam with a strain gauge at each end to every other side of the hexagonal penholder. But the penholders of ball-point pens generally on sale are usually made of plastic. Therefore, it easily bends with finger forces. Consequently, measurement will be impossible because of the interference from the finger forces. In order to prevent this interference, the portion held by the fingers is made of aluminium. The external appearance of the pen is shown in Figure 1.

The penholder consists of two parts. The left part, the penpoint side, is made of aluminium and its length is 62 mm. The right part, the tail side, is made of plastic with a length of 73 mm. The external size of the aluminium holder is similar to a ball-point pen and the section of the holder is hexagonal. In every other side of the aluminium holder, there is one slender hole. And one simple beam made of metal is attached to each hole. Consequently, the maximum outside diameter of the penholder is about 10.8 mm. Wherever subjects hold this pen within the surface of the simple beam, the force of the finger can be measured.

The cross section of the aluminium holder is shown in Figure 2. The tail side has a diameter of 5 mm and is inserted into the plastic penholder. The ink holder of a ball-point pen is inserted into the left hole of 4 mm diameter. This penpoint can easily be changed.

The side view of the simple beam is shown in Figure 3. The width of the simple beam is 4 mm. At the back of each end of the beam, one fulcrum is attached. The distance between the fulcra is 50 mm. The height of the fulcrum is 0.6 mm. One strain gauge is attached to the side of each fulcrum. Therefore, one beam has two strain gauges. The lead wires from the strain gauges are drawn into the inner hole of the aluminium holder and into the plastic one.

Whenever the subjects holds the pen, outputs from the strain gauges are generated by the finger forces. Because this output is generated continuously, the point at which the subjects begin to write a character cannot be recognised. In order to recognise the beginning of the writing, a wire of 0.5mm diameter was inserted into the ink holder so that the pointed top of the wire made contact with the ball point. When the ball point touches the metal foil attached to the writing paper which acts as an electrode, no electrical signal is generated. However, when the ball point is raised from the metal foil a signal is generated.

## *The calibration curve*

### How to draw the calibration curve

On the surface of each beam, seven loading points were established by dividing the surface into three portions on either side of the centre at 8mm intervals. The outputs from the gauges were recorded at each loading point as the load was increased from 0 to 4kgf in 0.5 kgf steps and then repeated with the load reduced from 4 to 0kgf in similar steps. Averages were taken after repeating this procedure a further two times. The resulting calibration curves for the penpoint-side and the tail-side gauges are shown in Figures 4 and 5 respectively. Figure 6 shows an example of how values within a 0.5kgf step can be obtained by a process of extrapolation.

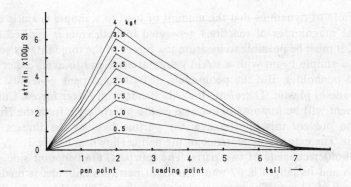

Figure 4. *An example of the calibration curves for the penpoint-side gauge*

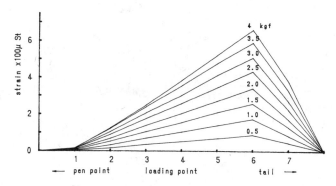

Figure 5. An example of the calibration curves for the tail-side gauge

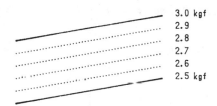

Figure 6. An example of the interpolation between two adjacent curves

## The usage of the calibration curve

When the load of magnitude **W** is applied to the simple beam at the point of $P_0$, both gauges are excited. Let A and B be the output strain from the penpoint-side and the tail-side gauge respectively. As seen from Figures 4 and 5, the shapes of every calibration curve are triangular. In Figure 4, the parallel line to the horizontal axis through A on the vertical axis is sure to intersect with some curves at one or two points. The same statement is true for the parallel line through B in Figure 5. From Figure 7, it can be said that the loading point for the penpoint-side gauge must be $P_1$ or $P_2$ and that the loading point for the tail-side gauge must be $P_3$ or $P_4$. But one of the following pairs, that is $\{P_1,P_3\}, \{P_1,P_4\}, \{P_2,P_3\}$ and $\{P_2,P_4\}$ must be the pair of $\{P_0,P_0\}$.

Based on this argument, both the magnitude of the load applied to the beam and the loading point can be obtained by the following steps:

Step 1. Obtain the points, namely $P_1$, $P_2$, $P_3$ and $P_4$, from the calibration curve for each magnitude of a load in Figures 4 and 5 by drawing parallel lines through A and B.

Step 2. Make the pairs of $\{P_1,P_3\}, \{P_1,P_4\}, \{P_2,P_3\}$ and $\{P_2,P_4\}$ for each calibration curve.

Step 3. Find the calibration curve which has the same pair and read the value of the magnitude of the curve.

Thus, both the real finger force and the real holding point can be measured.

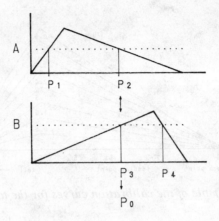

Figure 7. The use of the calibration curve

Table 1. The normal style

| finger | partial correlation coefficient ($\phi$=177) |
|---|---|
| middle finger | − 0.099 |
| forefinger | − 0.140 |
| thumb | 0.077 |

Table 2. The abnormal style

| finger | partial correlation coefficient ($\phi$=177) |
|---|---|
| middle finger | 0.110 |
| forefinger | 0.080 |
| thumb | − 0.091 |

## Application and discussion

Using this pen, eighteen right-handed male subjects were asked to draw a vertical line of 20 mm on the paper which was marked out at 2 mm intervals and on which the electrode was pasted. At this time, the subjects were directed to use two styles of holding the pen for confirming the roles of the fingers. One is to hold the pen with three fingers, namely, the forefinger, the middle finger and the thumb and so fix the penholder at the side of the palm. The other is to hold the pen with only three fingers without using the side of the palm. In this discussion, the former will be called the normal style and the latter the abnormal one.

Ten measured points were located on the drawn line 2 mm from the top. The degree of undulation of the drawn line was measured at each of ten points. In this report, the degree of undulation means the distance between the drawn line and the standard straight but invisible line through the top of the drawn line. There were 18 subjects and 10 measured points for each subject, so one hundred and eighty data were obtained. At the moment when the ball point parts from the electrode on the paper, the timer started automatically. Each subject's vertical line was recorded using a VTR with the time expression generated by the timer. Simultaneously, the output strains from the beams were recorded on a personal computer in order of the time. Using the time data as the common standard, the forces of the subject's three fingers at every ten measured points were collected.

Some subjects always hold a pen with a stronger finger force and some subjects always with a weaker finger force. Therefore, the finger forces were normalized for each subject to remove this effect.

The normalized forces of three fingers as the explanatory variables and the absolute value of the degree of undulation as the objective variable were applied to the multiple regression analysis for each of the two styles. The results of the normal style and that of the abnormal one are shown in Tables 1 and 2 respectively. The multiple correlation coefficient in the case of the normal style was 0.198 and significant at the 10% level. In the case of the abnormal style, it was 0.166 and significant at the 25% level. All partial correlation coefficients were significant at the 40% level at worst.

The directions of the correlations for the forefinger and the middle finger are opposite to that for the thumb as shown in Tables 1 and 2. Furthermore, the directions of the correlations for the normal and abnormal writing are opposite to each other.

From these results, the following may be concluded:
(1) a straight vertical line may be drawn by balancing the force of the thumb to the forces of the forefinger and the middle finger,
(2) the side of the palm is considered to act as a fulcrum to the pen in order to draw a straight vertical line.

Further studies on the roles of the fingers and the side of the palm in writing legible characters are necessary. The instrumented pen described in this paper would be of great help to such studies.

## *References*

Ishida, Y. 1982, A Study on a Manner of Holding a Ball-Point Pen to Write Legible Characters. In the Proceedings of the 8th Congress of the International Ergonomics Association, pp.250-251.

Ishino, R. 1981, A Textbook for Good Hand Writing (Kin En Sha, Tokyo)(in Japanese).

Iwata, K., Moriwaki, T. and Miyake, S. 1986, Dynamic Analysis of Hand Writing Motion. The Japanese Journal of Ergonomics, Vol. 22, No. 5, 223-229 (in Japanese).

Kiuchi, Y. 1984, An Explanation of Image Recognition (Nikkan Kougyo Shinbunsha, Tokyo)(in Japanese).

Oka, N. 1980, A Textbook Which Makes a Reader a Good Writer (Nippon Hohrei Yohshiki Hanbaisho, Tokyo)(in Japanese).

Ohsawa, K. 1985, The Science of a Character (The publishing bureau of Hohsei University, Tokyo)(in Japanese).

Sasanuma, Y. 1979, A Textbook Which Makes the Best Use of a Reader's Faults in Writing (Nippon Hohrei Yohshiki Hanbaisho, Tokyo)(in Japanese).

Tada, N. 1977, A Textbook for a Practical Penmanship (Nagaoka Shoten, Tokyo)(in Japanese).

Tomiya, E. 1977, An Easy Penmanship (Yuuki Shobou, Tokyo)(in Japanese).

Ueguchi, M. 1986, A Textbook Which Has the Knack of Writing a Good Hand (Shufu To Seikatsu Sha, Tokyo)(in Japanese).

# Part IV
# Ergonomics Methodology

# 27. ERGONOMICS FIELDWORK: AN ACTION PROGRAMME AND SOME METHODS

### E. Nigel Corlett

*Institute for Occupational Ergonomics*
*University of Nottingham,*
*Nottingham NG7 2RD, England*

**Abstract**

Although studies and research on specific ergonomic relationships are still necessary, looking at the whole person is now recognised to be important if the full benefit of these ergonomics improvements is to be obtained. Practical objectives include ensuring that the improvements are successfully introduced. If this objective is to be achieved the action sequence for an investigation must include bringing into the investigation those who will ultimately experience the changes. This requires a sequence of activities different from those usually followed in industrial studies, and an example of an appropriate sequence will be given. Fieldwork also requires techniques for data gathering and analyses which differ from the laboratory methods which practitioners often try to transfer to the field. A number of methods particularly concerned with posture, handling, repetitive work and the assessment of effectiveness are presented in this paper.

**Keywords:** *ergonomics intervention; posture; education in ergonomics; discomfort; physical workload; static workload; methodology; surveys*

## Introduction

In Britain, as in many other Western countries, early work in ergonomics was heavily influenced by its anatomical, physiological and experimental psychology roots. The investigation paradigm was that of the scientific laboratory whilst the word "perception" would be interpreted in relation to the visual or auditory sense.

The study of attitudes towards work and the effects of organisation on these, as well as on performance, were the domain of the occupational psychologist or industrial sociologist. Although pioneers (eg C S Myers, 1929) had attempted to link the parts together and look at the whole person at work, it was not until the 1950's that serious efforts were made to generate a coherent approach to the whole perspective of people at work.

Such an approach might be illustrated by Figure 1. Here it is proposed that there are two parallel approaches to creating better work situations. One is the classic ergonomics approach of matching the work demands to human capabilities, making it within someone's capacity to do it, i.e. making it possible. The second direction is to match the job and work situation to what people need to make them want to do it. This recognises needs and wants, attitudes and perceptions, and the roles of society, culture and organisation on influencing the ways people see their working lives.

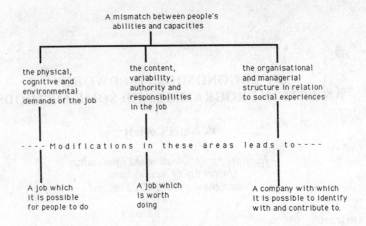

Figure 1. The mismatch

In the context of this paper these matters are germane to the introduction of change. The provision of a better-matching "ergonomic" piece of equipment is not sufficient to ensure its efficient use, and training is a necessary but not sufficient addition. "Technology transfer" has become a study area in its own right, but we must have a broad definition of technology if it is to embrace the introduction of new ideas about organisation or working practices. In this respect it is interesting to reflect on the industrial acceptance of ergonomics. For at least thirty years, in civil industry, it has been demonstrated that redesigned lathes, welders, presses, workbenches, tools and work activities would lead to safer, more accurate and better output from the users of these equipments. Yet these benefits have been ignored by all but the few pioneering managements sufficiently alert to see that the world was changing around them, and that the prejudices about the place of workers, developed in their youth, would not fit the working world of their middle age, or of their tomorrows. In consequence, the use of ergonomics in industry has advanced at a snail's pace, that is to say that the majority of managers see their workforce as an exploitable resource, below themselves in dignity and worth.

If the workforce is seen as of limited value, a cost to be minimised instead of a valued resource to be utilised, then there is no point in spending money on it. This is evident on a trip round too many factories. But society sometimes moves ahead, and demands some minimum standard. This most often happens in the fields of safety and health. Even where governments are reluctant to intervene, legislation gets introduced or the courts are used to bring pressure to bear on patently dangerous situations. In more advanced countries, eg Sweden, the legislation looked further than preventing people from being physically maimed, and set the stage for developments down the "worthwhile job" columns of Figure 1. But this route needs recognition of social partnership in industry, a condition far ahead of many nations which would be placed amongst the most advanced in industrialisation.

## Introducing change

However, the effect even of health and safety legislation is to focus sharply on the mismatches between people and their jobs. The UK Health and Safety Executive, as well as the US National Institute of Occupational Safety and Health, have used

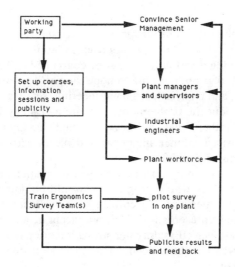

*Figure 2. Education and awareness programme*

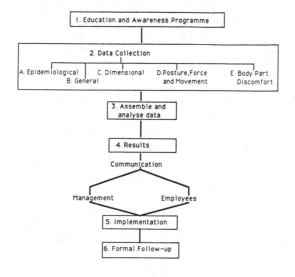

*Figure 3. The overall approach*

ergonomics to study and specify the safety standards they require, and publicise the fact that industrial safety problems require ergonomics investigation.

It is at this stage that a broader body of industry begins to take an interest in ergonomics. Most are looking for a "quick fix", but some will recognise that there are wider benefits to be gained if they embrace the subject more generously. But it is not enough just to get a book, or get a consultant, to answer a problem. Companies are complex cultures and a change in approach must permeate and gain support at many levels.

## An industrial case

The company which is the subject of this paper found itself facing a growing number of upper limb disorders (ULD). It is a large textile company, employing 38,000 people, and recent individual settlements (out of court) of ULD in other companies had reached £45,000. It was not possible to hope the problem would go away.

Personnel from safety, and from industrial engineering, set up a working party, both to inform themselves about the problem and prepare a strategy for dealing with it. Although things moved rather slowly at first, a "product champion" in the form of the company's head of industrial engineering began to drive the activity. The programme adopted is shown in Figures 2 and 3.

It will be seen that a primary target was senior management; without full commitment from the top the whole programme would be doomed. The company linked itself and its associated training college with the Institute for Occupational Ergonomics and prepared a presentation. This was accurate and well informed about the subject and its causes, as well as drawing strong attention to the likely costs if the proposed programme was not introduced. The costs and probable benefits of the programme were detailed.

The working party prepared the whole programme, and set it in train. With over 100 separate plants, the problem had to be tackled plant by plant. Each plant had to recognise the problem, realise the benefits and be able to follow through from the work of the 3 person Ergonomics Survey Team.

When the Survey Team was set up it received intensive training from Institute staff on methods for studying work, recognising situations where ULD may arise and using their results to determine and recommend changes. At the same time a 1-day course was set up for management, to inform them of the problem, its causes and consequences, and the contributions of ergonomics and the Survey Team to its minimisation.

A 2-day course, for all the company's industrial engineers, on ergonomics and its contributions in the context of classical industrial engineering - with particular reference to ULD - was also prepared. Part of each course was the information needed for work people, union personnel etc, and a programme of education and awareness was set up. In five months more than a quarter of the entire workforce had had presentations and information on the subject.

The Ergonomics Survey Team, responsible to Corporate Industrial Engineering, undertook a study at one company to test its methods. The results were startling in that 90% or so of the necessary changes to reduce the risks could be effected almost immediately and at little cost, whilst the plant's senior management were enthusiastic about the programme and its results. A typical plant required 3 weeks of the Survey Team's time, which was charged to the plant. The publicity given to the pilot plant's success - in spite of the charge for the Team - led to demands for their services which gave them a six-months forward programme in a few weeks.

Feedback from surveys has shown that the attack on ULD put forward by the company has been welcomed at all levels. As will be seen later, when the techniques used by the Ergonomics Survey Team are described, information gathered for ULD investigations can lead also to other ergonomic and organisational changes. The recognition of the whole nature of the work they were engaged in, as epitomised in Figure 1, has set the tone so that a steady programme of organisational development could be introduced.

## Procedure

The general procedure and data collected by the Survey Team are summarised in Figure 3. Five areas of data are listed horizontally across the figure. The contents and methods for each area are as follows:

*A. Epidemiological:*

For a period of at least three months prior to the survey study, all absences and reported sickness given on medical and personnel records are recorded. From these data an estimate of the contribution of ULD to sickness and absences as well as severity, likely critical jobs etc, can be estimated. Quantitative information to indicate current state, to justify where further studies should be done and to use in further assessment of intervention effectiveness is thus available. Names of personnel were **not** recorded.

*B. General:*

(i) A 1-page form for the plant manager, to assess the current state of loading of the plant, overtime and working conditions.

(ii) A 1-page form for the workforce, completed by each person studied, to gather attitudes to work, work support and cooperation. Answers were on a 4 point scale, giving negative to positive responses.

*C. Dimensional:*

Each workplace studied was measured, and the height and weight of the operator asked for and recorded. Photographs and videos were taken both for future analyses if needed and for educational purposes. Anonymity was provided.

*D. A Posture, Force and Movement recording form (PFM) was developed. On this was recorded:*

   (i) Posture of the hands, head and arms; need for reaching, bending or twisting; 8 questions.
   (ii) Physical stress; 3 questions on work rate and overtime.
   (iii) Force exertions; 3 questions on twisting, gripping or pinching.
   (iv) Equipment: questions on the design and use of the workplace, including handedness of equipment, seating and adjustability.
   (v) Repetitiveness; 3 questions.
   (vi) Design of tools; 3 questions.

*E. Body Part Discomfort:*

BPD form. This was based on the procedures of Corlett and Bishop (1976). Forms B,C,D and E were linked by a common code number.

The study sequence, briefly, was to discuss the programme with plant management, supervisors and union representatives. A day before the study the selected workpeople were visited, given a letter about the investigation and their cooperation requested.

On the study day each team member took a group of 3 or 4 employees. First each one was asked for information to complete their BPD diagram. Then one operator was observed and a PFM form completed for them. One to 1.5 hours after the start of the day, the BPD diagram for each operator was again completed, when the next operator's PFM form was completed. Just prior to the lunch break the BPD forms were again completed, and so on to the end of the shift. Dimensions, photographs and other data were gathered as needed in the time intervals between the BPD and PFM data gathering.

**Data assembly and analysis**

The key requirements for analysis were to assess what tasks were likely to cause ULD hazards, and what changes could be made to reduce those hazards. A further requirement was to identify areas for longer term improvement activities.

Form A — Epidemiological. Histograms of ULD and other incidence levels in relation to tasks done; relationships of incidence in relation to age and job experience.

Forms B(i) — The plant manager's data were linked to the PFM record, see D below.

and B(ii) — The employee's comments were analysed to give a profile, with average and range for each answer. Then B(ii) was linked with the following forms.

Forms C,D and E — The dimensions, PFM and BPD forms were linked to give complementary information. PFM was analysed to note extremes of posture or force, and their combinations. Repetition and other endangering factors were recorded. Criteria from the literature, e.g. Stevenson and Baidya (1987) and Putz Anderson (1988) were used to identify risk factors. BPD results were plotted across the working day.

Study of the BPD results in relation to PFM and workplace dimensional data demonstrated hazardous areas. With the other data, and discussions with industrial engineers as appropriate, an action plan was drawn up. This listed necessary actions, with their priorities and estimated costs. Time scales were agreed with local management for this wherever possible, and a future monitoring programme set up to monitor progress and results.

This comprehensive programme, driven through with close management control of its operation, shows clear evidence for dealing with ULD problems in the company. It also provides a vehicle for extending the impact of ergonomics to other aspects of work in the company. Where long repetitive tasks are experienced consideration for the wider aspects of job design could lead to reducing their impact without reductions in efficiency.

## Acknowledgments

I would like to acknowledge the support and contribution of Ms L. McAtamney, of the Institute, during the work programming described. Also the major contributions by the company, particularly Mr R.S. Benson, Corporate Industrial Engineer and Mr K. Meddings, Miss D. Brasil and Mr J. Bannel of the Ergonomics Survey Team. The work in the company, from awareness to follow-up, was entirely theirs. The support of Mr Tony Carter of the Harry Mitchell College is also gratefully acknowledged.

## References

Corlett, E.N. and Bishop, R.P. (1976) A technique for assessing postural discomfort. Ergonomics **19**, 175-182.

Myers, C.S. (1929) Industrial Psychology. Oxford University Press, Oxford, U.K.

Putz-Anderson, V. (1988) Cumulative Trauma Disorders. Taylor and Francis, London, U.K.
Stevenson, M.G. and Baidya, K.N. (1987) Some guidelines of repetitive work design to reduce the dangers of tenosynovitis. In Readings in R.S.I., (Ed. M.G. Stevenson) NSW University Press, Sydney, Australia.

# 28. FIELD TRIALS ON NETWORK MAINTENANCE SYSTEMS: AN ERGONOMIC APPROACH

## Rebecca Orring

*Swedish Telecom Networks*
*Sweden*

**Abstract**

An ergonomic assessment has been carried out as part of the Swedish Telecom field trials on systems for the remote testing of subscriber lines. In an interview study, the attitudes of the users and the effects of current task allocation on relations between maintenance staff and despatch operators have been investigated. In an experimental study to compare the usability of the four systems in the field trials, ease of learning, ease of use and user acceptance were evaluated.

**Keywords:** *methodology; telecommunications; user acceptance; usability; interviews; questionnaires; human-computer interface; maintenance*

## Background

In 1988, field trials were carried out to evaluate the performance of four subscriber line maintenance systems, in order to determine whether such equipment should be introduced at Swedish Telecom, and which of the four systems should be recommended for purchase and installation. The performance evaluation covered technical and economic factors, and also an assessment of ergonomic factors. The ergonomics assessment had two main purposes: the first was to investigate users' acceptance of this type of system, and the second was to determine which of the four systems was best from the point of view of the user. A secondary purpose of the study was to investigate various methods for evaluating systems usability, in order to establish a standard evaluation procedure for support systems for operation and maintenance of telecommunication networks.

The line-testing systems were installed in southern Sweden where they were in use between January and October 1988. Equipment installed at a number of exchanges enabled personnel at other locations to test various electrical characteristics of subscriber lines, including alternating and direct voltage, insulation resistance and capacitance between the two branches and between each branch and earth, and the current and resistance in the loop. From these measurements the type of terminal connected to the subscriber line and the balance in the cable pair can be calculated, and the values for each parameter are compared with reference values to assess the status of the line.

During the field study, terminals were installed at a despatch centre and maintenance centres in the area. Subscribers reported faults to despatch operators, who used the systems to test the condition of the line from the local telephone exchange to the customer. The test measurements could also be used by maintenance staff to diagnose and locate likely faults. Before the installation of the test systems, measurements were made by maintenance staff on site. The transfer of work tasks from maintenance staff (typically men with a technical background) to despatch operators (typically women without any technical training), could potentially affect working relationships between the two personnel groups.

Standard alphanumeric display screens were used for presentation of information for all four systems. One of the systems utilized a primarily menu-driven dialogue technique, one utilized function keys and a specially designed keyboard, while the other two were essentially command-driven.

Human factors aspects of line-testing systems have previously been studied by Tullis (1981), who in an experimental study compared how easily users could retrieve and interpret test results from alphanumeric, graphic and colour information displays. His results showed that a structured alphanumeric format was as effective as graphic formats for presentation of test data, but suggested that graphics could be useful as a training aid.

## User acceptance of the remote line-testing function

### Interviews with despatch centre staff

During a preliminary visit to the despatch centre, meetings were held with managers and the local working environment specialist, to discuss ergonomics aspects of the field trial and to collect information on work organization. Despatch office staff were interviewed on three separate occasions during the field trial, in order to identify changes in work content and procedures as a result of the introduction of the remote testing systems, and how these changes were experienced. A total of seven people were interviewed, although not all were interviewed on all three occasions. The interviews were informal and semistructured.

The first interviews with despatch centre staff were conducted in March, when the field trial had been running for nearly three months. Staff were asked to describe their work and demonstrate how they worked with the fault report system and line-testing system. Opinions concerning the systems, the working environment and work organization were also collected. The second interviews took place in May, approximately one month after the introduction of a new version of the fault report system, and further information about working with the line-testing system was collected. The final interviews were carried out in September, in order to summarize experience with the systems.

At the time of the first interview, despatch personnel spent relatively little time, approximately 10% of total work time, working with the fault report system. Fault reports from customers were usually recorded directly, generally during the conversation with the customer. A manual system was used in the exchange of information with maintenance staff. This information was later fed into the fault report computer system. At the time of the second interview, after the installation of the new fault report system considerably more time (up to 30% of total work time) was spent communicating with the computer, and less time in communication with maintenance staff. Despatch operators spent approximately 15 minutes per day working with the line-testing systems: thus the line-testing systems only resulted in a slight increase in the total time spent working with visual display units.

The line-testing system has not affected communication between despatch operators, but has had some effect on contact between despatch operators and maintenance staff. In the beginning of the trial period problems were caused by the fact that both despatch operators and maintenance staff were unsure about what measurements had been made

- the units of measurement were sometimes unfamiliar to both. Despatch operators reported that initially maintenance staff doubted the dependability of the test measurements, but that this distrust wore off fairly quickly.

The interviews show that changes in work content and workload as a result of the introduction of line-testing systems in the present work organization were regarded as positive. Although the extra task increased total workload, despatch personnel were able to fit in line-testing tasks when it suited them. The new task was seen as enriching their jobs, as well as giving more variety in work tasks. The opportunity to improve service to customers resulted in increased job satisfaction. Despatch operators considered that the line-testing system gave them a better overview of the whole process of which their jobs were a part. In addition to their function as communication links between customers and maintenance staff, the new systems enabled them to make a contribution to the solution of the customers' problems.

Each despatch operator was given some training for the systems she worked with. However it was generally agreed that more extensive training, in order to understand more about available functions, would have been useful. Difficulties in understanding the test results were also reported. A two-day course arranged by the manager of the despatch centre, to give operators some background information about what was being measured, was very much appreciated.

**Interviews with maintenance staff**

Interviews with all maintenance staff who came in contact with the systems (21 in total) took place towards the end of the field trial period, and focussed on how the new systems had affected the way they went about their jobs, and communication and cooperation with operators at the despatch office.

Maintenance staff were positive towards the new system, and saw the principal benefits as improvement in quality of service to customers and increased efficiency in maintenance work. For instance, measurement data from the test systems could be used to decide whether the fault was in the network, or in the customer's equipment, and which type of maintenance personnel would be required to mend the fault. Routine testing of subscriber lines could also help to identify problems in the network before they became noticeable to customers.

Several people mentioned that since they had to go to the exchange to collect maps and register information, they carried out new tests from the exchange to determine the distance to the fault. However, some maintenance staff had limited confidence in the ability of despatch operators to use the line-testing systems, and in many cases chose to repeat test measurements in order to obtain what they considered to be reliable results.

Voice response systems for two of the four systems make it possible to request line tests from a telephone. The telephone keypad is used to give commands, and test results are given by a recorded voice. Approximately half of the group had at some time tested the voice response systems, and it was generally agreed that they were easy to use.

## *Ease of learning*

An experimental test was carried out to determine how easily despatch office staff without previous experience of the systems could learn to use them. 16 despatch operators, not otherwise involved in the field trials, participated in this study. Each took part in a 4-hour training course in the use of one system. Four operators were thus trained to use each system. The participants then individually completed 10

Table 1. Mean time (in seconds) to complete a test measurement

| System | A | B | C | D |
|---|---|---|---|---|
| Mean time to complete task | 34 | 51 | 66 | 34 |

predefined tasks, consisting of performing tests on a series of numbers for which the most common types of fault had been simulated. These tests were video-recorded to facilitate analysis. Measurements of errors and times to complete the tasks were made.

The results showed that users could carry out tests of subscriber lines after only four hours of training. All participants succeeded in completing all tests, although especially in the first tests, time to complete the tasks varied considerably. In Table 1, the results from the first two tests have been excluded and the mean values are based on the times to achieve a test measurement for the final eight tasks. The letters A-D refer to the four systems included in the trials.

The participants were in many cases unable to interpret the results of the tests, and often expressed uncertainty about the type of fault diagnosed by the system.

## User evaluation questionnaire

A questionnaire, based on Shneiderman's forms for user evaluation of interactive systems (1987) was used to collect the opinions of despatch office staff. Shneiderman presents two versions of the form, one consisting of approximately 100 questions, and a shorter version consisting of approximately 25 questions. A selection was made from both forms of questions relevant for this study. These questions were translated, and seven-point rating scales were chosen instead of Shneiderman's ten-point scales.

The questionnaire was distributed in September to all staff using the test systems at the despatch centre. Each person completed one questionnaire for each of the systems he or she worked with. Eight questionnaires were returned for one of the systems, and seven for each of the remaining three systems.

The inexperienced despatch office staff also filled in a shorter version of the questionnaire. A summary of results for the four systems is presented in Figure 1. As in Table 1, the letters A-D refer to the four systems included in the trials.

Inexperienced (inexp) users' ratings are presented above the line, with experienced (exp) users' ratings below.

## Discussion and conclusions

### Methodological issues

This study is part of an attempt to establish suitable procedures for evaluating the human-computer interface for support systems for operation and maintenance of telecommunication networks. Some systems are developed within the organization, thus allowing testing of prototypes and design principles during the design process. In other cases, systems are purchased, and field trials are used to choose between a number of available products. These two approaches require different evaluation procedures.

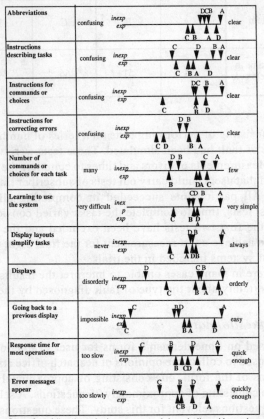

Inexperienced (*inexp*) users' ratings are presented above the line, with experienced (*exp*) users' ratings below

Figure 1. *Users' ratings of the system*

Three methods were used to collect data for this study: interviews, questionnaires, and experimental studies. In the experimental studies, video recording was used to facilitate analysis of results. While the interviews gave important background information concerning work content and work organization, this information was qualitative. The interviews addressed the issue of whether the new function was acceptable to despatch and maintenance staff. The quantifiable results obtained from the questionnaire and interview studies were of greater interest to the project group commissioning the work.

The interview studies were confounded by the unstable work situation at the despatch office. During the field trials, a new AXE-exchange was installed in the area, and the despatch operators primary support system was updated, resulting in considerable changes in work routines. Reorganization of areas of responsibility were planned, in which it was feared that this particular despatch office might cease to exist. These fears proved later to be well-grounded. Additionally, a policy recommending that contact with customers and contact with maintenance staff should be handled by separate work units, had recently been recommended by the central administration. In comparison with these changes, the effects of introducing the new line-testing system were seen as minimal. It was difficult to limit the scope of the interviews to opinions of the line-testing systems. Similar problems were encountered in the evaluation of a prototype system for network monitoring (Vainio-Larsson and Orring, 1990), where users were eager to discuss organizational and social issues while the purpose of the study was to evaluate the human-computer interface. While it is accepted that this organizational climate was not beneficial to the field study, it is inevitable that such things can happen in field trials. Evaluation methods must be robust enough to provide useful data even in the face of organizational turbulence.

Further work must be done on the design of the questionnaires for user evaluation of systems for operation and maintenance. The Shneiderman forms are a good base for further development, but efforts must be made to improve layout. Some questions must be reformulated, particularly those concerning error messages, which were found to be ambiguous, and those concerning visual attributes, which especially inexperienced users found difficult to answer.

## Swedish Telecom strategy

Swedish Telecom has recently adopted a policy for meeting working environment demands in the development of computer systems. Before the development of a new system is approved, an analysis of the effects of the new system on various aspects of the users' working situation must be carried out, and the results of this analysis must be examined by the working environment policy group. This means that studies of the kind reported here will become a standard part of the system development or system procurement process.

## Conclusions

Both despatch staff and maintenance staff were positive towards the line-testing systems, and it was generally considered that their use would lead to improved service to customers.

The experimental studies were useful for identifying problems in the human-computer interface, but even with small groups, the method was rather costly. However, the training of despatch staff not directly involved in the field trials was a useful means of spreading information about the system. Analysis of the video films was time-consuming, and the only data used in the analysis were time taken for the completion of each task. Had other data been retrieved from the video recordings (such as participants' comments and questions), this would have been acceptable, but in the given situation, manual recording of completion times would have been more efficient.

Experienced users gave two of the four systems in the field trial high ratings. Users unfamiliar with the systems were able to carry out measurement faster using these two systems than the other two. The new users gave one of the fast systems higher ratings than the other three systems. One particular system achieved the worst results in all comparisons.

Two of the four systems in the field trial were recommended for further consideration as a result of the comparison. Problems with the user interface are of minor importance in comparison with social and organizational issues associated with the introduction of the new system. Improvements in the presentation of results on the display screen will make it easier for operators to select relevant information. However, further training for despatch operators in the fundamentals of subscriber line-testing will be necessary to overcome difficulties in interpretation of the test results while maintaining the benefits of improved job design.

## References

Shneiderman, B. (1987), Designing the user interface: strategies for effective human-computer interaction, Reading, MA: Addison Wesley.

Tullis, T. (1981), An evaluation of alphanumeric, graphic and colour information displays, Human Factors, 23(5), 541-550.

Vainio-Larsson, A. and Orring, R. (1990), Evaluating the usability of user interfaces, To be published in Proceedings of Interact '90, 3rd IFIP Conference on Human-Computer Interaction.

# 29. THE USE OF SCALING TECHNIQUES FOR SUBJECTIVE EVALUATIONS

## Martin G. Helander and S. Mukund

*State University of New York at Buffalo*
*Buffalo, New York 14260*
*U.S.A.*

**Abstract**

Subjective evaluations of workplaces are of great importance in ergonomics because they are used for the identification of ergonomics problems as well as the evaluation of different design alternatives. The purpose of this study was to compare the different methodologies that have been developed for subjective evaluation of lighting, noise, vibration, climate, physical workload, mental workload, fatigue, comfort, and comprehensive evaluations of workplaces. There are three common ways of obtaining subjective evaluations: rating scales, interviews, and questionnaires. This study concentrated on the use of scaling techniques, such as paired comparisons, scaling by ranking, scaling by sorting, simple ratings (Lickert Scales) and anchored rating scales. These techniques sometimes introduce problems with validity of measurements, completeness, resolution and discrimination reactivity, utility and generalizability. Simple (Lickert) rating scales are used primarily for assessment of lighting, noise, vibration and climate. Anchored scales have been used for assessment of physical workload and comfort-discomfort. For mental workload, multidimensional procedures have been developed. Several design guidelines for rating scales are presented.

**Keywords:** *methodology; subjective evaluation; rating; scaling*

## Introduction

Subjective evaluations are essential for identification of ergonomics problems in the workplace. They are also used for comparative evaluations of job procedures and equipment to identify the best procedure or equipment. If it were not for people complaining or commenting, it would be difficult to evaluate and validate assumptions about the appropriateness of ergonomic design. The purpose of this study was to compare methods for subjective ratings that are commonly used for ergonomic evaluation of lighting, noise, vibration, climate, physical workload, mental workload, fatigue, comfort and comprehensive evaluations of workplaces. We are primarily interested in illustrating how different methods have evolved for assessment in different areas, and what types of scientific problems are associated with the use of different methods.

In summary the problems with subjective methods are the following:
  - How can ergonomists ask the right questions?
  - What are the problems with different techniques?
  - How can we avoid bias in answers?
  - How can we obtain high resolution and discrimination?
  - How can we ascertain validity of measurements, completeness,
    non-reactivity, utility, and generalizability?

There are three main techniques for obtaining subjective data: Interviews, questionnaires, and rating scales. Interviews and questionnaires are the most common. Questionnaires may be considered a refined type of interview and questions may be designed in several different ways: open-ended, close-ended, multiple choice, forced choice, and paired comparisons. Thus, questionnaires may also incorporate various types of scales. Several studies suggest that open-ended items provide more information than close-ended (Sinclair, 1975). However, they take more time.

## Potential problems in the use of scaling techniques

There are several potential problems in the use of rating scales. In many cases the human observer may not be very good at producing clear estimates. In risk perception, for example, individuals usually underestimate the probability of common risks and overestimate rare risks.

In addition, humans are not good observers of complex events such as multi-dimensional phenomena occurring at a high rate. One example is assessment of mental workload in a rapidly changing environment. This may partially be due to limitations in information processing and short-term memory, which would make it difficult to respond to rapid changes.

Poulton (1982) clarified that there are common biases in judgement that may affect ratings:

Halo effect - the response to a question is affected by previous responses.

Central tendency - the reluctance to choose extreme scale values, see Figure 1.

Leniency - the bias to consistently respond positively (easy rater) or negatively (hard rater).

Emotional content - the tendency to respond favourably or unfavourably to questions which invoke an emotional reaction.

Many of these biases may be overcome if an experienced rater is used or if subjects are trained. However, this is typically not provided in research studies.

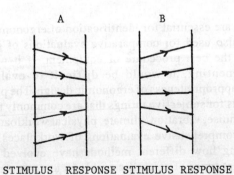

*Figure 1. Two common biases in rating: A. Underestimation of large sizes and differences, and B. Overestimation of small sizes and differences*

## Comparison of different scaling techniques

There are five different techniques that are commonly used for scaling in ergonomics research.

Paired Comparisons may be used to assess which of two stimuli (pair) is the greater. This technique has generally greater reliability than rating scales. However, all combinations of pairs must be assessed, and a large n, therefore, produces prohibitively many pairs.

Scaling by Ranking is used to rank order stimuli according to their magnitude. This technique may produce even greater reliability than "paired comparisons". It can also be used to rank several different attributes of an object.

Scaling by Sorting is used to sort objects in several piles. This technique is used for large n, and may be conceived as a mix of paired comparison and ranking. The endpoint piles may be labelled.

Simple Rating. Here stimuli are presented one at a time and rated along a scale. This technique is widely used and the Lickert scale is the most common.

```
never          sometimes         always
---------------------------------------
  1                5                9
```

Anchored Rating Scales. In this case numbers are assigned by comparison of the stimulus to a standard (anchor). It is easier to obtain interval properties than with ranking. Since it takes less time, this technique (as well as simple rating scales) is less elaborate to use than paired comparison.

There is a variety of anchors. For example, in a behaviourally anchored rating scale (BARS), "High Workload" may be described in the following way: Drive as many nails as possible in 5 minutes (Meister, 1985). Another example is "cross-modality matching", where the intensity of a stimulus may be indicated by pressing a hand dynamometer rather than by verbal response. Some types of anchors provide strong consistent association while other types of anchors may be drifting.

## Use of scaling techniques for assessment of different ergonomics problems

Below we will briefly review the types of scaling methods that have been used for assessment of different ergonomics problems.

Lighting. Ostberg and Stone (1974) provided a comprehensive overview of methodologies for glare discomfort evaluation from the beginning of the century. In later years, evaluations of the effect of lighting on VDT workstations have been common in the literature. Lickert scales are predominant. For example Stammerjohn et al. (1981) used a four point Lickert scale and Wibom and Carlsson (1986) a three point scale. Konz and Bennett (1987) used a semantic differential technique with seven point scales that were used to differentiate between adjective pairs.

Sound, Noise and Vibration. Classical psychophysics (e.g. Garner, 1954) has been employed to determine threshold values (using the method of limits) and discriminatory capacity (using the method of adjustment). Direct magnitude estimation is also common and has been used for estimation of loudness (Hellman and Zwislocki, 1965) and ride discomfort vibration (Leatherwood and Dempsey, 1976).

Leatherwood and Dempsey used a nine point unipolar interval scale to measure vibration discomfort (with endpoints 0 for zero discomfort and 9 for maximum discomfort). Subjects were regularly exposed to an anchor (comparison) stimulus. Hansson and Wikstrom (1981) used a simple nine point Lickert scale to evaluate vibration discomfort.

<u>Climate</u>. There is a standard seven point scale which has been developed by ASHRAE, see Figure 2.

|  | A. |  | B. |
|---|---|---|---|
| -3 | Cold | 1. | Cold |
| -2 | Cool | 2. | Cold/Cool |
| -1 | Slightly Cool | 3. | Cool |
| 0 | Neutral | 4. | Comfort/cool |
| +1 | Warm | 5. | Comfort |
| +2 | Hot | 6. | Warm Comfort |
| +3 | Very Hot | 7. | Warm |
|  |  | 8. | Hot/Warm |
|  |  | 9. | Hot |

*Figure 2. A - ASHRAE scale. B - Nine point scale (Rohles and Lavania, 1985)*

Rohles and Milliken (1981) used semantic differential scales with 71 pairs of bipolar adjectives to evaluate thermal comfort. Rohles and Lavania (1985) later developed a nine point scale to measure thermal comfort. They also used a "Normalized Certainty Scale" to let subjects estimate how certain they were of their estimate.

<u>Physical Workload</u>. This area has also been the subject of experimentation in psychophysics. Stevens (1957) proved that physical workload could be estimated as a ratio compared to a standard stimulus.

Borg (1962) developed the RPE scale (Rating of Perceived Exertion). The physical workload is estimated on a scale from 6-20, which incidentally multiplied by 10, corresponds to the approximate heart rate induced by the task. Ljunggren (1986) demonstrated that RPE may not produce results of better validity than regular magnitude estimation of the workload.

<u>Mental Workload</u> has been the topic of extensive research, primarily supported by the U.S. Air Force. The classic Cooper-Harper scale (1969) is a two level sequential menu-selection tree, with three alternatives at each level. It is used for estimation of pilot workload.

Phillip et al. (1971) compared "objective" workload for an air traffic control task with subjective assessments using Lickert scales. There were no significant correlations.

Reid et al. (1981) developed the SWAT scale (Subjective Workload Assessment Technique). In this technique, three dimensions of mental workload are assessed: 1) Proportion of time busy, 2) Complexity of task, and 3) Psychological stress. Each dimension is defined with a three point scale: low, medium and high. For the combined assessment there are three 3x3x3=27 combinations. Subjects are first trained to produce a rank ordering of the 27 cases in terms of mental workload. This may be accomplished by using examples (anchors) for each case. Subjects can then use the scale for assessments of workload.

Kilmer and Bateman (1988) found that SWAT had greater sensitivity to task difficulty than a Modified Cooper-Harper Scale - maybe because it has 27 levels.

Hart et al. (1984) demonstrated the use of the NASA TLX index. This incorporates nine dimensions: Task difficulty, time stress, performance, mental effort, physical effort, frustration, stress, fatigue and activity type.

Pairwise comparison was used by Shutte and Seidel (1988), who compared nine different components of workload.

Fatigue. Pearson (1957) used Gutman's method for generating words to describe different stages of fatigue. Judges sorted these words along a nine point scale where 1 = extreme well-being and 9 = extreme fatigue. Two equivalent thirteen point scales were then validated. One is illustrated in Table 1.

Table 1. Example of fatigue scale (Pearson, 1957)

1. Like I am bursting with energy.
2. Extremely peppy.
3. Very lively.
4. Very refreshed.
5. Quite refreshed.
6. Somewhat refreshed.
7. Slightly tired.
8. Slightly pooped.
9. Fairly well pooped.
10. Petered out.
11. Very tired.
12. Extremely tired.
13. Ready to drop.

Comfort. Shackel et al. (1969) developed the General Comfort Rating Scale - an eleven point interval scale. This was compared to 5 other methods to assess comfort:

- Body area comfort rating
- Chair feature check list
- Direct ranking
- Body posture change frequency
- Chair dimensions.

Corlett and Bishop (1976) used a seven point scale to rank body part discomfort from extremely comfortable to extremely uncomfortable.

The semantic differential technique was used by Brown et al. (1982) to measure "well-being" and "competence".

Comprehensive Evaluations. In recent years several methods have been developed for comprehensive assessment of ergonomics problems (Landau and Rohmert, 1989). Most of these are used by trained observers.

The AET method used five point scales to assess more than 100 variables. The JOB Profile developed by Renault automobile company evaluates 22 criteria using five point scales. The LEST Method assesses 16 criteria on thirteen point scales, and the

AVISEM Method evaluates 10 parameters on a five point scale. The GESIM Method evaluates 91 parameters using yes-no type answers. This is the only method that is actually used by workers themselves.

## Discussion

In summary, environmental factors including lighting, noise, vibration and climate have been assessed using primarily simple (Lickert) rating scales. For some type of assessments it was possible to present a comparison stimulus which was used as an anchor.

The assessment of physical workload has used anchored scales as well as simple rating scales. For mental workload, multidimensional assessment procedures have been developed.

For assessment of fatigue as well as comfort-discomfort verbally anchored scales have been developed. Finally comprehensive ergonomics evaluations use mostly Lickert scales.

The research on the Cooper-Harper scale for assessment of mental workload indicates that subjects provide much better, less affective responses if a two step process is used. Kappler and Pitrella (1988) obtained very promising results using this technique to assess parameters in diving performance. The stepwise process provided several advantages including: reduced load on short-term memory, less affective responses due to "less obvious" extreme alternatives, use of verbal authority at all levels of choice.

Several rules for design of rating scales may be used, see Table 2.

*Table 2. Principles in design for rating scales*

1. Continuous scales provide higher reliability and discriminability than category scales.
2. Horizontal scales are better than vertical.
3. A combination of numbers and verbal descriptors (anchors) is preferred.
4. Use brief and precise descriptors.
5. Use neutral (non-affective) descriptors.
6. Use psychologically scaled (rank-ordered) descriptors.
7. Use equally spaced descriptors.
8. Avoid use of negative numbers.
9. Preferred qualities should increase to the right (stereotype).
10. Avoid use of descriptors that call for evaluation.
11. Use eleven scale points, or more, up to the limit of available descriptors.
12. Minimize rater workload, e.g. by using two-stage scaling.

## Acknowledgments

This study was supported by a contract with NIOSH, Cincinnati, USA with Dr. Steven Sauter acting as contract monitor. The views are those of the authors, and do not necessarily represent policies and opinions of NIOSH. We gratefully acknowledge the help of Mrs. Patricia Waldron in typing the manuscript.

## References

Borg, G. (1962). Physical performance and perceived exertion. Copenhagen: Munksgaard.

Brown, I. D., Wastell, D. G., and Copeman, A. K. (1982). A psychological investigation of system efficiency in public telephone switchrooms. Ergonomics, 25, 1013-1040.

Cooper, G. E. and Harper, R. P. (1969). The use of pilot rating in the evaluation of aircraft handling qualities. NASA TN-D-5153, Moffett Field, CA: NASA, Ames Research Center.

Corlett, E. N. and Bishop, R. P. (1976). A technique for assessing postural discomfort. Ergonomics, 19, 175-182.

Garner, W. R. (1954). A technique and scale for loudness measurement. The Journal of the Acoustical Society of America, 26, 73-88.

Hansson, J. E. and Wikstrom, B. O. (1981). Comparison of some technical methods for the evaluation of whole-body vibration. Ergonomics, 24, 953-963.

Hart, S. G., Sellers, J. J. and Guthart, G. (1984). The impact of response selection and response execution difficulty on the subjective experience of workload. Proceedings of the Human Factors Society, Santa Monica, CA: Human Factors Society.

Hellman, R. P. and Zwislocki, J. (1965). Loudness function of a 1000 cps tone in the presence of a smashing noise. The Journal of the Acoustical Society of America, 36, 1618-1627.

Kappler, W.-D. and Pitrella, F. D. (1988). Evaluation of vehicle handling; design and test of a two-level sequential judgement rating scale. Proceedings of 23rd Annual Conference on Manual Control. Cambridge, MA: MIT.

Kilmer, K. J. and Bateman, R. (1988). A comparison of SWAT and Modified Cooper-Harper Scales. Proceedings of the Human Factors Society, Santa Monica, CA: Human Factors Society.

Konz, S. and Bennett, C. (1987). An evaluation of some VDT lighting variables. Proceedings of the Human Factors Society, Santa Monica, CA: Human Factors Society.

Landau, K. and Rohmert, W. (1989). Recent developments in job analysis. London: Taylor and Francis.

Leatherwood, J. D. and Dempsey, T. K. (1976). Psychophysical relationships characterizing human response to whole-body sinusoidal vertical vibration. NASA TN D-8188, National Aeronautics and Space Administration.

Ljunggren, G. (1986). Observer ratings of perceived exertion in relation to self ratings and heart rate. Applied Ergonomics, 17, 117-125.

Meister, D. (1985). Behavioral analysis and measurement methods. New York, NY: Wiley.

Ostberg, O. and Stone, P. (1974). Methods for evaluating discomfort glare aspects of lighting. Goteborg Psychological Reports, 4(4). Goteborg, Sweden: University of Goteborg.

Pearson, R. G. (1957). Scale analysis of a fatigue check list. Journal of Applied Psychology, 41, 186-191.

Phillip, U., Reiche, J., and Kirchner, J. H. (1971). The use of subjective ratings. Ergonomics, 14, 611-616.

Poulton, E. C. (1982). Biases in quantitative judgements. Applied Ergonomics, 13, 31-42.

Reid, G. B., Shingledecker, C., and Eggemeier, T. (1981). Application of conjoining measurement workload scale development. Proceedings of the Human Factors Society, Santa Monica, CA: Human Factors Society.

Rohles, F. H. and Lavania, J. E. (1985). Quantifying the subject evaluation of occupied space. Proceedings of the Human Factors Society, Santa Monica, CA: Human Factors Society.

Rohles, F. H. and Milliken, G. A. (1981). A scaling procedure for environmental research. Proceedings of the Human Factors Society, Santa Monica, CA: Human Factors Society.

Schutte, M. and Seidel, B. (1988). The strain-related scaling of different sequence sections of cash-work. Proceedings of the Ergonomics Society's Conference, London: Taylor & Francis.

Shackel, B., Chidsey K. D., and Shipley, K. (1969). The assessment of chair comfort. Ergonomics, 12, 269-306.

Sinclair, M. A. (1975). Questionnaire design. Applied Ergonomics, 6, 73-80.

Stevens, S. S. (1957). On the psychophysical law. Psychological Review, 64, 153-181.

Stammerjohn, L., Smith, M. J., and Cohen, B. G. F. (1981). Evaluation of workstation design factors in VDT operations. Human Factors, 23, 401-412

Wibom, R. I. and Carlsson, L. W. (1966). Work at video display terminals among office employees: Visual ergonomics and lighting. In: Knave, B. and Wideback, P.G. (Eds). Work with Display Units '86. Amsterdam, The Netherlands: Elsevier.

# Part V
Stress and Mental Workload

# 30. PSYCHOSOCIAL COMPONENTS OF ERGONOMIC MEASURES: METHODOLOGICAL ISSUES

Barbara G.F. Cohen[*], Kevin E. Coray[**]
and Chaya S. Piotrkowski[***]

[*]*Human Resources Technology*
*Internal Revenue Service*
*Washington, DC, U.S.A.*

[**]*Human Management Services, Inc., Severna Park*
*Maryland, U.S.A.*

[***]*Center for the Child, New York*
*New York, U.S.A.*

**Abstract**

Research on the effects of stress in the workplace consistently confirms the inextricability of physiological and psychological components from the outcomes of stress. Complicating the research analyses are many complexities of the measures themselves. This paper discusses the complexity of a measure of physical movement as an example of the measurement issues inherent in ergonomics research. In this self-report study, 625 female clerical employees completed a survey containing questions about working conditions, health and well-being. Physical work environment measures included the ability to move freely. In addition, measures of the psychosocial environment were created. Over 30 working conditions measures were obtained from responses to the questionnaire. The complexity of the Movement scale became apparent, in a set of psychometric analyses, in that the magnitude of its factor loadings placed it squarely within the Job Control dimension, a psychosocial construct, rather than in the Physical Environment dimension, an ergonomics construct, as would have been expected. Self-reported ergonomics measures appear to be sensitive to both physical and psychosocial conditions. Thus, construction and use of such measures must proceed with great care.

**Keywords:** *stress; methodology; questionnaires; offices; posture; subjective evaluation; psychosocial factors*

## Introduction

Occupational stress literature repeatedly demonstrates the confounding of physiological and psychological components found in stress consequences (e.g., French & Caplan, 1973; Sainforth & Smith, 1989). Complicating the research analyses are many complexities of the measurements themselves. Recognizing that controversial views of objective versus subjective measures exist, this paper addresses one example of the complexity in a self-reported measure of the physical environment. The Physical Movement Scale, developed to measure ergonomics aspects, was placed in the job control dimension. In addition, poor physical working conditions (e.g., lighting, space, privacy, etc., that were perceived as bothersome or inadequate) were related to organizational climate and excessive work load.

This supports findings of other similar dimensions related to occupational stress. For example, comparing several groups of clerical workers, Stellman et al. (1987) found one group was more satisfied with their physical environment characteristics but less satisfied with their jobs. They reported significantly less decision latitude, less skill utilization, and more task repetition in jobs not designed to promote much moving around. Kumashiro et al. (1989) found lack of control over work pace to be a common source of stress. This supports previous findings (e.g., Cohen et al., 1981; Karasek, 1981) that individuals' control over the way their work is performed is closely related to their stress, as well as to their job satisfaction. Dissatisfaction with organizational climate (e.g., Coch & French, 1948; Westlander, 1984), with job control (e.g., Johansson & Aronsson, 1984; Frese, 1987) and physical environment (e.g., Oldham & Fried, 1987; Smith et al., 1981) are significantly linked to occupational stress. On the other hand, good interpersonal relationships (e.g., Buck, 1972; French & Caplan, 1973) and skill utilization (e.g., Brook, 1973) are related to well-being and job satisfaction.

## Methods

### Data collection

Questionnaire data were collected from 635 women office workers from four U.S. office sites, two in the private and two in the public sector. A response rate of 71.7% resulted from 625 usable surveys. A nearly even split of the data came from the two sectors. However, organizations were selected for convenience rather than through systematic sampling. Subjects voluntarily completed the questionnaire.

### Sample characteristics

Information processors comprised the largest single category of subjects (40.7%), followed by secretaries (35.1%), clerks (20.6%) and managers (3.5%). Average age was approximately 37. The women had been on the job about two years (median), were mostly Caucasian (71.7%), and were mostly married (44.6%). 39% had completed high school, but another 59.2% had at least some college education. About a third had children living at home.

### Measures

The questionnaire had a large number of items tapping a wide spectrum of working conditions, health and well-being, family stressors, and coping strategies. The first step in managing this data was to subject a priori scales from published studies used in the questionnaire to internal reliability analyses. Scales with internal consistency reliabilities of 0.64 or better were retained. Items not in a priori scales were inspected for or psychometrically improved to ensure adequate variability and normality. Items with excessive missing data or limited variability were eliminated. These remaining items, along with psychometrically sound items from pre-existing scales which did not meet the internal consistency criterion, were factor analyzed in two groups to form new internally consistent scales. The first group consisted of items pertaining to physical aspects of work while the second pertained to the psychosocial work environment. These factor analyses guided construction of additional internally consistent scales.

*Pre-existing measures*

Five scales from the *Work Environment Scale* (Insel & Moos, 1974) met the internal consistency criterion set in this study. These were Staff Support, Task Orientation, Work Pressure, Clarity and Innovation. Scales from Caplan et al. (1975) included Variance in Work Load, Role Ambiguity-Environment, Role Conflict, Underutilization of Abilities, Participation, and Job Future Ambiguity. Non-Support from Boss came from Haynes et al. (1978).

*Physical aspects of the work environment*

The initial factor analysis of items relating to physical aspects resulted in measures of Lighting Adequacy, Noisiness, Lack of Space, Poor Air Quality, and Movement. Movement is a three item scale measuring the extent to which a person must maintain one position for extended periods of time, i.e. job constraints on mobility or static posture. The alpha for Movement was 0.61. It was retained for the analysis of physical aspects of the working environment because of its ergonomics relevance.

*Psychosocial work environment*

The second factor analysis resulted in the construction of internally consistent measures assessing organizational and task variables such as workload, co-worker relations, rule orientation and task autonomy, work recognition and reinforcement, skill utilization, desire to stay in one's position, and influence over equipment.

## Results

To better understand the constructs inherent in the ergonomics and working conditions scales, because the total number of measures of the work environment was still unwieldy (34 measures of working conditions), and because many of the measures were redundant (high intercorrelations), a series of successively refined second-order factor analyses of the scales was conducted. The final solution, based on 31 scales and exhibiting considerable stability, contained seven rotated factors. The first six factors in the final solution reflected meaningful dimensions of the work environment that are similar to those reported in the organizational psychology literature, offering some construct validity to the measures created in this study. The dimensions are: (I) organizational climate, (II) workload, (III) physical environment, (IV) interpersonal relations, (V) skill utilization, and (VI) job control. The scales defining each factor are partially reported in Cohen et al. (1987). Additional refinement of the factors was subsequently reported in Piotrkowski et al. (1987).

To relate these dimensions to each other, unit-weighted indices were created using factor loadings of 0.40 as a minimum criterion for including scales in an index. Individual scales within an index were standardized and then averaged. In this manner, six global indices were created. No scale was allowed to be part of more than one index. The final unit weighted indices of working conditions are described in Table 1.

One unexpected finding was that Movement, a scale initially conceptualized as reflecting ergonomics factors, had only a relatively modest loading (-0.21) on the physical environment factor. Instead, it loaded (-0.66) with other scales tapping control over one's job tasks and the rule orientation of the organization. The following items comprised this scale: (1) Do you hold your arms in one position for long periods of time when performing your job? (2) Does your job allow you to change positions and sit and stand when you want? (3) At work, do you sit or stand in the same position for several hours? It is noted that item 2 differs from the others because of the insertion "when you want". Subsequently the dimension of control was introduced and complicated the measure. Thus the ability to move around appears to be experienced as an aspect of control. In other words, what had appeared to be an ergonomics or physical aspect of the environment was perceived by the women in our sample as a psychosocial phenomenon. As a result, Movement was placed into the measure of the Lack of Control Index.

The Poor Physical Conditions Index was correlated both with the Work Load Index [$r = -0.24$, $p. < 0.001$] and to the Organizational Climate Index [$r = -0.32$, $p. < 0.001$]. In other words, poor physical working conditions were related to a poor organizational climate and excessive work load.

*Table 1. Descriptions of indices measuring dimensions of the work environment and scales making up each index, with factor loadings*

---

I. *Organizational Climate Index*: The extent to which rules and roles are clear, efficiency and new ideas are encouraged, management is supportive of and appreciative of office workers. A high score is a better climate.

Measures in index: Staff Support (.75); Unappreciation (-.71); Clarity (.66); Nonsupport from Boss (.63); Innovation (.62); Task Orientation (.62); Role Ambiguity--Environment (-.48); Participation (.46).

II. *Work Load Index*: The extent to which frequent increases in work demands do not occur, do not conflict with each other, work pressure does not dominate the job milieu and underload occurs. A high score reflects low demands while a low score reflects high demands.

Measures in index: Work Load (.80); Variance in Work Load (.72); Work Pressure (-.70); Underwork (-.68); Role Conflict (-.50).

III. *Poor Conditions Index*: The extent to which the physical environment (air quality, lighting, noisiness, space and privacy) is perceived as bothersome or inadequate. A high score is a less adequate environment.

Measures in index: Lack of Space (.74); noisiness (.69); Poor Air Quality (.67); Lighting Adequacy (-.59).

IV. *Skill Utilization Index*: The extent to which employees can use their skills and education on the job and the extent to which they wish to stay in their current position. A high score is more skill utilization; a low score is less skill utilization.

Measures in index: Skills Unused (-.78); Underutilization of Abilities (.75); Stay in Job (.56).

V. *Lack of Control Index*: The extent to which the organization uses rules to control workers; workers cannot exercise decision latitude in doing their job tasks; and workers must maintain static positions during their work. A high score is less control; a low score is more control.

Measures in index: Lack of Autonomy (.70); Movement (-.66); Rule Orientation (.62).

VI. *Interpersonal Tensions Index*: The extent to which co-workers are a source of stress; sexual harassment exists; co-workers are not helpful, supportive or accepting; workers do not feel they are treated with dignity and respect. A high score is more tension, less support, more harassment.

Measures in index: Stress from Co-workers (.66); Sexual Harassment (.64); Treated with Dignity (-.44); People Get Along (-.51); Acceptance (-.42).

---

## Discussion

Understandably, many researchers and managers experience more confidence and comfort with evaluation of the physical environment's effects on the workers than with the measurement of the psychosocial aspects. Chairs, work surfaces, lighting, etc. can be measured with reliable instruments like rulers, meters, and other objective, quantitative devices. Yet what appear to be precise measurements, on closer scrutiny, can be quite complex. Controversies abound regarding optimal ranges of workstation components' angles, heights, etc. as well as how measurements are to be made (Dainoff & Dainoff, 1986), and even as to what postures are preferable (Mandal, 1985; Kroemer, 1983).

There were two interesting findings in this study. First, the Physical Movement Scale illustrates the complexity of what had appeared to be a straightforward measurement of an ergonomics aspect but was experienced as being related to organizational climate and control over job tasks. The second finding was that poor physical working conditions were related to poor organizational climate and excessive workload. Interviews conducted with a subsample of respondents at each site provide examples of the complexity and intertwining of these issues.

Interviews indicated that, at one site, a tense organizational climate was created by a newly initiated two-tiered system: a day shift comprising regular workers paid by a system tied to the amount of claims they processed and an evening shift of part-timers paid an hourly rate to do the easier cases. Instead of the varied case load they had prior to this new system, the regulars had to maintain their rate via processing only the difficult, time-consuming claims. In addition, many indicated that their workload was further increased because they had to correct errors made by the inexperienced evening shift.

This resulted in many of the women feeling constant pressure to sit at their desks and work at their fastest pace to complete what they perceived an inordinately heavy workload. Often they skipped their breaks, and sometimes even lunch, in an attempt to maintain their previous pay status. Thus the lack of mobility reported seemed due more to an organizational climate that promoted fear of diminished pay and job dissatisfaction than to various ergonomics factors. In summary, the highly stressed organization had employees who remained seated for long periods with their arms outstretched keying information as fast as they could. At the same time they were dissatisfied with both the increased workload and the decreased opportunities to accelerate their pace by working easier cases.

Smith et al. (1987) noted work pressure intensifies the perception of work expended, creating psychological fatigue. This psychologically exacerbated fatigue was likely to be experienced by the women in this study who worked under constant pressure to maintain their fastest pace. Moreover, working without getting up and moving about for long periods increased their physiological strain. As static body posture influences noradrenaline output, it may be assumed that these workers produced higher amounts. Johansson et al. (1978) found high noradrenaline output for workers who remain in the same position and significantly lower amounts in those who moved about. More positive, calmer feelings are also correlated with mobility (Johansson & Aronsson, 1984).

It is recognized that the physical work environment contributes to employees' vulnerability to other stressors such as interpersonal tensions, heavier workload, lack of job control and an unsupportive environment (Smith et al., 1981). Less obvious but equally important is the role that psychosocial factors play in the meaning and measurement of physical ergonomics aspects. Hence, the seemingly simple measure of how much employees move about in the course of their workday is an example of a rather complex measure of workers' perceptions and reactions to their organizational climate as well as to their physical environment.

Results of this study suggest that, despite the complexities of using self-report measures, the solution is not to exclude or avoid self-report methods, but to be keenly aware of the fact that measures of physical work environment are intricately bound up with other aspects of organizational climate and workload.

## Acknowledgements

Support for this research was provided under National Institute for Occupational Safety and Health (NIOSH) contract number 86-71437. Opinions expressed are those of the authors and are not to be construed as necessarily reflecting the official view or endorsement of NIOSH or of the U.S. Department of Health and Human Services.

## References

Brook, H., 1973, Mental stress and work. The practitioner, 210, 500-506.

Buck, V., 1972, Working Under Pressure. (London: Staples).

Caplan, R.D., Cobb, S., French, J.R.P., Jr., Harrison, R.V., & Pinneau, S.R., 1975, Job Demands and Worker Health. HEW Publication No. (NIOSH) 75-160.

Coch, L. & French, J.R.P., 1948, Overcoming resistance to change. Human Relations, 11, 512-532.

Cohen, B.G.F., Smith, M.J., & Stammerjohn, L., Jr., 1981, Psychosocial factors contributing to job stress of clerical VDT operators. In G. Salvendy & M. Smith (Eds.) Machine Pacing and Occupational Stress. (London: Taylor & Francis), 337-345.

Cohen, B.G.F., Piotrkowski, C.S., & Coray, K.E., 1987, Working conditions and health complaints of women office workers. In G. Salvendy, S.L. Sauter, & J.J. Hurrell, Jr. (Eds.) Social, Ergonomic and Stress Aspects of Work with Computers. (Amsterdam: Elsevier), 365-372.

Dainoff, M.J. & Dainoff, M.H., 1986, People and Productivity: A Manager's Guide to Ergonomics in the Electronic Office. (Chichester, UK: John Wiley & Sons Ltd).

French, J.R.P. & Caplan, R., 1973, Organizational stress and individual strain. In A.J. Marrow (Ed.) The Failure of Success. (New York: AMACOM), 30-66.

Frese, M., 1987, A concept of control: Implications for stress and performance in human-computer interaction. In G. Salvendy, S. Sauter & J.J. Hurrell, Jr. (Eds.) Social, Ergonomic and Stress Aspects of Work with Computers. (Amsterdam: Elsevier), 43-50.

Haynes, S.G., Levine, S., Scotch, N., Feinleib, M., & Kannel, W.B., 1978, The relationship of psychosocial factors to coronary heart disease in the Framingham Study. American Journal of Public Health, 20, 133-141.

Insel, P.M. & Moos, R.H., 1974, Work Environment Scale. (Palo Alto, CA: Consulting Psychologists Press).

Johansson, G., Aronsson, G., & Lindstrom, B., 1978, Social psychological and neuroendocrine stress reactions in highly mechanized work. Ergonomics, 21, 583-599.

Johansson, G. & Aronsson, G., 1984, Stress reactions in computerized administrative work. Journal of Occupational Behaviour, 5, 159-181.

Karasek, R., 1981, Job decision latitude, job design, and coronary heart disease. In G. Salvendy and M. Smith (Eds.) Machine Pacing and Occupatonal Stress. (London: Taylor & Francis), 45-56.

Kroemer, K.H.E., 1983, Engineering anthropometry, work space and equipment to fit the user. In D.J. Oborne & M.M. Gruneberg (Eds.) The Physical Environment at Work. (Chichester, U.K.: Wiley).

Kumashiro, M., Kamada, T., & Miyake, S., 1989, Mental stress with new technology at the workplace. In M. Smith & G. Salvendy (Eds.) Work with Computers: Organizational, Management, Stress, and Health Aspects. (Amsterdam: Elsevier), 270-277.

Mandal, A., 1985, The Seated Man. (Copenhagen: Dafnia Publications).
Oldham, G. & Fried, Y., 1987, Employee reactions to workspace characteristics. Journal of Applied Psychology, 72, 75-80.
Piotrkowski, C.S., Coray, K.E., & Cohen, B.G.F., 1987, The relationship of working conditions to health and well-being among women office workers. NIOSH Technical Publication No. 86-71437, Cincinnati, OH.
Sainforth, P.C. & Smith, M.J., 1989, Job factors as predictors of stress outcomes among VDT users. In M.J. Smith & G. Salvendy (Eds.) Work with Computers: Organizational, Management, Stress, and Health Aspects. (Amsterdam: Elsevier), 233-241.
Smith, M.J., Cohen, B.G.F., & Stammerjohn, L.W., 1981, An investigation of health complaints and job stress in video display operations. Human Factors, 23, 387-400.
Smith, M.J., Carayon, P., & Miezio, K., 1987, VDT technology: Psychosocial and stress concerns. In B. Knave & P. Wideback (Eds.) Work with Display Units 86. (Amsterdam: North-Holland), 695-712.
Stellman, J., Klitzman, S., Gordon, G., & Snow, B., 1987, Work environment and the well-being of clerical and VDT workers. Journal of Occupational Behaviour, 8, 95-114.
Westlander, G., 1984, Office automation, organization factors and psychosocial aspects of health with special reference to word processing. In Reports from the Department of Psychology, University of Stockholm, Stockholm, Sweden.

# 31. MULTIVARIATE ANALYSIS OF MENTAL STRESS AND SINUS ARRHYTHMIA

## Kyung S. Park and Dhong H. Lee

*Department of Industrial Engineering*
*Korea Advanced Institute of Science (Technology)*
*Seoul, Korea*

**Abstract**

In many studies of the relationship between mental load and sinus arrhythmia (SA), the effect of minor motor actions inevitably accompanying light work has been disregarded. In this study mental and physical loads were respectively scored by information processing and finger tapping rates. According to a multivariate regression analysis of SA scores on mental and physical load components, an increase in mental load decreased SA, but did not increase heart rate (HR) significantly. An increase in physical load decreased SA and increased HR significantly.

**Keywords:** *mental workload; physical workload; heart rate; sinus arrhythmia; regression analysis*

## Introduction

With the advance of industrial technology and the automation of manufacturing processes, tasks requiring mental effort will gradually prevail. Consequently the evaluation of the mental load of a task may be useful for job design or work measurement.

In previous studies on choice reaction tasks, some researchers (Kalsbeek and Ettema, 1963; Kalsbeek, 1968; Wartna et al., 1971) report that an increase in mental load decreases sinus arrhythmia (SA) without changing the heart rate (HR); while others (Blitz et al., 1970; Ettema and Zielhuis, 1971; Mulder et al., 1972) report that it also increases the HR. This inconsistency might have resulted from disregarding the effect of minor motor actions (such as finger tapping) accompanying the 'mental' tasks used (thus inadvertently allowing their colinearity), because conventional univariate ANOVA only tests the equality of mean SA scores for different mental loads. However, minor motor actions not reflected in metabolic rate can influence SA, since chemoreceptors for muscle cell activities are directly coupled with HR through neural pathways (Luczak, 1979).

We investigated the effects of minor motor actions accompanying mental load on the SA scores by multivariate regression analysis. High-mental/low-physical and low-mental/high-physical tasks were included to reduce their colinearity.

## Method

Thirty tasks from 5 stimulus-response modes (mental counting, 4-choice with single tapping, 2-choice combinations of either single or double tapping and, finally, simple tapping) and 6 stimulus-presentation rates (0, 20, 40, 60, 80, 100 per min) were designed (Table 1). A personal computer was used to present a combined visual and auditory stimulus and to record and check the keying responses of the 12 subjects (Figure 1).

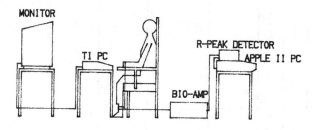

Figure 1. Layout of apparatus

While the seated subject performed 30 randomly sequenced tasks over 5 days, electrocardiographic (ECG) data were collected (3 minutes for each task); RR (interbeat) intervals were measured by an R-peak detector and timer using an Apple II PC (accuracy = 0.5 msec).

## Results

The mental load was scored by the actual information processing rate (M bits/min), the physical load by the finger tapping rate (P taps/min) (See Table 2 and Figure 2). The figure suggests that there are large intersubject variations. To reduce individual differences in levels of mean and standard deviation for RR intervals, it is desirable to normalize each score by an individual reference value. The mean and standard deviation of RR intervals were normalized by an individual resting value to obtain the "percentage change above resting level" in HR(%μ) and SA(%σ).

One of the problems in regression analysis is to find the best set of independent variables to be employed for describing the dependent variable. For this purpose, various functions of mental and physical load variables (M, $M^2$, ln M, P, $P^2$, ln P, and M∗P) which may conceivably be related with the dependent variable, were selected as the potential independent variables. For both %μ and %σ, we obtained all possible regression equations involving every possible combination of the potential independent variables taken singly, in pairs, in triples, etc., using "all possible regression procedures" in the SAS statistical package. Only a pair of variables M and P turned out to be significant:

$$\begin{vmatrix} \%\mu \\ \%\sigma \end{vmatrix} = 10^{-3} \begin{vmatrix} 0.2189 \\ -86.18 \end{vmatrix} + \begin{vmatrix} -0.03538 & -0.3441 \\ -4.558 & -0.7030 \end{vmatrix} \begin{vmatrix} M \\ P \end{vmatrix}, \quad \begin{aligned} &(M<121 \text{ bpm}) \\ &(P<118 \text{ tpm}) \end{aligned}$$

$$R^2 = \begin{vmatrix} 0.05973 \\ 0.3558 \end{vmatrix}, \quad p = \begin{vmatrix} 0.000017 \\ 0.000001 \end{vmatrix}, \quad \text{sample} = 360.$$

Variations of %μ and %σ explained by M and P are small (low $r^2$) but the linear relationships between (%μ & %σ) and the entire set of (M & P) are significant (p.<0.0001). It can be concluded that: Mental load decreases SA; Physical load (minor motor actions) increases HR and decreases SA.

Table 1. Stimulus-response modes of tasks

| Stimulus-response mode (i) | Stimuli visual | Stimuli auditory | Response key | Response for a stimulus mental | Response for a stimulus physical | Scores for a correct response mental (bit) | Scores for a correct response physical (tap) |
|---|---|---|---|---|---|---|---|
| mental counting (i=1) | red blue | 250 Hz 2000 Hz | - | mental counting of total number of blue | - | 1 | - |
| four-choice reaction (i=2) | white red blue black | 262Hz 523Hz 1047 Hz 2095 Hz | (I) (J) (N) (K) | selection of response | single tapping of response key | 2 | 1 |
| two-choice reaction (i=3) | red blue | 250 Hz 2000 Hz | (J) (K) | selection of response | single tapping of response key | 1 | 1 |
| two-choice reaction with double tapping response (i=4) | red blue | 250 Hz 2000 Hz | (J) (K) | selection of response | double tapping of response key | 1 | 2 |
| simple tapping response (i=5) | blue | 2000 Hz | (K) | - | single tapping of response key | - | 1 |

| stimulus-response mode | | target stimulus-presentation rate (stimuli/min) | | | | | | |
|---|---|---|---|---|---|---|---|---|
| | | 0 (j=1) | 20 (j=2) | 40 (j=3) | 60 (j=4) | 80 (j=5) | 100 (j=6) | |
| mental counting (i=1) | mental load (bits/min) | 0 | 19 | 40 | 60 | 77 | 95 | |
| | physical load (taps/min) | 0 | 0 | 0 | 0 | 0 | 0 | |
| four-choice reaction (i=2) | mental load (bits/min) | 0 | 38 | 64 | 81 | 95 | 106 | |
| | physical load (taps/min) | 0 | 20 | 33 | 42 | 50 | 57 | |
| two-choice reaction (i=3) | mental load (bits/min) | 0 | 21 | 36 | 49 | 58 | 67 | |
| | physical load (taps/min) | 0 | 22 | 37 | 50 | 61 | 69 | |
| two-choice reaction with double tapping response (i=4) | mental load (bits/min) | 0 | 19 | 32 | 42 | 49 | 54 | |
| | physical load (taps/min) | 0 | 39 | 65 | 85 | 100 | 111 | |
| simple tapping response (i=5) | mental load (bits/min) | 0 | 0 | 0 | 0 | 0 | 0 | |
| | physical load (taps/min) | 0 | 22 | 41 | 58 | 71 | 84 | |

Table 2. Mean scores of mental and physical loads over 12 subjects during $Task_i(j)$ $(i = 1-5, j = 1-6)$

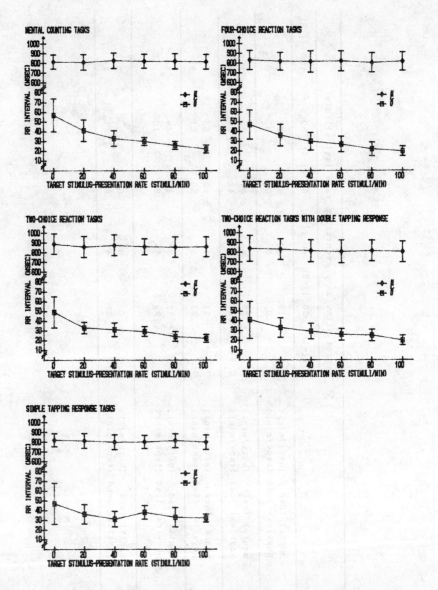

Figure 2. Mean and +/-1 SD of mean($\mu$) and standard deviation ($\sigma$) for RR intervals, as a function of stimulus-response mode and target stimulus-presentation rate

## Conclusions

As the results show, mental and physical load scores explained only a small part of the variations of HR and SA. The result may be sharpened by including other stress variables (such as emotional, environmental, subjective factors) and extra physiological strain variables (such as EEG, EOG, GSR, and respiratory pattern).

For mental and physical load evaluations of light work, the following multivariate inverse regression function between equivalent M, P and observed $\%\mu$, $\%\sigma$ may be relevant:

$$\begin{vmatrix} \hat{M} \\ \hat{P} \end{vmatrix} = \begin{vmatrix} 16.0808 \\ 30.1514 \end{vmatrix} + \begin{vmatrix} 52.5123 & -76.6032 \\ -146.1822 & -17.6765 \end{vmatrix} \begin{vmatrix} \%\mu \\ \%\sigma \end{vmatrix}.$$

This pair of regression functions separates mental and physical load components from physiological variables ($\%\mu$, $\%\sigma$) and converts them into equivalent information processing and finger tapping rates respectively.

## References

Blitz, P.S., Hoogstraten, J., and Mulder, G., 1970, Mental load, heart rate and heart rate variability, Psychologische Forschung, 33, 277-288.

Ettema, J.H., and Zielhuis, R.L., 1971, Physiological parameters of mental load, Ergonomics, 14, 137-144.

Kalsbeek, J.W.H., 1968, Measurement of mental work load and of acceptable load, possible applicatons in Industry, International Journal of Production Research, 7, 33-45.

Kalsbeek, J.W.H., and Ettema, J.H., 1963, Scored regularity of the heart rate pattern and the measurement of perceptual load, Ergonomics, 6, 306.

Luczak, H., 1979, Fractioned heart rate variability. Part II: Experiments on superimposition of components of stress, Ergonomics, 22, 1315-1323.

Mulder, G., and Mulder-Hajonides Van Der Meulen, W.R.E.H., 1972, Heart rate variability in a binary choice reaction task: An evaluation of some scoring methods, Acta Psychologica, 36, 239-251.

Wartna, G.F., Danev, S.G., and Bink, B., 1971, Heart rate variability and mental load: A comparison of different scoring methods, Pfluger's Archiv, 328, 262.

# 32. STRESS MOODS AND STRESSORS OF WORKERS IN KOREAN HEAVY INDUSTRIES

Yoo-Jin Seo[*] and Masaharu Kumashiro[**]

[*]Department of Industrial Engineering
University of Kyung-Nam, Masan, Korea

[**]Department of Ergonomics, IIES
University of Occupational and Environmental Health, Japan

### Abstract

The purpose of this paper was to identify the stress moods experienced by workers in Korean heavy industries, and to examine their effects on overall job stressors. To evaluate the effects of stress moods in the three companies, the stress arousal checklist (SACL) was applied to the investigation of the stress moods in blue collar and white collar workers. This study was conducted with 490 subjects and an anonymous questionnaire was distributed to each division. In rating the feeling of stress, the frequency of representing the feeling of stress as presented in the SACL was used as a stress score ranging from 0 to 17. From our analysis, we came to three conclusions. Firstly, the score of surveyed stress moods was related not only to the presence of a trade union but also "Working pace distribution", "Anxiety for other issues except the original duty", "Working conditions", "Human relations", and "Working attitude" ($p. < 0.0001$). Secondly, workers working for a company without a trade union complained more often than workers with a trade union, regardless of the difference between white collar and blue collar workers. Thirdly, blue collar workers complained more than white collar workers. Based on the results of this study, we recommend a study into methods which would allow top managers and/or personnel (or works) managers to apply the results from the SACL to solve the problems raised by the relationship between labour and management, and to reduce the conflicts.

**Keywords:** *stress; questionnaires; trade unions; arousal; industrial relations; factor analysis; heavy manufacturing industries*

## Introduction

These days, companies in Korea are using logical management themselves and so material, equipment and methodology problems along with others are being solved quickly. However, there are still many problems relating to labour-management relations (Ministry of Economic Planning, Korea, 1988). The rise in the number of industrial disputes shows how serious these are: from 276 in 1986 to 3,749 in 1987, and then to 1,873 in 1988 (Ann, 1989). Though it seems to be declining considerably, there has so far been no fundamental solution to the problem. Because managers, who are in a position to solve these problems smoothly, have in the past depended solely on their experience, they have often made problems worse, rather than better. Furthermore, to maintain smooth industrial relations, mutual recognition and cooperation between labour and management are needed (Seo & Cheon, 1989). It is necessary therefore to collect and analyse data for level and frequency of discontent, dissatisfaction and complaints. The results should then be applied to personnel management or labour services.

The discontent of blue collar and white collar workers can roughly be divided into two groups; one is discontent of the physical system and the other is of the psychoneurotic system (Kumashiro et al., 1989). The former can easily be solved by adapting work conditions, the design of the workplace, and work-rest schedules. However, the effect on the psychoneurotic system, which is a problem of mental stress, is quite a difficult problem to solve.

Mackay et al. (1978) produced a stress arousal check list (SACL) to observe emotional irregularities induced by living and working environments. It was further studied by Cox et al. (1982) and King et al. (1983). Cruickshank (1984) raised questions on the use and composition of the SACL. In response, Cox & Mackay (1985) presented another study on the effectiveness of the SACL to justify it as a method for the measurement of self-reported stress and arousal. Several validation studies on the SACL have been reported (Burrows et al., 1977; King et al., 1983; Hollingworth, 1988; Kumashiro et al., 1989). The SACL is specifically designed to evaluate the sense of stress and the sense of arousal, which are two distinct types of emotions: negative hedonic tone or stress ("tense", "worried", and "uncomfortable") and arousal ("alert", "awake", and "aroused") (Cox et al., 1982; Watts et al., 1983).

We used the SACL to investigate the stress mood of Korean workers. In this study, we compared the stress mood of workers from three companies and explored its associated factors.

## Method of survey

We chose three representative heavy industrial corporations in Korea according to the presence of a trade union and the degree of problems: company "A" had 385 white collar and 264 blue collar workers and no trade union; company "B" had 787 white collar and 3,198 blue collar workers who were suffering from severe problems between management and labour with a trade union; company "C" had 297 white collar and 513 blue collar workers with a trade union experiencing a few problems.

An initial sample of 100 persons was taken from each company and category of worker totalling 600 persons. In March 1989, an anonymous questionnaire was distributed to each division, then collected through company personnel staff. The return rates, comparing white collar to blue collar workers, were: company "A", 69% : 95%; company "B", 98% : 96%; company "C", 92% : 73%. The average return rate was 86.3% : 88.0%. Though the total return rate for the entire sample was 87.2%, the possible data for analysis for white collar workers were on average 81% and 82.3% for blue collar workers, totalling 81.6% of the original sample of 600.

The questionnaire comprised two scales: one was the translated Korean version of the SACL, the other consisted of 24 items concerning work-related conditions and other issues. In rating the sense of stress, the frequency of representing the sense of stress, as presented in the SACL, was used as a stress score ranging from 0 to 17.

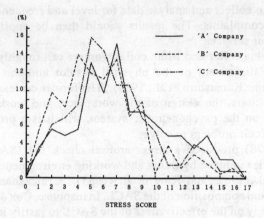

Figure 1. Percentage distribution of stress scores in Korean heavy industry workers

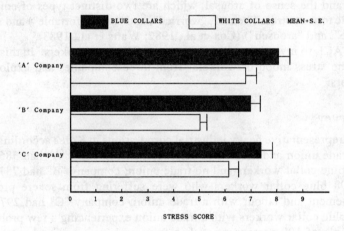

Figure 2. Comparison of the stress scores between blue collar and white collar workers in Korean heavy industry workers

## Results and discussion

### Stress score

From answers to the anonymous questionnaire, the total average stress score of the 490 respondents (243 white collar workers and 247 blue collar workers) working for heavy industrial companies was found to be 6.67 +/-3.51 (M +/-SD). The average stress scores for each company were: 7.54 +/-3.71 (M +/- SD) for company "A" with 155 respondents (62 white collar and 93 blue collar workers); 5.98 +/-3.28 for company "B" with 194 respondents (98 white collar and 96 blue collar workers); 6.62 +/-3.42 for company "C" with 141 respondents (83 white collar and 58 blue collar workers). The stress scores for the three companies were tested by ANOVA (p.<0.001). Meanwhile, in respect of the mode for stress score, the mode of workers in company "B" was 4.00, 5.00 for company "C", but the mode for company "A" was 7.00.

Figure 1 illustrates the stress score distribution for white collar and blue collar workers. We found that the average stress score for workers in company "A" (without a trade union) was higher than that of "B" and "C" (with a trade union). Figure 2 illustrates the distribution of workers and managers for each company. The average stress scores for managers and workers were: 6.80 +/-3.08 and 8.06 +/-4.02 for company "A"; 5.03 +/-2.53 and 6.97 +/-3.67 for company "B"; 6.15 +/-3.39 and 7.38 +/-3.35 for company "C". The stress scores, for white collar and blue collar workers, were tested by ANOVA (p.<0.001). We found that blue collar workers complained more often than white collar workers.

Factor analysis of 24 work-related items concerned with work was used to explore the stressors that exist in the workplace. In Table 1, 24 items concerned with work are classified into 8 factors: working conditions, expectations in the company, human relations, trust in the company, anxiety for other issues except original duty, work arrangement, working attitude, working pace distribution. The stress scores quoted by the SACL as dependent variables, and 8 factor scores as independent variables were then subjected to multiple regression analysis. Five factors yielded significant association as shown in Table 2.

From Table 2, we find that the factor "Working pace distribution", "Anxiety for other issues except the original duty", "Working conditions", Human relations", and "Working attitude" should be examined further.

Table 2. *Stepwise multiple regression analysis of occupational stressors against SACL stress scores*

| Step | Stressor variable | Multiple R | $R^2$ | $R^2$ Change |
|---|---|---|---|---|
| 1. | Working pace distribution | .394 | .153 | .153 |
| 2. | Anxiety for other issues except the original duty | .493 | .239 | .086 |
| 3. | Working conditions | .531 | .276 | .037 |
| 4. | Human relations | .557 | .303 | .027 |
| 5. | Working attitude | .564 | .309 | .006 |

Overall F=34.96  P<0.0001

Inter-individual conflict can be divided into substantive and emotional conflict. Emotional conflict cannot be solved through negotiations or other problem solving methods. However, the substantive conflict can be solved through the intervention of a third party (Wofford, 1977). If workers were to decrease their substantive conflict through their trade union activities, they would show low stress scores. However, if workers were not able to solve their substantive conflict, not only would it be transferred to emotional conflict but emotional conflict would also increase, resulting in a combined high score. Analyzing this data, as in Figure 1, we observed that the stress score for company "A" workers, not having a trade union, is higher than for company "B" workers and that is thought to be caused by the presence of a trade union. The stress score for company "B" workers was lower than "C" workers, despite the fact that company "B" was seriously suffering problems, as opposed to company "C" which rarely experienced problems.

Table 1. Factor analysis of stressors among in heavy industry workers

| Items | Loading |
|---|---|
| **Factor 1: Working conditions** (Eigenvalue = 2.44; Pct. of variance = 10.2%) | |
| 1. Safety notion against work place | 0.77 |
| 2. Suitability for working environment | 0.69 |
| 3. Fairness of work time and relax time | 0.64 |
| 4. Wages | 0.61 |
| **Factor 2: Expectations to the company** (Eigenvalue = 2.30; Pct. of variance = 9.6%) | |
| 1. Long work history for present company | 0.75 |
| 2. Intimacy for present company | 0.73 |
| 3. Certainty of one's role in work place | 0.55 |
| 4. One's own opinion reflected to the company | 0.49 |
| **Factor 3: Human relations** (Eigenvalue = 1.82; Pct. of variance = 7.6%) | |
| 1. Smooth human relations with peers | 0.83 |
| 2. Smooth human relations with coworkers | 0.79 |
| 3. Smooth human relations with bosses | 0.48 |
| **Factor 4: Trust to the company** (Eigenvalue = 1.69; Pct. of variance = 7.0%) | |
| 1. Degree of desire to change company | -0.68 |
| 2. Degree of satisfaction of promotion | 0.60 |
| 3. Satisfaction of the type of duty | 0.49 |
| **Factor 5: Anxiety for other issues except the original duty** (Eigenvalue = 1.61; Pct. of variance = 6.7%) | |
| 1. Anxiety for future plans after retirement | 0.67 |
| 2. Others conflicts except the original duty | 0.66 |
| 3. Anxiety for health | 0.60 |
| **Factor 6: Work arrangement** (Eigenvalue = 1.59; Pct. of variance = 6.6%) | |
| 1. Monotony of work | 0.74 |
| 2. Responsibility of work | -0.64 |
| 3. Existence of inconsistent work | 0.52 |
| **Factor 7: Working attitude** (Eigenvalue = 1.38; Pct. of variance = 5.8%) | |
| 1. Suitability for ability and duty | 0.77 |
| 2. Extra pressure from overtime work and holiday work | 0.67 |
| **Factor 8: Working pace distribution** (Eigenvalue = 1.37; Pct. of variance = 5.7%) | |
| 1. Work smoothness | 0.78 |
| 2. Distribution of work pace | 0.55 |

## Conclusions

From our analysis, we conclude:
1. The score for stress mood is related not only to the presence of a trade union but also to "Working pace distribution", "Anxiety for other issues except original duty", "Working conditions", "Human relations", and "Working attitude".
2. Workers in a company without a trade union complained more than workers with a trade union, regardless of the difference between white collar and blue collar workers.
3. Blue collar workers complain more often than white collar workers.

As a result of this study, we recommend a study into methods which would let a top manager and/or personnel (or works) manager apply the results from the stress arousal check list to solve the problems raised by the relationship between the labour force and management, and which would lead to a reduction in conflicts.

## References

Ann, C.S., 1989, Industrial Relations in After 6.29 Declaraton, Nippon Roudou Kyokai, Japan, 354, 45-51.
Burrows, G.C., Cox, T., and Simpson, G.C., 1977, The Measurement of Stress in a Sales Training Situation, Journal of Occupational Psychology, 50, 45-51.
Cox, T., Mackay, C., and Page, H., 1982, Simulated Repetitive Work and Self-Reported Mood, Journal of Occupational Behaviour, 3, 247-252.
Cox, T., and Mackay, C., 1985, The Measurement of Self-Reported Stress and Arousal, British Journal of Psychology, 76, 183-186.
Cruickshank, P.J., 1984, A Stress and Arousal Mood Scale for Low Vocabulary Subjects, British Journal of Psychology, 75, 89-94.
Hollingworth, C., 1988, Job Satisfaction and Mood: An Exploratory Study, Work and Stress, 2, 225-232.
King, M.G., Burrows, G.D., and Stanley, G.V., 1983, Measurement of Stress and Arousal: Validation of The Stress/Arousal Adjective Checklist, British Journal of Psychology, 74, 473-480.
Kumashiro, M., Kamada, T., and Miyake, S., 1989, Mental Stress with New Technology at the Workplace, In: Work with Computers: Organizatonal, Management, Stress and Health Aspects, edited by M.J. Smith and G. Salvendy, Elsevier, 270-277.
Mackay, C., Cox, T., Burrows, G.C., and Lazzerini, A.J., 1978, An Inventory for the Measurement of Self-Reported Stress and Arousal, British Journal of Social and Clinical Psychology, 17, 283-284.
Ministry of Economic Planning, 1988, An Economic White Paper, Seoul, Korea, 415. (In Korean)
Seo, Y.J., and Cheon, M.B., 1989, Worker Participation and Industrial Relations, Thesis of Labor Welfare, University of Kyung-nam Korea, 8, 39-70. (In Korean)
Watts, C., Cox, T., and Robson, J., 1983, Morningness-Eveningness and Diurnal Variations in Self-Reported Mood, Journal of Psychology, 113, 251-256.
Wofford, J.C., et al., 1977, Organizational Communication, McGraw-Hill, 230-241.

## 33. A STUDY ON THE HYGIENE AND ERGONOMICS OF MEDICAL STAFF IN A HOSPITAL IN CHINA

Tianlin Li[*], Xiufen Zhang[**], Ailan Feng[**], Lijuan Zheng[**], Zhengxiang Wu[**], Chunfa Zhang[**] and Zhenying Fu[*]

[*]*Department of Applied Physiology*
*Institute of Occupational Medicine, CAPM*

[**]*Harbin Institute of Industrial Hygiene and Occupational Diseases, China*

**Keywords:** *stress; health and safety; hospitals; medical staff; work experience*

## Introduction

Although there are hundreds of publications on medicine, hygiene and labour protection, there are few reports on the health of medical workers in hospitals (Izmerov and Kaptsov, 1984). Medical staff come into contact with patients suffering from different kinds of illness. Their work is arduous and they have little rest during work. It is very often necessary for them to work overtime. In addition, in their workplaces there may be physical or chemical factors that have hazardous effects on them. How and to what degree do these factors influence the health of medical staff? So far, there has been no detailed study (Bates and Moore, 1975). To ascertain the actual health situation of medical personnel, and to provide evidence for protecting and improving their health, we carried out the following research.

## Method of study

A middle-sized comprehensive hospital was selected and a reviewing method used. We investigated working conditions, such as the normal work situation, illness, morbidity, working time for medical staff and compared the morbidity rate between medical workers and odd-jobmen. The above information was collected, estimated and analysed, and some of the results were classified, ordered and analysed. To prevent the confounding effects of age, care was taken to balance the groups.

## Results and discussion

(1) Firstly, we compared morbidity among staff members of different ages in certain hospitals in 1988. The results show that morbidity among medical staff increases with increasing age. Three kinds of diseases are most common, in the following order, diseases of the circulatory system, the digestive system and the respiratory system.

The morbidity of respiratory diseases in the 20-year old age group appeared highest and was highly significant (p.<0.01). The morbidity of diseases of the circulatory system was highest in the 30-year old age group. This high morbidity indicated that the health of medical staff was far from good.

In the 40-year old age group the morbidity of diseases of the digestive system was significantly higher than for other age groups (p.<0.001).

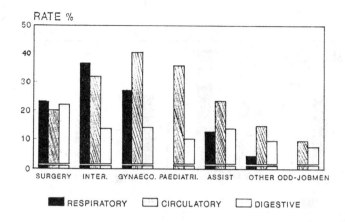

Figure 1. The standardized morbidity of common diseases in different departments

(2) We compared the morbidity among medical staff of different clinical experience. The situation was almost the same as for age, i.e. the morbidity of medical staff rose with increase in clinical experience. Diseases of the circulatory system were highest.

Among those people whose clinical experience was less than 10 years, the morbidity of respiratory diseases was the highest (20.8%). In the group with 20 years experience, the morbidity of digestive diseases was highest (14.3%).

The morbidity for diseases of the musculoskeletal system among people whose experience is more than 20 years was 6.14% (for 20 years of experience) and 13.5% (for 30 years of experience) respectively. It follows that the greater the experience, the higher the morbidity of the musculoskeletal system.

(3) Estimation and analyses of the standardized morbidity of staff members from different departments in certain hospitals in 1988 (Figure 1) showed that the morbidity of respiratory diseases in the departments of internal medicine and gynaecology were highest (37.13% and 27.09%). Perhaps this was because of the heavier work and the greater contact with patients making it more likely for them to become infected. The morbidity of respiratory diseases is above 20% of the total.

The morbidity of circulatory system diseases in the departments of gynaecology and paediatrics was highest (40.5% and 36.77%). The reason for this phenomenon has been mentioned above. The morbidity of circulatory system diseases in the department of internal medicine and the surgical department was higher than 20%. It seems that the morbidity of circulatory system diseases in most departments was usually higher, so some work must be carried out to prevent this cardiovascular disease. The total morbidity would decline if the morbidity of cardiovascular diseases declines.

According to Figure 1, the morbidity of digestive system diseases in the surgical department was obviously higher than other departments. Perhaps this was because the working time in the surgical department was relatively longer, their work and living habits were not regular. Hence, we should try to help them to arrange better working and resting schedules so that the morbidity could be limited.

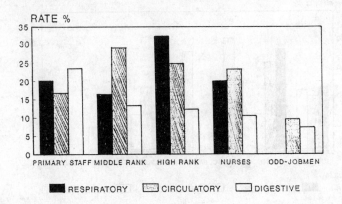

Figure 2. The comparison of diseases for standardized morbidity of diseases between medical staff and odd-jobmen

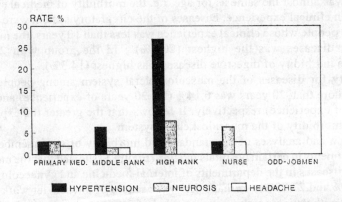

Figure 3. The comparison of the morbidity of three kinds of diseases between medical staff and odd-jobmen

(4) Figure 2 shows that the morbidity of diseases of the circulatory system among middle rank medical staff was highest (29.2%), the morbidity of diseases of the digestive system for primary medical staff was highest. On the contrary, the morbidity of the odd-jobmen was lower, it was usually less than 10%. This shows that the tasks performed by medical staff were heavier and may be the cause for the decreased capability of preventing diseases. So they can more easily become ill. Therefore we should pay more attention to the health of medical staff in hospitals.

(5) We compared the most commonly occurring diseases of medical staff and odd-jobmen. It is known that there are 18 kinds of diseases that commonly occur among medical staff, but for the odd-jobmen, there are only 4 kinds. The morbidity for medical staff was 1.2%-5.0%. Each of the 18 kinds of diseases showed certain morbidity. The morbidity for odd-jobmen was 1.8-3.5%. This shows that there is an

evident difference between the health of medical staff and of odd-jobmen. So it is clear that, in general, the health situation of medical staff is worse than of the odd-jobmen in the same unit (Formanek et al., 1989).

(6) The comparison of three kinds of disorders between the medical staff and the odd-jobmen (see Figure 3) shows that in higher rank staff the rates of hypertension and neurosis were 26.9% and 7.7%. The morbidity of hypertension for middle rank staff was 6.3%; nurses mainly suffered from neurosis (6.2%). But none of the odd-jobmen suffered from any of these three diseases in three years. This may be because the work of medical staff was always intense, they were under nervous and mental strain, for long periods (Vincent and Tatham, 1976).

## Summary

1. The results of the investigation show that morbidity increases with increasing age and clinical experience. Ordinarily, the morbidity for people over 40 years of age and whose experience is more than 20 years is highest.

2. Diseases of the circulatory, respiratory and digestive systems are most common. The morbidity of the respiratory system is relatively high in the departments of internal medicine, gynaecology and obstetrics. The morbidity of the digestive system is relatively high in the surgical department. And the rate of hypertension among high rank medical staff is much higher than others. These results can provide the scientific bases for the protection of and improvement in the health of medical staff working in hospitals.

3. The results also show that the morbidity of medical staff in hospitals is much higher than that of the odd-jobmen working in the same unit. It points to the need for some protective measures, such as decreasing the workload and shortening the working time per day (Pond, 1969; Waring, 1974).

## References

Bates, E. M. & Moore, B. N. Stress in hospital personnel. Medical Journal of Australia, 15, 765-767, 1975.

Formanek, J. et al. Interdisciplinary investigation of the work of geriatric nurses. The Proceedings of the 3rd Conference on "Work and Health of Health Care Workers" Georgia, USSR, 2-5 October 1989.

Izmerov, N. F. & Kaptsov, V. A. Physician heal thyself. World Health Forum, 5, 273-275, 1984.

Pond, D. A. Doctors' mental health. New Zealand Medical Journal, 69, 131, 1969.

Vincent, M. O. & Tatham, M. R. Psychiatric illness in the medical profession. CMA Journal, 115, 293-296, 1976.

Waring, E. M. Psychiatric illness in physicians: a review Comprehensive Psychiatry, 15, 519, 1974.

# 34. RELATIONSHIP OF SUBJECTIVE SYMPTOMS OF FATIGUE TO ATTITUDE TO LIFE AND STRESS

Yukio Hiraoka, Junko Tanaka, Masahide Oda, Hisanori Okuda and Hiroshi Yoshizawa

*Department of Hygiene, Hiroshima University School of Medicine*
*1-2-3 Kasumi, Minami-ku, Hiroshima 734, Japan*

## Abstract

A survey was carried out on 2108 employees of 32 small-to-medium-sized enterprises in Hiroshima Prefecture. The questionnaire items related to working conditions, stress, daily life, the subjective symptoms of fatigue and other factors. It was found that persons reporting numerous complaints of subjective symptoms of fatigue lead a more unhealthy life, they were less enthusiastic and active at work, felt less pleasure and satisfaction about family life and work, and experienced a lot of stress. Thus, it can be inferred that considerable attention should be directed to persons reporting numerous complaints of subjective symptoms of fatigue in order to alleviate their stress.

**Keywords:** *stress; fatigue; mental fatigue; mental health; questionnaires; home life*

## Introduction

We have reported (Okuda and Tanaka, 1986; Okuda and Tanaka, 1988; Okuda et al., 1989; Tanaka, 1988; Tanaka, 1989) that physical fatigue tends to lead to mental fatigue, and that mental fatigue accumulates and can become chronic. It has also been determined that stress is a cause of mental fatigue. As it has become very difficult for workers in Japan and other industrialized countries to avoid stress, investigation into countermeasures to stress would appear to be of great importance.

In the present study, some relationships between subjective symptoms of fatigue and some situations connected with job attitudes and attitudes to life and with stress were investigated.

## Subjects and methods

### Subjects

A survey was carried out in 32 small-to-medium-sized enterprises in Hiroshima Prefecture. Of 2108 questionnaires given out, 1844 were returned, 1808 of which were usable, making the survey 85.8% effective.

### Contents of the questionnaire

The questionnaires consisted of 27 questions about working conditions, stress, and daily life, and subjective symptoms of fatigue. The questionnaire has been recommended by the Japanese Association of Industrial Health (Tanaka et al., 1989). The questions are classified into three groups, I, II and III (Yoshitake, 1975). In addition, the Self-Rating Depression Scale (SDS) (Association for the Study of Industrial Fatigue, 1970; Fukuda and Kobayashi, 1973; Zung, 1965; Sarai, 1976) developed from Zung and Breslow's "Seven desirable habits of good health" was applied at the same time. This relates to factors such as life habits and conditions, diseases related to stress, and physical conditions.

**Period and methods of the survey**

The survey was conducted on weekdays from 1 to 14 September, 1989. Questions on subjective symptoms of fatigue, bedtime of the previous day, and behaviour between rising and going to the office were answered on reaching the office. Other questions could be answered at any time.

**Analytical methods**

Responses to questions on the subjective symptoms of fatigue were compared, and divided into three groups; group A, in which the number of complaints was 0, group B, in which there were from 1 to 3, and group C, in which there were 4 or more complaints. The questionnaire returns were further divided by sex, so that there were 6 sets of answers for each item, which were totalled, and considered separately.

## Results

### Average age, type of occupation and working conditions

The number of subjects and average age of each group are shown in Table 1. Average ages for both males and females in group C were lower than those in groups A and B.

Table 1. Number of subjects and average ages

|  | Male | | | Female | | |
| --- | --- | --- | --- | --- | --- | --- |
|  | Grp. A | Grp. B | Grp. C | Grp. A | Grp. B | Grp. C |
| Subjects | 121 | 377 | 601 | 71 | 242 | 396 |
| Average age | 40.3 | 39.1 | 37.7 | 37.0 | 37.3 | 34.5 |

Table 2. SDS scores

|  | Male | | | Female | | |
| --- | --- | --- | --- | --- | --- | --- |
|  | Grp. A | Grp. B | Grp. C | Grp. A | Grp. B | Grp. C |
| Average | 37.7 | 39.6 | 43.6 | 38.6 | 41.8 | 45.3 |
| Subjects | 100 | 340 | 537 | 58 | 211 | 358 |

Production workers constituted the majority of the subjects (males 44.4% and females 32.1%), followed by, in males, technical and research workers (15.6%), office workers (15.5%), and executives (10.5%). In females, the order was office workers (30.2%), technical and research workers (18.5%), and saleswomen and service workers (14.4%). The number of nominal days-off and real days-off in a month and overtime in a week were not appreciably different among the three groups. Overtime in a week was 8-9 hours in males and 2-3 hours in females. Overtime in both males and females was the longest in group C, followed by groups B and A.

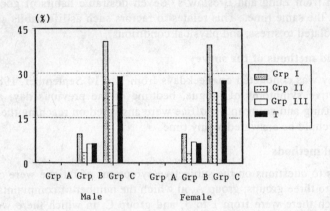

*Figure 1. Complaint rates of subjective symptoms of fatigue*

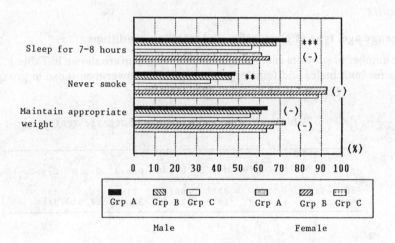

*Figure 2. Habits regarding good health - A*

## SDS scores and the rate of subjective symptoms of fatigue

SDS scores in group C were the highest, followed by group B, and then group A in both males and females (see Table 2). SDS scores were inversely related to the level of mental health. Thus, persons in group C showed poorer mental health.

As for the complaint rates of subjective symptoms of fatigue, for both males and females in group B, group I complaints were most commonly reported followed by groups III and II (see Figure 1). This pattern of subjective symptoms was designated "general type" by Yoshitake (1975). The complaint rate of males and females in group C was highest in group I, followed by groups II and III (Figure 1). This pattern was designated "mental work type or night-shift type" by Yoshitake. For group A, it was impossible to identify the type of fatigue since the complaint rate was 0.

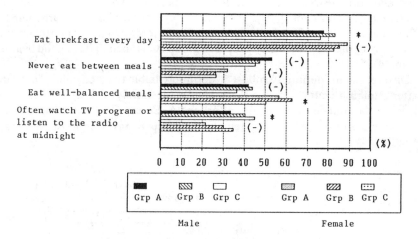

*Figure 3. Habits regarding good health - B*

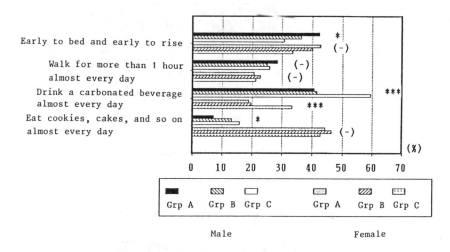

*Figure 4. Habits regarding good health - C*

## Health conditions

Regarding diseases in the year preceding the questionnaire, differences in the rate of stomach and duodenal ulcers, asthma, angina pectoris, and depressive psychosis among groups A, B, and C could not be determined. Low back pain, gastrointestinal neurosis, and other neuroses were noted more frequently in both males and females in group C.

For Breslow's "Seven desirable habits of good health", differences in the three groups regarding "sleep for 7-8 hours" and "never smoke" in males were statistically significant at the 0.1% and 1% levels respectively (Figure 2). Many males in group A maintained these desirable habits of good health, while females showed a significant difference. In males, differences in the three groups for "eating breakfast every day", "often watch TV programme or listen to the radio at midnight", "early to bed and early to rise", and "eat cookies, cake and so on almost every day" were statistically significant at the 5% level. Differences for "drink a carbonated beverage almost every day" in both males and females were statistically significant at the 0.1% level. In all these items, many people in group A claimed to have good health (Figures 3 and 4).

Differences in "definitely robust", "definitely liable to catch cold", "definitely puffy after climbing stairs" (at the 0.1% level), "definitely lack of exercise" (at the 1% level) in both males and females of groups A, B, and C were statistically significant (Figures 5 and 6). Males and females in group A gave the most answers for good health.

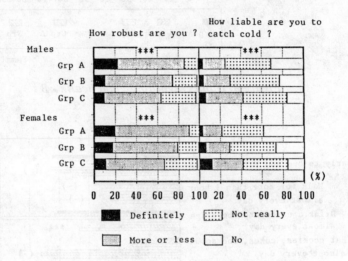

Figure 5. Health condition - A

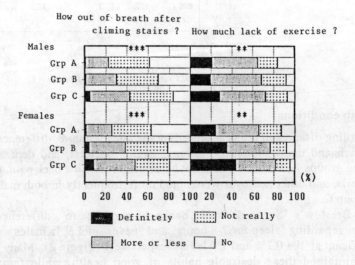

Figure 6. Health condition - B

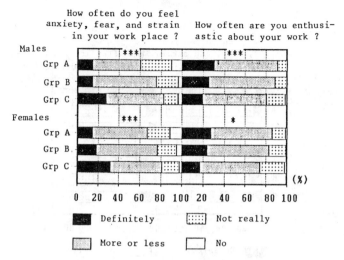

Figure 7. Working conditions

Table 3. Major stress factors at work and their frequency

| | Male | | | | Female | | | |
| | Grp. A | Grp. B | Grp. C | SD | Grp. A | Grp. B | Grp. C | SD |
| --- | --- | --- | --- | --- | --- | --- | --- | --- |
| Too much pressure to work harder | 40<br>33.9 | 125<br>33.7 | 283<br>47.9 | *** | 21<br>30.0 | 86<br>36.3 | 170<br>43.5 | * |
| Human relations in the work place are not good | 22<br>18.6 | 82<br>22.1 | 178<br>30.1 | ** | 15<br>21.4 | 82<br>34.6 | 157<br>40.2 | ** |
| Working hours are too long | 20<br>17.0 | 52<br>14.0 | 160<br>27.1 | *** | 8<br>11.4 | 30<br>12.7 | 90<br>23.0 | *** |

Upper;real number, Lower;percentage. SD means significant difference
*<0.05, **<0.01, and ***<0.001, as evaluated by the chi-square test.

## Workplace and family life

Persons enthusiastic about work and who feel less anxiety, fear, and strain in the workplace were most numerous in both males and females in group A, followed by groups B and C (Figure 7).

Differences in "too much pressure to work harder", "human relations at work are not good", and "working hours are too long" in both males and females in the three groups were statistically significant (Table 3). Thus, many persons in group C had too much work pressure, bad human relations, and worked too long hours. They were followed by groups B and A.

Regarding family life, males who were enthusiastic about family life and felt less anxiety, fear, and strain were most numerous in group A and differed statistically significantly at the 0.1% level from groups B and C (Figure 8). Females experiencing less anxiety, fear, and strain were most numerous in group A and the difference was statistically significant at the 5% level but there was no significant difference in "how often are you enthusiastic about your family" among the three groups (Figure 8). Thus, persons in group C felt less constructive, willing, and happy about family life. Most persons in group A felt more satisfied with family life and work than those in groups B and C (Figure 9). The difference was statistically significant at the 0.1% level.

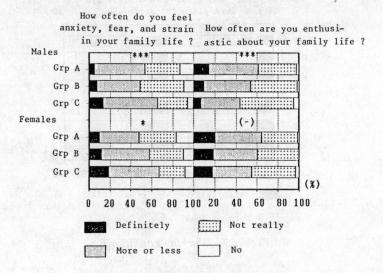

*Figure 8. Family life conditions*

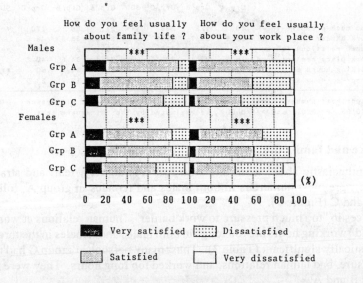

*Figure 9. Daily life conditions*

Regarding happiness, except when working, many males and females of group A answered "having a good time with the family" and "playing sports". Many in group C answered "smoking" and "vulgar amusements or gambling" in males and "owning one's house" in females.

## Discussion

From the surveys of workers which we have carried out, daily life appears to cause a greater frequency in subjective symptoms of fatigue than working conditions. In the present study, persons with numerous complaints of subjective symptoms of fatigue not only had a more unhealthy life style, but poorer mental health. They were less enthusiastic and active at work, felt less pleasure and satisfaction about family life and work, and experienced a lot of stress. Thus, it can be inferred that considerable attention should be directed to persons having numerous complaints of subjective symptoms of fatigue to avoid stress.

The results of this study demonstrate that an enthusiastic attitude towards family life and towards work leads to reduced stress.

## References

Association for the Study of Industrial Fatigue 1970, Report on "the investigation of subjective symptoms" of industrial fatigue, Digest of Science of Labors, 25, 12-62.

Fukuda, K. and Kobayashi, S. 1973, A study on a self-rating depression scale, Folia Neurologica et Psychiatrica Japanese, 75, 673-679.

Okuda, H. and Tanaka, J. 1986, Subjective fatigue among workers in small and medium-sized enterprises, The Journal of the Hiroshima Medical Association, 39, 1816-1827.

Okuda, H. and Tanaka, J. 1988, Subjective fatigue among workers in small and medium-sized enterprises. An investigation over four days in one week, The Journal of the Hiroshima Medical Association, 41, 2065-2077.

Okuda, H., Yamamoto, M. and Tanaka, J. 1989, Subjective fatigue among workers in small and medium-sized enterprises. Relevance of marital status and whether wives are co-workers, The Journal of the Hiroshima Medical Association, 42, 1800-1809.

Sarai, K. 1976, The epidemiology of depression, Folia Neurologica et Psychiatrica Japanese, 81, 778-853.

SAS Institute 1982, SAS user's guide (1982 edition), North Carolina SAS Institute Inc.

Tanaka, J. 1988, Fatigue of contemporary workers, Hiroshima Medical Journal, 36, 1037-1051.

Tanaka, J. 1987, A study on subjective symptoms of fatigue in gainfully occupied couples, Japanese Journal of Industrial Health, 29, 486-493.

Tanaka, J., Hiraoka, Y., Imada A. and Okuda, H. 1989, Relation of subjective symptoms of fatigue to work conditions and daily life of VDT workers, Hiroshima Medical Journal, 37, 665-680.

Yoshitake, H. 1975, Industrial fatigue, Rodokagaku Sosho, 33, 22-35.

Zung, W.W.K. 1965, A self-rating depression scale, Archives of General Psychiatry, 12, 63-70.

# 35. THE LINK BETWEEN STRESS AND ATTITUDE TOWARDS LIFE

Masahide Oda, Junko Tanaka, Yukio Hiraoka,
Hisanori Okuda and Hiroshi Yoshizawa

*Department of Hygiene, Hiroshima University School of Medicine*
*1-2-3 Kasumi, Minami-ku, Hiroshima 734, Japan*

### Abstract

In order to investigate the relationship between enthusiasm and stress and stress reactions, a survey was conducted via a questionnaire on 2,108 workers in Hiroshima Prefecture. 1,844 of these (87.5%) were returned, and 1,795 (85.2%) were usable. The results showed that the group who was enthusiastic about work, family and others showed the least reaction to stress. In comparison with the other groups, this group felt stress in more or less the same way but felt more happiness and pleasure. It is considered that an enthusiastic attitude toward life raises the level of happiness in life, and this tends to either eliminate stress or to weaken undesirable stress reactions.

**Keywords:** *stress; home life; fatigue; mental fatigue; questionnaires; job attitudes*

## Introduction

It has been shown in our previous investigations into fatigue that living conditions as well as working conditions are relevant to worker fatigue (Okuda and Tanaka, 1986; Okuda and Tanaka, 1988; Okuda et al., 1989). It has also become clear that there is a tendency for worker fatigue to shift from physical to mental fatigue, and for the mental fatigue to accumulate. Furthermore, there is a possibility that this mental fatigue, together with adult diseases, may be the cause of the 'deaths from overwork' which have recently been in the news. Therefore, in order to deal with these problems, it seems necessary to develop ways of dealing with stress.

The modern worker is continuously exposed, not only at his place of work but in various places and situations, to pressures which can subject him to stress. If these causes of stress cannot be avoided, then attention must be focused on resistance to stress (the ability to suppress outbreaks of stress generated by these pressures) and methods of getting rid of stress. In the belief that a man's resistance to stress depends on his attitude toward life (his enthusiasm or commitment), we have conducted an investigation into the connection between stress and attitude toward life. We now present our findings.

## Method

### Subjects of the investigation

A survey was carried out on workers at 32 small-to-medium-sized businesses in Hiroshima Prefecture, with questionnaires being distributed to 2,108 workers. 1,844 of these (87.5%) were returned, and 1,795 (85.2%, 1,091 from males, 704 from females) were usable.

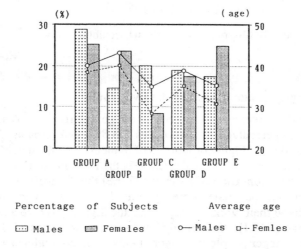

Figure 1. *Percentage of subjects and average age in each group*

## Content of the questionnaire

The questionnaire covered the thirty headings proposed by the 'investigation of subjective symptoms' (Association for the Study of Industrial Fatigue, 1970) by the fatigue research group of the Japan Association of Industrial Health. Additional headings covering areas relevant to the assessment of stress such as frequency and intensity of occurrences of stress at work, at home, and elsewhere, and frequency and intensity of feelings of enthusiasm and happiness were incorporated into the questionnaire, as well as some general questions about the time of going to bed the night before the investigation, the time of getting up on the day itself, and activities prior to arriving at work. Furthermore, in order to measure mental health, the twenty headings of the Zung 'Self-rating Depression Scale' (SDS) were included in the questionnaire (Fukuda and Kobayashi, 1973; Zung, 1965; Sarai, 1976). However, it must be said that this paper bases its conclusions about stress on the complaints of anxiety and insecurity.

## Dates and procedure of the investigation

The survey was conducted on working days from 1-14 September 1989. The 'investigation of subjective symptoms' and the items concerning the time of going to bed and activity prior to arrival at work were to be completed when the subject arrived at work, while the other items could be completed at any time.

## Method of analysis

The subjects were assigned to one of the following five groups, according to their enthusiasm (or commitment) or lack of it in three areas of life:

Group A  (balanced type) enthusiastic about work, family, and other things

Group B  (home type) enthusiastic about work and family, but not about other things

Group C  (family-ignoring type) enthusiastic about work and other things, but not about family

Group D    (work-centered type) enthusiastic about work, but not about family or other things

Group E    (work-hating type) not enthusiastic about work.

## Results

### Number and average age in each of the five groups

Figure 1 shows the percentage of subjects in each of the five groups and their average age. The largest percentage of both males (28.8%) and females (25.3%) were in Group A. Group B had a higher proportion of females (23.7%) than of males (14.6%), Group C, a higher proportion of males (20.1%) than of females (8.4%), and Group E, a somewhat higher proportion of females (25.0%) than of males (17.6%). In other words, more females than males were committed to their families, and more females than males disliked their work. Group B had the highest average age for both males (43.2) and females (40.1), and Group C had the lowest (males 35.0, females 28.3), indicating that younger people were more likely to be not enthusiastic about family life.

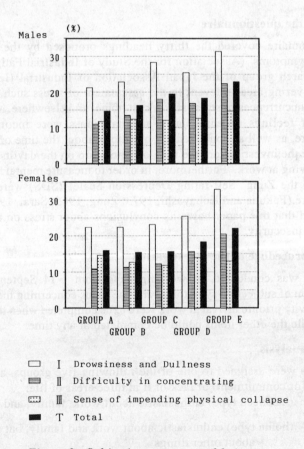

Figure 2. Subjective symptoms of fatigue

## Subjective symptoms of fatigue

From the results of the 'investigation of subjective symptoms' the rate of complaint about subjective symptoms of fatigue (hereafter referred to as T) was calculated. These symptoms were also studied as three separate groups, and the rate of complaint about each group was calculated.

*Table 1. Subjective symptoms of fatigue*

|  |  | GROUP A | GROUP B | GROUP C | GROUP D | GROUP E |
|---|---|---|---|---|---|---|
| males | I (%) | 21.2 | 22.8 | 28.3 | 25.4 | 31.4 |
|  | II (%) | 11.0 | 13.5 | 17.9 | 16.8 | 22.3 |
|  | III (%) | 11.7 | 12.4 | 12.1 | 12.6 | 14.6 |
|  | T (%) | 14.6 | 16.2 | 19.4 | 18.3 | 22.8 |
|  | II／T | 0.75 | 0.83 | 0.92 | 0.92 | 0.98 |
|  | TYPE | I | II | II | II | II |
| Females | I (%) | 22.4 | 22.4 | 23.2 | 25.5 | 32.9 |
|  | II (%) | 10.9 | 11.3 | 12.3 | 15.7 | 20.5 |
|  | III (%) | 14.6 | 12.7 | 11.8 | 14.0 | 13.0 |
|  | T (%) | 16.0 | 15.5 | 15.7 | 18.4 | 22.1 |
|  | II／T | 0.68 | 0.73 | 0.78 | 0.85 | 0.93 |
|  | TYPE | I | I | II | II | II |

Each group contained 10 of the 30 headings of the 'subjective symptoms of fatigue' investigation: Group I being the group of symptoms of 'drowsiness and dullness', Group II, the group of symptoms of 'difficulty in concentrating', and Group III, the group of symptoms of 'a sense of impending physical collapse'. These names for the groups are taken from Yoshitake (1975), who classifies types of fatigue on the basis of the order of frequency of complaints about each of the three groups of symptoms as follows:

I-dominant type (general type)
  in which Group I > Group III > Group II

II-dominant type (mental work type, night-shift type)
  in which Group I > Group II > Group III

III-dominant type (physical work type)
  in which Group III > Group I > Group II

where > = greater than

Figure 2 and Table 1 show the subjective symptoms of fatigue for each group. They indicate a tendency for T-complaint rate, in both males and female, to be lower in Group A and Group B (males A 14.6% and B 16.2%, females A 16.0% and B 15.5%), and higher in Group E (males 22.8%, females 22.1%).

With regard to the type of mental fatigue, the results indicate a tendency for Group A males and females, and Group B females, to be I-dominant, and for all other groups to be II-dominant. With regard to male/female differences, the results for Group C indicate a higher rate of complaints of fatigue and a more definite tendency toward mental fatigue among males than among females.

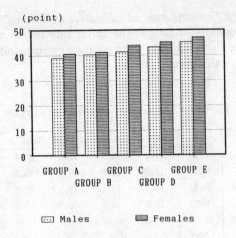

Figure 3. SDS scores

Table 2. SDS scores

|  | GROUP A | GROUP B | GROUP C | GROUP D | GROUP E |
|---|---|---|---|---|---|
| Males | 38.9 | 40.2 | 41.4 | 43.2 | 45.1 |
| Females | 40.6 | 41.1 | 43.9 | 45.1 | 46.9 |

Groups A and B, the groups showing enthusiasm for both work and family life, had a low rate of complaint of fatigue for both males and females. With regard to the type of fatigue, the same two groups, except for the Group B males, were general type (I-dominant), not manifesting a tendency toward mental fatigue. Groups C, D, E, on the other hand, the groups lacking enthusiasm for work or lacking enthusiasm for family life, had a high rate of complaint about fatigue and a tendency toward mental fatigue.

### The Zung self-rating depression scale

Figure 3 and Table 2 show symptoms measured on the Zung scale. For both males and females, the score rises gradually all the way from Group A to Group E, an indication that Groups A and B (the groups showing enthusiasm for both work and family life) have a higher level of mental health, and both Group E (the group with no enthusiasm for work) and Group D (the group enthusiastic about work but not enthusiastic about family life and other things) have a lower level of mental health. The scores for Group E (the group with no enthusiasm for work) are particularly high (45.1 for males, 46.9 for females), coming close to the Zung scores for groups of patients with symptoms of neurosis (Fukuda and Kobayashi, 1973).

### Stress and happiness at the workplace

#### Stress

See Figures 4 and 5, which tabulate, respectively, the frequency and intensity of stress (anxiety, insecurity, etc.) at the workplace for each group.

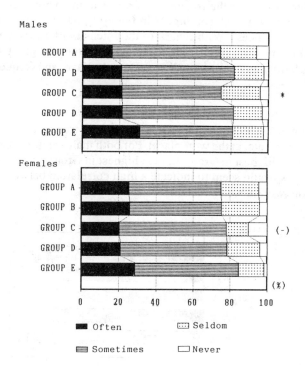

*Figure 4. Frequency of stress at workplace*

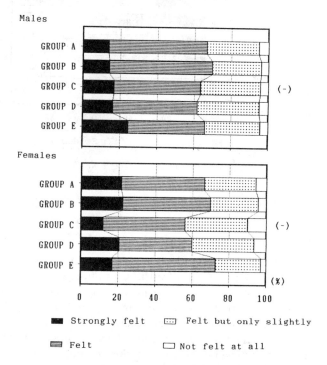

*Figure 5. Intensity of stress at workplace*

There was a tendency for the frequency of stress at the workplace among males to be lower in Groups A and C, and higher in Group E, but among females there were no significant differences. With regard to the intensity of stress among both males and females, there were no significant differences between the groups. It would seem that there was hardly any correlation between enthusiasm and frequency or intensity of feelings of stress.

## Happiness

See Figures 6 and 7 which tabulate, respectively, the frequency and the intensity of happiness, pleasure, etc. at the workplace. For both males and females, the frequency and intensity were highest in Group A, next highest in Groups B, C, and D, and lowest in Group E. This would seem to indicate a high correlation between enthusiasm and happiness at the workplace.

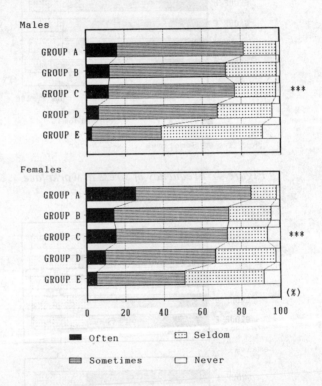

Figure 6. Frequency of happiness, pleasure etc at the workplace

## Discussion

Selye says that whether an individual, in given circumstances, feels eustress (stress in a good sense, implying a degree of fulfillment of the whole person) or distress (an unhappy condition of fretful anxiety or mental anguish) depends to a considerable extent on that person's attitude toward life, and how he faces up to life (Selye, 1974). Our current studies however, indicate that all of our groups, with or without enthusiasm for life, feel stress in more or less the same way. Additionally, the studies indicate that where the level of enthusiasm does make a difference it is in regard to happiness or pleasure. From this, it may be conjectured that there is a link between enthusiasm and happiness, and between happiness and freedom from stress. It seems that happiness reduces the subjective symptoms of fatigue and raises the level of mental health. But if we take it that there is a partial correlation between subjective symptoms of fatigue and the level of mental health on one hand, and stress reactions on the other, then to increase happiness and pleasure should be an effective strategy to eliminate or counter stress. We expected to find some correlation between the level of enthusiasm and resistance to stress, but in the answers from our five groups about

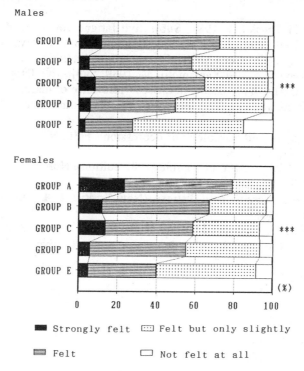

Figure 7. *Intensity of happiness, pleasure etc at the workplace*

whether or not they felt stress, there was no evidence that the level of enthusiasm made any appreciable difference. It seems that the point about resistance to stress is not whether the stress can be felt or not, but whether there are means to eliminate it and to deal with the stress reaction. It would seem that the means required is a combination of enthusiasm for life, and happiness, which increases that enthusiasm. Group A, both males and females, showed the least stress reaction, and this indicates the importance of a balanced spread of enthusiasm over the three areas of life, namely work, family, and others. Group B showed the least stress reaction after Group A, and this indicates the importance of enthusiasm spread over at least the two areas of work and family. However, although we expected to find that enthusiasm for life would increase resistance to stress and reduce the occurrence of stress, our investigation of stress, as measured by complaints of feelings of anxiety or insecurity, did not provide any definite indication of a connection between enthusiasm and stress. Our conclusion is that rather than reducing the causes of stress, an enthusiastic attitude to life raises the level of happiness in life, and this tends to eliminate stress or to weaken undesirable stress reactions.

## References

Association for the Study of Industrial Fatigue. 1970, Report on the 'investigation of subjective symptoms' of industrial fatigue. Digest of Science of Labor, 25(6), 12-62.

Fukuda, K. and Kobayashi, S. 1973, A study on a Self-Rating Depression Scale. Folia Neurologica et Psychiatrica Japanese, 75(10), 673-79.

Okuda, H. and Tanaka, J. 1986, Subjective fatigue among workers in small and medium-sized enterprises. The Journal of the Hiroshima Medical Association, 39, 1816-27.

Okuda, H. and Tanaka, J. 1988, Subjective fatigue among workers in small and medium-sized enterprises - an investigation over four days in one week. The Journal of the Hiroshima Medical Association, 41, 2065-77.

Okuda, H., Tanaka, J. and Yamamoto, M. 1989, Subjective fatigue among workers in small and medium-sized enterprises - relevance of marital status and whether wives are co-workers. The Journal of the Hiroshima Medical Association, 42, 1800-09.

Sarai, K. 1976, The epidemiology of depression. Folia Neurologica et Psychiatrica Japanese, 81, 778-853.

SAS Institute. 1982, SAS user's guide (1982 edition). North Carolina: SAS Institute Inc.

Selye, H. 1974, Stress without distress. (New York: J. B. Lippincott Co.) (Japanese translation 1986).

Yoshitake, H. 1975, Industrial fatigue. Rodo Kagaku Sosho, 33, 22-35.

Zung, W.W.K. 1965, A self-rating depression scale. Archives of General Psychiatry, 12, 63-70.

# 36. AN EXAMINATION OF THE MENTAL WORKLOAD OF DESIGN WORK IN OFFICES

## Takeshi Aoyama and Mamoru Umemura

*Department of Management Science*
*Science University of Tokyo*
*1-3 Kagurazaka, Shinjuku-ku, Tokyo 162, Japan*

**Abstract**

In this paper, the social problems of industrial and occupational stress and fatigue were studied. Psychological methods were used to examine the degree of mental workload. The task chosen for the study of mental workload was "Design Work in Offices". Each subject was examined for one continuous week. It is hoped that the results from this study can be used as a basis for future studies into overall workload assessment.

**Keywords:** *mental workload; heart rate; offices; designers; commuting*

## Introduction

Recently in Japan, the public issue of industrial stress and fatigue has received much attention. Death caused by overwork (karoshi), which is the worst case, has been frequently discussed. It can be argued that the mental pressure required to accomplish tasks is increasing due to the complexity of society, the fractionalization of technology, the increase of information and the diversification of value systems. In this study it was decided to examine the degree of work stress and fatigue from the viewpoint of workload in compulsory work in everyday life. Up to now, most studies have been concerned with industrial stress and fatigue originating from muscular work, although recently there have been many studies concerned about perceptual work. However, there have been few studies which have examined the mental work covering the whole period of compulsory working hours.

In this paper, "Design Work in Offices" was selected as an example of mental work and it was measured by means of continuous heart rate (HR) monitoring during actual work time and commuting time continuously for one week.

## Methods

This study was a field examination. Continuous HR monitoring was selected as the instrument to measure work stress because of its relative unobtrusiveness.

Four subjects, all in their early thirties, were studied, their personal details are given in Table 1. They were mainly engaged in design work in offices and all were judged to be healthy by the company's medical examination. None of them complained about their health during the period of the experiment.

The examination period was from July 17 to August 28 1989, and each subject was examined for one week, excluding Saturday and Sunday.

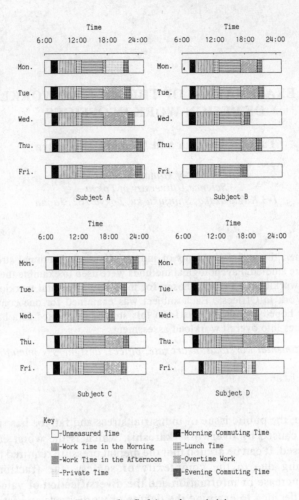

Figure 1. Subjects' activities

Table 1. Subjects' details

| Subject | Sex | Home Background Environment | Commuting Time (one way) | Shift |
|---|---|---|---|---|
| A | M | Bachelor | About 70 min. | Daytime |
| B | M | Wife and a Child | About 60 min. | Daytime |
| C | M | Bachelor | About 70 min. | Daytime |
| D | M | Wife and a Child | About 60 min. | Daytime |

Continuous HR was recorded without a break from when the subjects left home in the morning until they returned in the evening. During the study, subjects did not go on overnight business trips and they acted as normal during leisure and working time. Each subject decided for himself whether or not to work overtime depending on the content and progress of the design work, and private circumstances. However, often they were required to work overtime unwillingly on unexpected work.

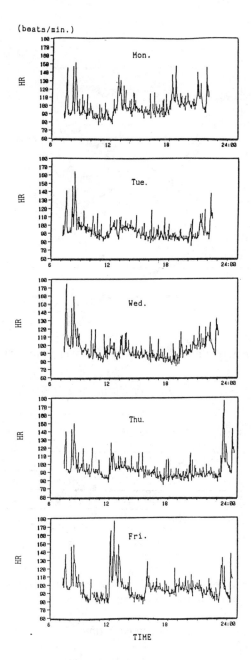

*Figure 2. Sequential variation in HR (Subject A)*

## Results

### Activity of subjects during the examination period

The activity of subjects is classified by time periods shown in Figure 1. Figure 2 shows the sequential variation in HR for one subject (subject A).

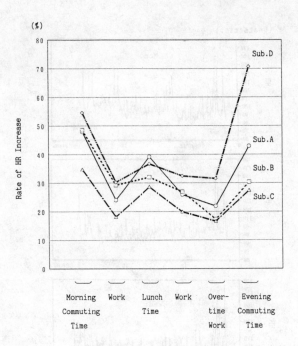

Figure 3. *Rate of HR increase (%) over time periods*

Table 2. *Rate of HR increase (%)*

| Time Bloc<br>Subject | Morning<br>Commuting<br>Time | Work | Lunch<br>Time | Work | Over-<br>time<br>Work | Evening<br>Commuting<br>Time |
|---|---|---|---|---|---|---|
| A | 47.9 | 24.0 | 39.2 | 25.9 | 21.9 | 42.9 |
| B | 48.5 | 29.2 | 32.0 | 26.9 | 17.3 | 30.3 |
| C | 34.7 | 18.1 | 28.6 | 19.9 | 16.6 | 27.4 |
| D | 54.5 | 30.2 | 36.7 | 32.4 | 31.5 | 70.7 |

## Heart rate

The HR results indicate remarkable differences among individuals. The rate of the HR increase was computed by the following equation -

$$\text{Rate of HR increase (\%)} = \frac{(\text{mean HR in each time block}) - (\text{HR at rest})}{(\text{HR at rest})} \times 100$$

For resting HR, we used the lowest HR recorded from each subject:
- Subject A : lowest HR = 74 (beats/minute)
- Subject B : lowest HR = 80 (beats/minute)
- Subject C : lowest HR = 71 (beats/minute)
- Subject D : lowest HR = 60 (beats/minute)

Results are shown in Table 2 and Figure 3.

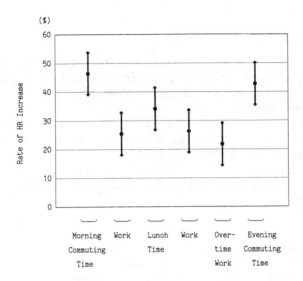

*Figure 4. Estimates (95% confidence interval) of rate of HR increase (%) over time periods*

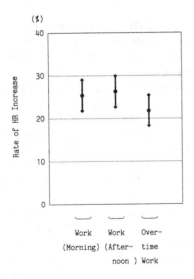

*Figure 5. Estimates (95% confidence interval) of rate of HR increase in design work time*

Figure 4 gives estimates (95% confidence interval) of the rate of HR increase in all the time periods. These time periods are divided into three types, design work time (in the morning, afternoon and overtime), restricted hours excluding design work time (lunch time) and the time necessary to go to the office (commuting time). It took from between 60 to 70 minutes for each subject to commute between home and the office, but each commuting route was different. They were free to spend their time at lunch. Therefore, we estimated the rate of HR increase only during the time they did similar design work to each other, as shown in Figure 5.

## Discussion

### Estimation method of workload in mental work

At present there are numerous methods used to estimate mental workload (for example critical fusion frequency, electroencephalogram, electric impedance of the skin, skin temperature, blood pressure, survey of subjective symptoms, sinus arrhythmia, HR etc). The reasons it was decided to use HR in this study were as follows:

(1) HR is reliable (Lacey, 1967; Meshkati, 1988). Meshkati (1988) in particular gives a full account of the use of HR in the examination of mental workload.

(2) Continuous records of data can be obtained.

(3) The method does not disturb the subjects or interrupt their work.

(4) The measure can be used not only during design work time, but also during compulsory work time.

### Design work

The design work times chosen to examine were as follows:
(1) Design work in offices involving mental work.
(2) Basic work incidental to design work (development etc).
(3) Meetings, discussions, etc.
Each subject was occupied primarily with item (1).

### Design work time

In design work periods, the rates of HR increase were nearly equal in each time period, 25% in the morning, 26% in the afternoon, and 21% during overtime work. There were no significant differences among work time periods, and the degree of HR variation during design work time was somewhat more stable than during commuting time. In design work time periods, the increase in mean HR was 17 beats/minute and design workers had equal workloads throughout.

According to the Handbook of Industrial Fatigue (Endo, 1988), "There is no standard method of estimating HR concerned with mental work under the present state of studies". Therefore, for reference purposes, we propose to compare the HR of mental work with that of static muscular work. Hashimoto and Endo (1973) reported, "The upper limit increase of HR in static work is 10 beats/minute", but HR in the design work in our examination greatly exceeded that value.

It can be observed from Figure 5 that the degree of rate of HR increase in the morning is approximately the same as that in the afternoon, but is least in overtime. Consequently, it may be considered that during overtime work, external stress due to conferences, meetings and reception of callers is less. On the other hand, as working time passes by, the subjects gradually became fatigued and too tired to react to their workload immediately. During this time period, each subject's rate of HR increase was always between 20 to 30%.

### Commuting time

During the morning commuting period, the rate of HR increase was 46%, mean HR was 104 beats/minute, and the HR increase was 33 beats/minute. During evening commuting time the rate of HR increase was 43%, mean HR was 102 beats/minute, and the HR increase was 31 beats/minute. In these commuting periods, the mean rate of HR increase was 45%, and the mean HR increase was 32 beats/minute. In spite of the differences in their commuting routes, each subject's HR increase rate was almost of the same value, but it was obviously different from that in design work time. This may be because each subject commutes under the same rush hour conditions in the train and subway, etc.

Hashimoto and Endo (1973) report that "The limit of HR increase, with continued dynamic work for many hours, was 30 beats/minute". Comparing our results with the studies mentioned above, it may be considered that commuting aggravates strain.

**Lunch time**

The rate of HR increase at lunch time was 34%, a little larger than the active design work period before and after lunch. It could be suggested that their internal body organs become active while moving to lunch shops and having lunch. The rate of HR increase of subject A was larger than others probably because this subject went out of the office at lunch time and took exercise.

## Conclusions

From the results of this study, the following conclusions can be reached concerning design workload:
(1) During design work time, the rate of HR increase was 25% in each time period, and subjects had similar workloads for many successive hours.
(2) During commuting time, the rate of HR increase was 45%. This was the biggest increase during compulsory activities.
(3) There were no changes in the rate of HR increase by day of the week over the one-week examination period.

The necessity to eliminate factors that cause work stress and fatigue is increasing. Mental work is different from muscular work in that there is no way to quantify it. Establishment of counterplans (effective use of work ability and advancement in overall quality of work life) are lagging behind, compared to other fields.

In design work time, it is necessary to reduce the rate of HR increase by improving the designer's surroundings, work content, and by lightening the workload by the reduction of work time (especially overtime). In commuting time, it is necessary for us to change our circumstances drastically by the reduction of commuting time and the administrative improvement of crowded trains during the rush hour. The rate of HR increase could be reduced by avoiding the rush hour, as in a flexitime system, in the future.

## References

Endo, T. 1988, Handbook of Industrial Fatigue, Japan Association of Industrial Health, (Rodo Kijun Chosakai, Tokyo), 178-181.
Hashimoto, K. and Endo, T. 1973, Vision of Function of Human Body, (Japan Publication Service, Tokyo), 124-137.
Lacey, I. 1967, Somatic response patterning and stress. Some revisions of activation theory, In Psychological Stress: Issues in Research, M.H.Apperley and R.Trumball (Eds) (Appleton-Century-Crofts, New York), 14-37.
Meshkati, N. 1988, Heart rate variability and mental workload assessment, In Human Mental Workload, P.A. Hancock and N. Meshkati (Eds) (North-Holland, Amsterdam), 101-115.

# 37. A STUDY ON THE EVALUATION OF SIMPLE WORKLOAD BY A THERMAL VIDEO SYSTEM

### Yoshinori Horie

*College of Industrial Technology*
*Nihon University, Japan*

**Keywords:** *mental workload; skin temperature; critical fusion frequency*

## Introduction

Many methods have been tried to evaluate the criteria of an optimum working environment for man at work since labour science began. Some are objective or subjective, some continuous or intermittent etc. Evaluation of learning and skill acquisition is thought to be an important factor when deciding on the appropriate method. It goes without saying, it is very important to measure man's workload in a simple but effective way. It is important, however, not only for experimenters to be able to estimate a subject's load under the lesser stress caused by unrelated factors, but also for the subject to be able to work as usual.

The experimental study, therefore, was conducted by measuring skin surface temperatures which is thought to be a useful index for evaluating the variations in physiological functions.

In the research field, measuring skin temperatures by the thermal video system (TVS) is a comparatively new methodology which makes it possible to take surface skin temperatures moment by moment as consecutive distributed pictures without touching the body. The relative strength of emitted infrared rays from the skin is then converted into electric signals which can be indicated as an image on a CRT display after being amplified. The data can also be processed through the various kinds of image processing techniques.

In the experiment, the Critical Flicker Fusion Frequency (CFF) value was also measured. CFF is frequently used by ergonomists in Japan because the measuring method is very easy and data obtained are effective for estimating man's workload.

## Outline of the experiments

In order to avoid the unnecessary influence of outer environments, all the environmental conditions were kept constant at a temperature of $20^{\circ}C$, humidity of 60%, and the air current from 0.1 to 0.3 meters per second. Other infrared rays were excluded so they would not affect the results.

Subjects were male college students. They were prohibited from smoking for three hours before starting the experiments, so that they would not be affected by nicotine which is apt to cause fluctuations in blood volumes.

The equipment used for the experiment was made by the Nippon Avionics Co., Ltd. and called the TVS-4000 series. The measurements of the subjects' skin surface temperatures were taken by the TVS with a resolution of $0.2^{\circ}C$, and under conditions of radiation rate equivalent to $E = 0.98$.

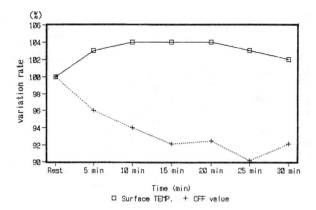

*Figure 1. The variations in surface body temperature and CFF value (N = 7), results from the first experiment.*

## The first experiment

The part of the body measured was the palm for the reasons that the spot is raked, easy to measure, and big fluctuations of blood volume can be observed.

The experimental procedure was that the main task of measuring the CFF value, with the right hand, started after 20 minutes of rest and continued for 30 minutes of work at 5 minute intervals. At the same time the subject was told to put his left hand on the platform for the skin temperature to be measured through the TVS. No other task was given during the experiments.

## The second experiment

The task adopted for the second experiment was simple visual work. Twenty-one healthy male college subjects, aged from 20 to 24 years, were used. All subjects were right handed and their visual and other physical functions were normal. Prior to the experiment, all subjects were asked to enter and stay in the experimental room which was at a temperature of about 20°C for one hour to acclimatize themselves to the thermal environment of the room.

The task of trying to add a pair of randomized one digit numbers from 3 to 9 was given to all subjects. They were told to input only the last digit of the added number into the computer using the ten keys with the right hand. Subjects had to continue the task for 30 minutes continuously. During each experimental session, various functions of the subjects were measured every 5 minutes. Evaluation methods were as previously mentioned - surface body temperature (using the left hand), critical flicker fusion frequencies (CFF) (using the right hand), - and the environmental conditions of the experimental room represented by the temperature, humidity, and air current were measured to make clear the relationship between subjects and their working environment.

## *Results from the experiments*

### The first experiment

Mean skin temperatures increased continuously for 20 minutes immediately after starting work, and then decreased. The continuous and gradual deteriorations in CFF

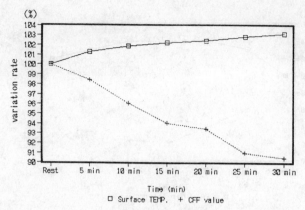

*Figure 2. The variations in surface body temperature and CFF value (N = 21), results from the second experiment.*

variations for the subjects were also seen for 25 minutes (Figure 1).

From these results, it is assumed that the deterioration in the level of the cerebral cortex was seen to be mainly due to the occurrence of feelings of monotony. At the end of the experiments, the significant activation of arousal level for the subjects was clearly seen to reflect an end spurt effect. There was a -0.72 correlation coefficient, at the 5% significance level, observed between temperatures of averaged area and CFF values.

### The second experiment

The results obtained from the 2nd experiment were quite similar to those results of the 1st experiment. Namely, the tendency of skin surface temperatures to increase was observed through the experiment. The deterioration in the level of the cerebral cortex activity was also seen clearly owing to the increase in the feelings of monotony caused by the neuro-sensory mental work associated with the VDT task.

This was clearly confirmed by the results obtained from the CFF values (Figure 2). There was a -0.97 correlation coefficient observed at the 1% significance level, between surface skin temperature and CFF values.

## Conclusion

In conclusion, the following can be deduced from the results obtained from the two experiments; The surface skin temperature data recorded by the TVS can be validated as an objective physiological index, since all the data closely relate to the variation in CFF - a good index with which to evaluate the psychophysiological load of man at work. Because a comparatively short period of 30 minutes was adopted as the duration for these experiments, longer periods of working would be needed for future experiments in order to be able to apply the results to real industrial situations.

Information obtained by the TVS is effective and useful for evaluating workload. However, there might still be many problems to be solved if it is intended to use the system more effectively in the future, such as the cost of processing, technological problems, for example improving process graphics, how to measure or how to analyze the data quantitatively and so on. Once these problems have been solved, it can be said that the TVS measuring method would be one of the simplest and most effective methods for evaluating physiological variation in man at work.

# Part VI
# Workplace Evaluation

# 38. DIFFERENCES IN WORKLOAD AND THE INFLUENCE OF AGE ON PRESS AND WELDING WORKERS

Koki Mikami[*], Soichi Izumi[*] and Masaharu Kumashiro[**]

[*]Department of Industrial Engineering
Hokkaido Institute of Technology, Japan

[**]Department of Ergonomics, IIES
University of Occupational and Environmental Health, Japan

## Abstract

The subject of this investigation was the main work, press and welding work, in the manufacture of automobile parts, and the purpose was to contribute to work management in the workplace from the human viewpoint. The differences in workload and the influence of age on press and welding workers were investigated. The investigation showed that the following considerations are important to work management in this industry: 1) Exchanging existing workers with workers who can do more than one kind of work as a result of their work experience. 2) Introducing combined work systems under which the middle- or advanced-aged workers and younger workers compensate for each other. 3) Improving work to compensate for reduced psychophysiological functions in order to best utilize middle- or advanced-aged workers.

**Keywords:** *physical workload; ageing; fatigue; work experience; job design; critical fusion frequency; press work; welding; job attitudes*

## Introduction

Automobile manufacturers occupy an important position in Japanese industries, and are supported by many automobile parts manufacturers. Automobile parts manufacturers operate diversified small-quantity production and "Just In Time" production corresponding to the production lines of their parent manufacturers, and measures to use the labour force flexibly under such production systems are indispensable for the work management (Nakane, 1986). In addition, Japan is rapidly turning into a society composed largely of the aged (Ida, 1990), and countermeasures to the ageing of workers are also necessary for work management (Koide, 1990).

The subject of this investigation is the main work, press and welding work, in an automobile parts manufacturer, and the purpose is to contribute to work management in the workplace. The differences in workload and the influence of age on press and welding workers, who play an important role in the automobile parts industry, were investigated from the viewpoint of psychophysiological functions, feelings of fatigue and workers' behaviour.

## Methods

### Differences in workload between press and welding workers

Six workers in the press workshop and seven in the welding workshop were selected as subjects. They were about the same age and had about the same duration of work experience. Their average ages were 29.5 in the press workshop and 30.4 in the welding workshop. The average durations of work experience were 9.5 years in the press workshop and 9.9 in the welding workshop.

The main task assigned to the press workers was preparing materials, setting moulds, pressing, watching with vigilance and carrying materials, moulds and products. The main task assigned to the welding workers was welding by hand, operating machines, filing off welds, adjusting sizes, inspecting and carrying materials and products. The working hours were from 8:00 to 10:00, from 10:15 to 12:00, from 12:45 to 15:00 and from 15:15 to 17:00. The actual working time was 7 hours and 45 minutes. The holidays were every Sunday and Monday (the statutory holidays were not holidays) in line with those of the parent manufacturer.

The following were examined: Target-Aiming Function (TAF) (Takakuwa, 1982), Critical Fusion Frequency (CFF), Multiple Choice Reaction Time (MCRT), and Subjective Feelings of Fatigue (SFF) which consisted of 30 items (defined by the Japan Association of Industrial Health). Heart rate and workers' behaviour were recorded during the working hours. Sound levels, temperature and humidity were also measured. Moreover, job consciousness, working conditions and health conditions were investigated through questionnaires.

### Differences in workload with age

Five younger workers, average age 22.2, and five middle- or advanced-aged workers, average age 54.2 in the press workshop, and five younger workers, average age 24.6, and five middle- or advanced-aged workers, average age 51.2 in the welding workshop, were selected as subjects.

The following were examined: CFF, MCRT, near point accommodation, ocular accommodation time, short-term memory ability, crural circumference, tapping ability, the body's centre of gravity, grip strength and lung capacity. Workers' behaviour was also recorded during working hours.

## Results and discussion

### Differences in workload between press and welding workers

Work environment. The average noise levels were 91.8 +/-3.5db(C) in the press workshop and 81.4 +/-2.4dB(C) in the welding workshop. Illumination levels in the work areas ranged from 70 lx to 1000 lx in the press workshop and from 16 lx to 2460 lx in the welding workshop. The average temperature was 32.3 +/-0.5 °C in the press workshop and 33.7 +/-0.5 °C in the welding workshop. The average humidity was 73.3 +/-1.0% in the press workshop and 66.8 +/-2.2% in the welding workshop.

Aspects of workers at work. The rates of actual work activity, excluding personal allowances were 96.4% in the press workshop and 97.5% in the welding workshop. Comparison between the morning and the afternoon showed that the rates for the main task decreased 4% in the afternoon in both workshops. In addition, the variations in the rates of the personal allowances with the lapse of time showed higher appearance rates in the press workshop than those in the welding workshop. In both workshops

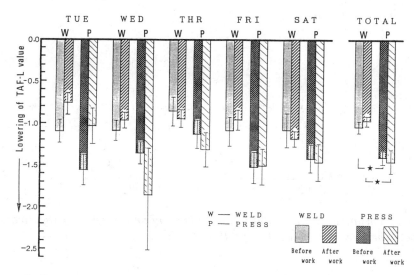

Figure 1. *Variations in TAF-L values for 5 days where * = significant difference between the two workshops at the 1% level*

the personal allowances had a tendency to increase gradually in the afternoon compared to the morning.

With regard to working postures, the total appearance rates of both the 'standing straight' posture and the 'bending a little forward with the knees extended' posture were high, 92.6% in the press workshop and 78.4% in the welding workshop. In the press workshop there was no large difference between the postures in the morning and those in the afternoon. On the other hand, in the welding workshop the total appearance rates of the 'bending a little forward' postures and the 'bending forward with the back bent' postures decreased by 9% in the afternoon and the rate of the standing straight posture increased by 10%.

Behaviour of workers off duty. The average sleeping hours per day were 6.9 +/-0.8 hours for the press workshop workers and 7.2 +/-0.8 hours for the welding workshop workers. The workers' main activities after work were "relaxing at home", and no difference was observed between the two workshops.

Changes in psychophysiological functions. As shown in Figure 1, there was no significant lowering of TAF-L values in either workshops after work in any of the mean values, namely, the mean values for each weekday and the mean values for the week. However, the TAF-L values in the press workshop were low for both before and after work for each weekday except the value after work on Tuesday, which was the first day after the two days holiday of the week, as compared to those in the welding workshop. Comparison of the TAF-L mean values for the week between the two workshops showed significantly low values both before and after work in the press workshop.

With regard to the CFF mean values for each weekday and for the week, there were no significant decreases after work in either workshop. As Figure 2 shows, there was hardly any CFF variation recorded with the lapse of time in the two workshops, but the values at every tested time were low in the welding workshop compared to the press workshop, and there was a remarkable difference recorded between the two workshops.

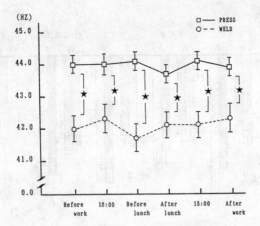

Figure 2. Variations in CFF values with the lapse of time, where * = significant difference between the two workshops at the 1% level

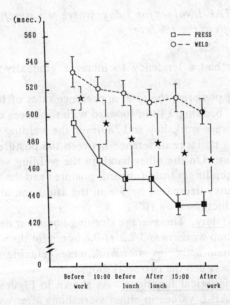

Figure 3. Variations in mean reaction time in multiple choice reaction time, where ▫● = significant at the 5% level and ■ = significant at the 1% level as compared with the value before work, and ☆ ★ = significant difference between the two workshops at the 5% and 1% levels, respectively

MCRT had a tendency to shorten with the lapse of time in both workshops as shown in Figure 3. At every tested time, MCRTs were shorter in the press workshop than those in the welding workshop, and a remarkable difference was observed between the two workshops.

As Figure 4 shows, there were no marked increases in heart rate with the lapse of time either in the press workshop or in the welding workshop, but the heart rate of the welding workers decreased less during the lunch break than that of the press workers.

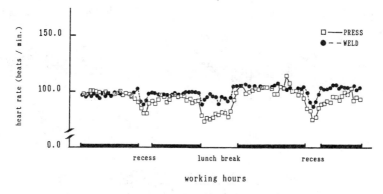

Figure 4. Variations in heart rate with overtime

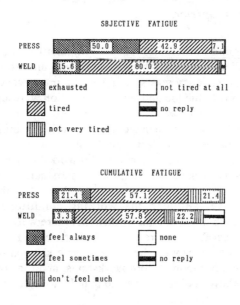

Figure 5. Complaint rates (%) of subjective and cumulative fatigue

Subjective feelings of fatigue (SFF). The mean complaint rates of the T category of SFF were 18.0% before work and 14.6% after work in the press workshop, and 25.4% before work and 27.8% after work in the welding workshop. The first category, "dullness and drowsiness", was the dominant pattern observed both before and after work in the two workshops. The items with over 50% complaint rates after work were "dullness over the whole body", "feel unsteady in standing" and "want to lie down", which were only observed in the welding workshop.

Feelings of fatigue and cumulative fatigue. Feelings of fatigue and cumulative fatigue showed high complaint rates in both workshops as shown in Figure 5. The complaint rates of feelings of fatigue were 92.9% in the press workshop and 95.6% in the welding workshop. The complaint rates of cumulative fatigue were 78.5% in the press workshop and 71.7% in the welding workshop.

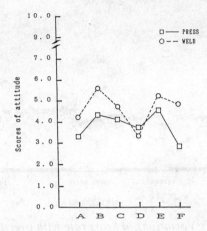

Figure 6. Workers' attitudes, where A = management, B = work, C = human relations with co-workers, D = wages and life, E = working conditions and F = human relations with superiors

Job consciousness and workers' attitudes. With regard to job consciousness, in both workshops the workers had little job satisfaction and interest in the job, and reported feeling weary. This tendency was stronger in the press workshop. In addition, the workers' frustration with the job and sense of isolation were also stronger in the press workshop.

Workers' attitudes consist of "management", "work", "working conditions", "wage and life", "fellow workers" and "superiors" (Saegusa, 1962). As Figure 6 shows, comparison of these six items between the two workshops showed an inharmonious pattern in the press workshops, where their job satisfaction and morale tended to be lower. In particular, the score of "superiors" (2.8) was low there.

As mentioned above, although the working conditions were bad with high temperatures and much noise in both workshops, there were no substantial decreases in psychophysiological functions recognized after work. However, the variations in the personal allowance rates with the lapse of time showed that workers' desire for rest increased in the afternoon in both workshops, and the complaints of feelings of fatigue and of cumulative fatigue suggested that the workers were experiencing heavy workload in both workshops.

Comparing the press workshop and the welding workshop, the complaint rates of SFF were higher both before and after work in the welding workshop, which suggests that the welding workers were more fatigued. In addition, the heart rate showed a difference in the effect of rest between the two workshops. On the other hand, workers' negative attitude to work was stronger in the press workshop, and their dissatisfaction with their superiors was clearly recognized. Furthermore, there were remarkable differences recognized in the function levels of TAF, CFF and MCRT between the two workshops, and it was seen that these differences resulted from the difference in workers' experience at work.

Therefore, in order to reduce workers' consciousness of heavy workload in both workshops, it is necessary to improve the work itself, to take their work experience into account to change existing workers with workers who can do more than one kind of work for flexible production systems. In addition, superiors should give their subordinates proper instructions and advice so that the above improvement, advice and change may help improve workers' morale.

Table 1. Comparison of psychophysiological function levels between middle- or advanced-aged groups and younger groups

| Items of examination | PRESS | | WELD | |
|---|---|---|---|---|
| | before work | after work | before work | after work |
| 1 near point accommodation(Diopter) | ↓★ | ↓★ | ↓★ | ↓★ |
| 2 contraction time | (-) | (-) | ↓★ | (-) |
| 3 relaxation time | (-) | (-) | (-) | (-) |
| 4 critical fusion frequency (CFF) | ↓★ | ↓★ | ↓★ | (-) |
| 5 reaction time based on the difference in the distance condition of movement (MCRT) | | | | |
| 1) reaction time (distance;40cm) | ↓★ | (-) | (-) | (-) |
| 2) response time (distance;0cm) | (-) | ↓★ | (-) | ↓★ |
| 3) movement time ( 1)-2) ) | (-) | (-) | (-) | (-) |
| 6 short-term memory | | | | |
| 1) correct response (%) | ↓★ | (-) | ↓★ | ↓★ |
| 2) revival time (sec) | (-) | (-) | (-) | (-) |
| 7 the body's center of gravity | | | | |
| 1) total shift | (-) | (-) | (-) | (-) |
| 2) lateral shift | (-) | (-) | (-) | (-) |
| 3) anteroposterior shift | (-) | (-) | (-) | (-) |
| 8 crural circumference | (-) | (-) | (-) | (-) |
| 9 tapping ability | ↓★ | (-) | (-) | ↓★ |
| 10 graping power | ↓★ | ↓★ | (-) | ↓★ |
| 11 breathing capacity | (-) | ↓★ | (-) | (-) |

NOTE) ↓★ indecates a significant lowering of the psychophysiological function in the middle- or advanced-aged group at 5% level as compared with the younger group.

## Differences in workload in relation to age

### Levels of psychophysiological functions and age

Comparing levels of psychophysiological functions between the middle- or advanced-aged groups and the younger groups shows (Table 1), there was significant lowering of six functional levels before work in the press middle- or advanced-aged group, namely, near point accommodation, CFF values, reaction time of MCRT, correct response of short-term memory, tapping ability per 20 second and grip strength and of four function levels in the welding middle- or advanced-aged group, namely, near point accommodation, contraction time, CFF values and correct response of short-term memory. On the other hand, there were no significant differences observed in relaxation time, response time and movement time of MCRT, revival time of short-term memory, the body's centre of gravity, crural circumference and lung capacity before work between the middle- or advanced-aged group and the younger group in both workshops.

*Psychophysiological function levels and working hours*

The rates of the actual activity not including the personal allowance were 83.3% in the press middle- or advanced-aged group, 76.3% in the press younger group, 87.3% in the welding middle- or advanced-aged group, and 88.5% in the welding younger group. As regards the changes in psychophysiological functions with the lapse of time, the mental reaction time of the welding middle- or advanced-aged group after work showed a significant lowering compared to that before work, but there were no lowerings of the other psychophysiological functions after work in the four groups.

As mentioned above, it was suggested that the levels of some psychophysiological functions, which are closely related to work, lowered considerably with age in both workshops, but there were no substantial differences in the workload between the middle- or advanced-aged groups and the younger groups. Therefore, the necessity to improve work to compensate for the lowering of psychophysiological functions and the possibility of making use of middle- or advanced-aged workers with long work experience were suggested.

## Conclusions

In workshops of this kind, diversified small-quantity production and "Just In Time" production are indispensable, and this investigation has shown that the following are important in the work management of this manufacturing industry:

1) Exchanging existing workers with workers who can do more than one kind of work as a result of their work experience and approaches to introducing combined work systems under which middle- or advanced-aged workers and younger workers compensate for each other.

2) Improving workers' morale in this kind of simple, repetitive work, in particular proper instructions and advice from their superiors are important.

3) Improving work to reduce heavy workload and compensating for the lowering of psychophysiological functions in order to best utilize middle- or advanced-aged workers.

## References

Ida, A. 1990, The aged and ergonomic job design, The Japanese Journal of Ergonomics, 26(1), 17-20.

Koide, I. 1990, Application of ergonomics to industrial health, The Japanese Journal of Ergonomics, 26(1), 20-24.

Nakane, J. 1986, A problem awaiting solution for the new frontier in manufacturing, Bulletin of the System Science Institute Waseda University, 17, 267-288.

Saegusa, M. 1962, Study of Employee Psychology, (Japan Management Association, Tokyo, Japan), 23-238.

Takakuwa, E. 1982, Evaluation of Fatigue and the Function of Maintaining Concentration (TAF), Hokkaido University Medical Library Series 14, Hokkaido University School of Medicine, Sapporo.

# 39. ERGONOMICS PROBLEMS AMONG KITCHEN WORKERS IN NURSERIES

Eiji Shibata[*], Yuichiro Ono[*], Jian Huang[*], Naomi Hisanaga[*], Yasuhiro Takeuchi[*], Midori Shimaoka[**] and Shuichi Hiruta[**]

[*]Department of Hygiene
Nagoya University, School of Medicine
Japan

[**]Research Center of Health, Physical Fitness and Sports
Nagoya University, Japan

### Abstract

To investigate the ergonomics problems among kitchen workers in nurseries, a questionnaire survey and health examination were conducted on kitchen workers in 20 private nurseries. In addition, the dimensions of the kitchen and installations were measured in 10 of these facilities. Repetitive use of the arms and fingers and keeping the trunk flexed for a certain amount of time appeared to impose musculoskeletal loads on the kitchen workers. Workers' views of the suitability of kitchen installations were thought to be linked to musculoskeletal symptoms. These symptoms also seemed to be affected by the number of lunches, age, and difficulty in taking a short rest.

**Keywords:** *workplace; musculoskeletal disorders; ageing; working hours; nurseries; kitchens*

## Introduction

Food preparation poses various ergonomics problems, although some regulations do exist with regard to the number of workers and the dimensions of kitchens in primary schools in Japan. However, few regulations apply to kitchens in nurseries, where the number of workers is small and workspace is less than in primary schools. Many workers complain of musculoskeletal symptoms and work stress in the workplace. Severe cases of musculoskeletal disorders have also been reported. However, the factors behind the complaints have not, so far, been sufficiently analyzed. Moreover, there is no settled opinion on the ergonomics standards for designing kitchens in nurseries. Cooking tasks in nurseries were suspected of creating specific musculoskeletal load due to repetitive movements and to the strained posture from inadequate ergonomic conditions. This study was carried out to identify the factors causing the trouble by analyzing working conditions including ergonomic factors which could cause workers to assume strained postures.

## Materials and methods

A questionnaire survey and health examination were conducted on kitchen workers in 20 private nurseries in Nagoya City. The questionnaire focused on working conditions and physical complaints. The questions on working conditions included workers' views on the height of kitchen installations, tasks which added to their load, preparation of special lunches for babies or allergic children, the total number of lunches served, the number of kitchen workers and the difficulty in taking a short rest. The questions on physical complaints covered health conditions and musculoskeletal symptoms such as stiffness and pain in the neck, shoulders, back, arms, fingers and lower extremities, cold arms and lower extremities, numbness in the arms, fingers and lower extremities. Thirty-eight workers responded to the questionnaire. They were all female, aged 36 +/-11 years, who had worked an average of 7 +/-5 years.

The 38 workers who responded to the questionnaire underwent physical examinations. Muscle tenderness in the arms, neck, shoulders, upper back and lower back was checked. In 10 of the nursery schools surveyed, we measured the dimensions of the kitchens and installations, and the luminous intensity of the workplaces.

## Results

The main tasks indicated by workers as imposing loads on them were cooking special lunches for babies or allergic children (50%), washing dishes and cutlery (42%) and cutting up food (37%) (Table 1). Among the kitchen installations, the highest cupboard shelf (66%), the rice cooking equipment (45%) and gas burners (26%) were too high, while the lowest cupboard shelf (55%) and sink bottom (47%) were too low (Table 2).

Table 1. Main tasks which workers found heavy and/or stressful (n = 38)

|  | No. | % |
|---|---|---|
| Cooking special lunches for babies or allergic children | 19 | 50 |
| Washing dishes and cutlery at sinks | 16 | 42 |
| Cutting up foods | 14 | 37 |
| lifting and carrying dish baskets | 14 | 37 |

Table 2. Workers' opinions of kitchen installation suitability (n = 38)

|  | Too High | | Suitable | | Too Low | | no response |
|---|---|---|---|---|---|---|---|
|  | No. | % | No. | % | No. | % | |
| Gas burners | 10 | 26.3 | 25 | 65.8 | 2 | 0.1 | 1 |
| Rice cooker | 17 | 44.7 | 15 | 39.5 | 5 | 13.1 | 1 |
| Cooking table | 4 | 10.5 | 29 | 76.3 | 4 | 10.5 | 1 |
| Rim of sink | 2 | 0.1 | 30 | 78.9 | 5 | 13.2 | 1 |
| Sink bottom | 3 | 0.1 | 16 | 42.1 | 18 | 47.3 | 1 |
| The highest cupboard shelf | 25 | 65.8 | 10 | 26.3 | 1 | 0.0 | 2 |
| The lowest cupboard shelf | 0 | 0 | 14 | 36.8 | 21 | 55.2 | 3 |

Table 3. Height (cm) of workers in terms of their opinions of the suitable height of kitchen installations

|  | No. | Workers who felt too low | No. | Workers who felt suitable or too high | p |
|---|---|---|---|---|---|
| Sink bottom | 18 | 157.0±5.4 | 16 | 153.0±4.0 | * |
| Lowest cupboard shelf | 21 | 155.3±5.8 | 14 | 154.0±3.3 | NS |

*: Significantly different between groups (p<0.05)
NS: not significant

|  | No. | Workers who felt too high | No. | Workers who felt suitable or too low | p |
|---|---|---|---|---|---|
| Gas burners | 10 | 152.3±4.2 | 25 | 155.7±4.7 | NS |
| Rice cooker | 17 | 153.8±3.6 | 15 | 155.8±5.8 | NS |
| Highest cupboard shelf | 25 | 154.5±5.2 | 10 | 155.4±4.5 | NS |

Table 4. Multiple regression analysis of symptoms in the low back and lower extremities

| Valuable | Parameter estimate | Standardized estimate |
|---|---|---|
| Age | 0.1317 | 0.4091* |
| Personal height | 0.0334 | 0.2754 |
| Difficulty in taking rest | 1.3997 | 0.3811§ |
| Lunches served | 0.0284 | 0.4362* |
| Lunches served for allergic children | 0.0384 | 0.1031 |
| Days of cooking special lunches for allergic children | 0.1508 | 0.0671 |
| Intercept | -12.9536 | 0 |
| Multiple correlation coefficient |  | 0.6545 (p<0.05) |

§ p<0.10, * p<0.05

The workers who replied that gas burners or rice cooking equipment were too high, complained of significantly more musculoskeletal problems than those who felt them to be suitable. The workers who found the sink bottoms too low were significantly taller than those who did not (Table 3). Multiple regression analysis suggested that symptoms in the lower back and lower extremities were attributable to the total number of lunches, age, and difficulty in taking a short rest (Table 4).

Nurseries with a large number of lunches tended to have a large kitchen area and fewer lunches per worker. Some automatic equipment such as a rotating cooking facility, a dishwasher and an elevator for carrying dishes was available (Table 5).

The ratios of the height of the main kitchen installations to 27 workers' body heights were 0.29-0.46 for sink bottoms, 0.52 for kitchen tables, 0.64 for cooking pots on gas burners, 0.33 or 0.67 for rice cooking equipment, respectively.

The width of passages in most workplaces was less than 1 m. Only three passages were equal to or more than 1 m.

Table 5. Details of 10 workplaces (A-J) where kitchen installations were measured

|  | A | B | C | D | E | F | G | H | I | J |
|---|---|---|---|---|---|---|---|---|---|---|
| No. of children | 32 | 46 | 46 | 45 | 46 | 60 | 60 | 104 | 106 | 120 |
| No. of workers | 2 | 3 | 3 | 3 | 3 | 3 | 3 | 3 | 3 | 4 |
| Children per worker | 16.0 | 15.3 | 15.3 | 15.0 | 15.3 | 20.0 | 20.0 | 34.7 | 35.3 | 30.0 |
| Area ($m^2$) | 18.5 | 19.6 | 16.3 | 13.3 | 16.4 | 27.2 | 28.0 | 27.5 | 35.0 | 36.4 |
| Area per worker | 9.2 | 6.5 | 5.4 | 4.4 | 5.5 | 9.1 | 9.3 | 9.2 | 11.7 | 9.1 |
| No. of special lunches for babies | 6 | 9 | 9 | 9 | 10 | 10 | 10 | 12 | 11 | 12 |
| Rotating caldron | - | - | - | - | - | + | + | + | + | + |
| Dishwasher | - | - | - | - | - | - | - | + | - | + |
| Carrying elevator | - | - | - | + | + | + | + | - | + | + |
| Floor Material* | S | S | W | C | S | W | S | W | C | C |
| Floor warmer | + | - | - | - | - | - | - | - | - | - |

*S: synthetic resin, W: wood, C: concrete

## Discussion

In most nurseries, though fewer lunches are served than in primary schools, special lunches for allergic children and babies are cooked besides regular lunches for non-allergic children. Cooking special lunches for allergic children by itself does not involve specific muscle work. But it can cause mental tension, because mistakes in selection of material would cause allergic reactions in children. On the other hand, cooking special lunches for babies requires the mashing and straining of materials, which helps babies digest foods.

Repetitive use of the arms and fingers and keeping the trunk flexed for certain lengths of time appeared to impose musculoskeletal loads on the kitchen workers.

The positioning of rice cooking equipment, sink bottoms, the highest and lowest shelves in cupboards were found to be unsuitable by more than 44% of the workers. Workers' views of the adequacy of the kitchen installations appeared to have some relation to their body height. Pekkarinen and Anttonen (1988) evaluated furniture and equipment in relation to individual elbow height in a survey of 11 kitchens. The worktables with cutting boards were found to be too high for 33% of the workers, and the kitchen equipment too high for 34-80% of the workers. They pointed out that the working level should be adjusted to a suitable one by using thinner cutting boards or platforms etc.

From the results of multiple regression analysis, the suitability of kitchen installations was thought to be linked to the musculoskeletal symptoms, though a specific relationship between them was not clearly seen. These symptoms also seemed to be affected by the number of lunches served, age and the difficulty of taking a short rest.

The ratios of height of sink bottom to worker height were mostly lower than the recommended height of 0.40-0.45 that was proposed in view of perceived exertion, surface EMG of the back muscles, and metabolic load following experimental results by Ichimune (1981). The ratio of kitchen tables to height was almost the same as the recommended height of 0.53 proposed by the Committee for Human Measurement (1980) of The Japan Ergonomics Research Society, though the ratios were 0.46-0.51 in some workplaces. The mean ratio of the upper end height of cooking pots with a 20 cm depth on gas burners compared to workers' body heights was 0.64. The ratio of the height of the elbow joint of body height of the mean Japanese is supposed to be 0.62. Therefore, some workers have to work with their hands elevated above elbow level, and in some cases the height of hands is shoulder height. As regards the height of the grips of the rice cooker, the mean ratio was 0.33 when the cooker was located under the table. In this case, the task of lifting the cooker, which weighs about 15 kg on to the table was observed. On the other hand, when the cooker is located on the table, the ratio was 0.67, which was higher than elbow level. The highest and lowest shelves in the cupboard were not within the standard range proposed by the Committee for Human Measurement.

Grandjean (1973) recommended the minimum room for movement in front of the sink, cooker or working surface to be 40 cm, and passage between the wall and table where the hands are free to be 60 cm. Therefore, the minimum width for passages in kitchens is considered to be 100 cm. Most of the passages in the workplaces we measured were narrower than this recommendation. The standard luminous intensity in workplaces proposed by Japanese Industrial Standards is 200-500 lx. The luminous intensity in the workplaces we measured was in the main nearer the lower limit of the standard.

Huang et al. (1988) compared two school lunch service centres, in which the number of meals prepared per worker was almost the same, but where there was a markedly different prevalence of musculoskeletal complaints. One of the centres had an automatic dishwasher and conveyor, which provided some biomechanical improvement. Some of the workplaces with a large number of children in the present study were automated and this could reduce musculoskeletal load among workers. On the other hand, the number of lunches including special ones per worker tended to be large in the workplaces with a large number of children. School lunch workers are reported to have more musculoskeletal symptoms the more lunches they prepare (Oze, 1984; Ono, 1988). These factors should be taken into account in any attempt to make ergonomic improvement in nursery kitchens.

## References

Committee for Human Measurement. 1980, The Measurement of Man, 3rd edn (Japanese Press Service), 76.

Grandjean, E. 1973, Ergonomics of the Home. (Taylor & Francis, London), 144-146.

Huang, J., Ono, Y., Shibata, E., Takeuchi, Y. and Hisanaga, N. 1988, Occupational musculoskeletal disorders in lunch centre workers. Ergonomics, 31, 65-75.

Ichimune, H. 1981, The desirable depth of the kitchen sink (Part 5). Kaseigakuzassi, 32, 628-631.

Ono, Y. 1988, Health problems among cooks providing a school lunch service. Shinrodokagakuron (Rodokeizai-sha), 593-609

Oze, Y. 1984, Studies on health hazards among cooks providing a school lunch service. Report 2. An analysis of factors associated with the development of health hazards. Japanese Journal of Industrial Health, 26, 425-437.

Pekkarinen, A. and Anttonen, H. 1988, The effect of working height on the loading of the muscular and skeletal systems in the kitchens of workplace canteens. Applied Ergonomics, 19, 306-308.

# 40. WORKLOAD OF WORKERS IN SUPERMARKETS

## Koya Kishida

*Takasaki City University of Economics*
*Japan*

**Abstract**

The workload of workers employed in a supermarket chain was investigated during the busiest weeks in December 1988. The subjects were 172 employees, 50 male and 122 female. Several surveys were conducted: time study, job structure, subjective feelings of fatigue, local fatigue symptoms, daily-life time structure. Workers in the food department reported a high number of complaints of symptoms of fatigue. The job structure of managers was clearly different from that of salespersons. Section chiefs complained most of subjective fatigue. The complaints of subjective feelings of fatigue among part-timers was less than those of managers. The off-duty time of part-timers was generally over 960 minutes, while section chiefs in the main had under 600 minutes of off-duty time. Managers continue to feel more tired than other employees. They suffered from cumulative fatigue caused by long working hours.

**Keywords:** *fatigue; posture; discomfort; physical workload; workplace; working hours; supermarkets; job analysis; home life*

## Introduction

Occupational cervicobrachial disorders (OCD) among cashiers have become an important problem in Japan since 1969. Many researchers (Nishiyama et al., 1973; Nakaseko et al., 1975; Itani, 1976; Ohara et al., 1982a; Ohara et al., 1982b), have studied the problems of cashiers. Some improvements in working conditions, for example shortened operating time, adoption of worker rotation systems and changes from mechanical cash registers with heavy keytouches to electronic ones with light keytouches, have been introduced into supermarkets. Itani (1976) pointed out that, compared with other female workers at the same supermarkets, cashiers complained more frequently of OCD and general fatigue. Thus, it seemed that improvements in their working conditions and hours would be an important factor in reducing occupational hazards.

In recent years, working conditions in supermarkets have gradually changed. A working system with two days off per week has been implemented by several large supermarket chains. After introduction of the two-days-off system, many part-timers were hired by supermarkets. Most of them were middle-aged female workers in their 40's and 50's, resulting in new problems in supermarket management.

## Method

### Test subjects and locations

The workload of workers employed in a supermarket chain was investigated during the busiest weeks in December 1988. (In Japan, the term supermarket is applied to large chain stores which include departments for clothing and household appliances in addition to food and household utensil sections.)

The subjects were 172 employees (50 male, 122 female). They worked at two stores, one with 103 employees and the other with 69. Of these, 106 were part-timers, all of them female. The duty time for full-timers was nine hours including a one-hour break. The duty hours for part-timers varied, but half worked six hours or more. The two stores normally open from 10:00 to 19:00. However, one store was open to 19:30 during December. Many full-timers were on duty from 9:00 to 20:00. Several managers were on duty from 8:00 to 21:00.

### Survey items

Several surveys were conducted. A time study was conducted to analyze work content and working postures during various kinds of work. In this survey, five workers were surveyed each day for twelve days. Thus a total of sixty workers was surveyed. A job structure survey was also conducted using the method developed by the Railway Labour Science Research Institute (Soma et al., 1965). This survey was performed as an attempt to analyze job structures through employee perceptions. The job structure survey consisted of twenty-four items categorising eight factors. Each item was graded on a five-point scale. The eight factors were: managerial abilities I & II, intellectual manipulations, motor abilities, mechanical manipulations, sociability, cooperativeness, and objective clarity.

Subjective feelings of fatigue and local fatigue symptoms were surveyed before and after work for twelve days using the method developed by the Industrial Fatigue Research Committee of the Japan Association of Industrial Health (Kogi et al., 1970; Saito et al., 1970).

Finally, a daily-life time structure survey was carried out for twelve days. Workers were asked to record the times that they departed for work and returned home each day, as well as the times that they arrived at work and left work. They were also asked to record the times at which they went to bed and got up (Kishida et al., 1985).

## Results and discussion

### Job structure

Figure 1 shows the job structure of four departments. Job structure, as perceived by the employees, showed almost the same pattern, especially in three departments; food, clothing and home products. The job structure of the management department showed a slightly different pattern from the other three departments, but was still similar. These results indicate that the employees working in the first three departments were salespersons and, thus their perceived job structures were virtually identical. Most of the workers in the management section formerly worked in one or other of the other three departments, and so their perceived job structure was similar.

Figure 2 shows the results of the job structure survey of cashiers compared with that of salespersons. The job structure of cashiers was different from that of salespersons, particularly in sociability and mechanical manipulations, because cashiers must talk to customers and use a cash register.

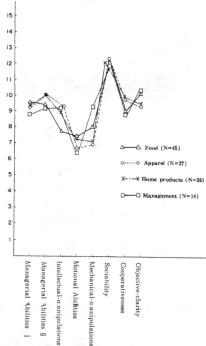

*Figure 1. Job structures by job status in all four departments*

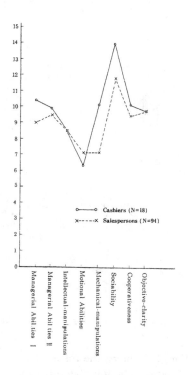

*Figure 2. Job structures of cashiers and salespersons*

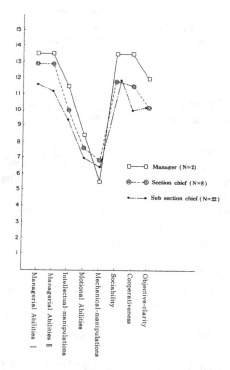

*Figure 3. Job structures by job status*

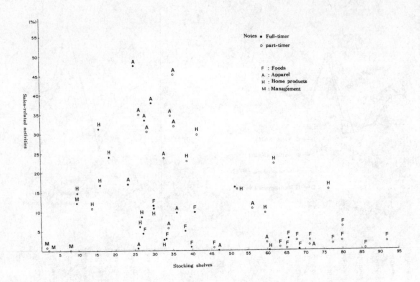

*Figure 4. Results of the time study (sales-related activities vs stocking shelves)*

Figure 3 shows the results of the job structure survey of managers compared with job status. The job structure of store managers and section chiefs showed high scores in factors relating to management, such as managerial abilities I (task management) and managerial abilities II (manpower management).

Store managers showed the highest scores in managerial abilities I, II, sociability, co-operativeness and objective-clarity, and showed the lowest scores in mechanical-manipulations. The job structure of managers was clearly different from that of salespersons.

**Time study**

Figure 4 shows the relationship between the times spent stocking shelves and those conducting sales-related activities. The results showed that salespersons in the clothing department spent most time making sales, while salespersons in the food department mainly stocked shelves.

Figure 5 shows the relationship between times spent standing in one location and those spent walking by workers in supermarkets. The results showed that salespersons in the clothing department mainly worked standing, while salespersons in the food department walked more frequently than in the other departments. This is because stockrooms in the food department were more distant.

Figure 6 shows the relationship between time spent standing in one location and time spent in uncomfortable postures such as squatting, bending, stretching etc. Salespersons in the food department frequently squatted, bent over and stretched. These last postures were used mainly when workers stocked the shelves with food.

The median for standing for all departments was 54.5% of working hours and $Q_3$ was 72.9%. Many workers in the clothing department exceeded the value of $Q_3$. This indicates that the dominant posture in the clothing department was standing in one location.

The $Q_3$ values for squatting and bending were 7.6% and 5.3%, respectively. These values were higher for salespersons in the food departments.

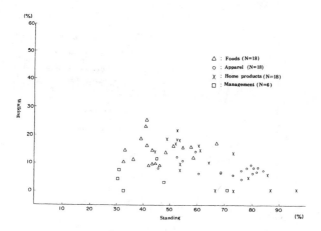

Figure 5. Relationship between standing and walking

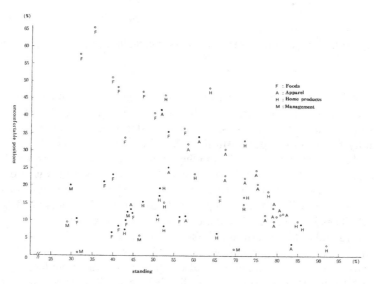

Figure 6. Relationship between standing and uncomfortable postures

## Subjective feelings of fatigue

Figure 7 shows the subjective feelings of fatigue of workers in each department (based on a student's t test).

Complaints by cashiers of drowsiness and dullness were significantly higher than in the other three departments. In addition, total complaints of subjective feelings of fatigue by cashiers was the highest. Complaints of projection of physical disintegration of salespersons in the food department were the highest. The home products department had the lowest complaints of the three groups for subjective fatigue feelings.

Table 1 shows the results of an analysis of 30 subjective fatigue symptoms for each department (based on a Chi-square test).

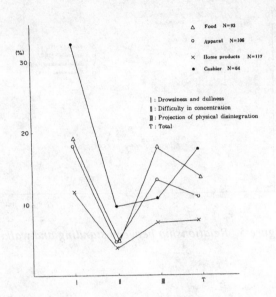

Figure 7. Subjective feelings of fatigue by department and job category (after work)

Subjective fatigue symptoms of cashiers, such as 'get tired of the whole body' 'feel the brain muddled', 'become drowsy', 'feel strained in the eyes', 'want to lie down', 'unable to have interest in things' and 'unable to straighten up in a posture', differed significantly from those reported in the other three departments. 'Feel thirsty' was complained of significantly less than in the other departments. These results showed that cashiers engaged in less body movement, but their workload was still very high.

Workers in the food department reported a high number of complaints of symptoms of fatigue, including 'legs feel heavy', 'feel stiff in the shoulders' and 'suffer low back pain'. These results confirm the results of the time study.

Figure 8 reveals that section chiefs complained most of subjective fatigue. Complaints of drowsiness among section chiefs increased from 18.9% before work to 30.2% after work. The total number of complaints of subjective feelings of fatigue among section chiefs increased from 11.8% before work to 19.2% after work. The complaints of subjective feelings of fatigue among part-timers was less than that of managers. These results show that the higher an employee's job status, the greater is the workload.

Table 2 shows the results of an analysis of the 30 subjective fatigue symptoms by job status (based on a Chi-square test).

Subjective fatigue symptoms of section chiefs, such as 'feel heavy in the head' 'get tired of the whole body', 'feel the brain muddled', 'become drowsy', 'feel difficulty in thinking', 'unable to have interest in things' and 'no energy', differed significantly from those of part-timers.

The characteristic complaint of section chiefs was muddled thinking. This complaint was voiced by 36.4% of section chiefs after work. This value was the highest complaint in all job levels and departments.

Local fatigue symptoms showed that low back pain (46%) and shoulder stiffness (26%) were also common complaints by section chiefs.

Table 1. Percentage of workers complaining of the thirty subjective fatigue symptoms by department and job category (after work)

|  | Food | Home products | Apparel | Cashier |
|---|---|---|---|---|
| Feel heavy in the head | 10.8 | 15.1 | 6.7 | 9.4 |
| Get tired of the whole body | 11.8 | 15.1 | 8.1 | 43.8 |
| Legs feel heavy | 47.3 | 34.9 | 30.2 | 40.6 |
| Give a yawn | 10.8 | 17.0 | 6.7 | 14.1 |
| Feel the brain muddled | 12.9 | 11.3 | 10.7 | 29.7 |
| Become drowsy | 26.9 | 18.9 | 8.1 | 57.8 |
| Feel strained in the eyes | 44.1 | 52.8 | 33.6 | 67.2 |
| Become rigid or clumsy in motion | 10.8 | 2.8 | 5.4 | 12.5 |
| Feel unsteady in standing | 6.5 | 1.9 | 3.4 | 7.8 |
| Want to lie down | 9.7 | 12.3 | 6.7 | 42.2 |
| I : Drowsiness and dullness | 19.2 | 18.2 | 12.0 | 32.5 |
| Feel difficult in thinking | 5.4 | 3.8 | 3.4 | 7.8 |
| Become weary of talking | 3.2 | 5.7 | 4.7 | 4.7 |
| Become irritable | 7.5 | 3.8 | 2.0 | 12.5 |
| Unable to concentrate attention | 3.2 | 0.9 | 3.4 | 7.8 |
| Unable to have interest in things | 2.2 | 4.7 | 2.7 | 17.2 |
| Become apt to forget things | 5.4 | 6.6 | 8.1 | 1.6 |
| Apt to make mistakes | 3.2 | 1.9 | 2.7 | 4.7 |
| Feel uneasy about things | 8.6 | 5.7 | 3.4 | 4.7 |
| Unable to straighten up in a posture | 0.0 | 1.9 | 2.0 | 15.6 |
| No energy | 4.3 | 8.5 | 5.4 | 18.8 |
| II : Difficulty in concentration | 4.3 | 4.3 | 3.8 | 9.5 |
| Have a headache | 8.6 | 13.2 | 4.7 | 6.3 |
| Feel stiff in the shoulders | 64.5 | 35.8 | 23.5 | 37.5 |
| Suffer low back pain | 46.2 | 22.6 | 17.5 | 25.0 |
| Feel oppressed in breathing | 2.2 | 4.7 | 1.3 | 3.1 |
| Feel thirsty | 35.4 | 34.9 | 20.1 | 20.3 |
| Have a husky voice | 12.9 | 17.0 | 4.7 | 9.4 |
| Have a dizziness | 2.2 | 1.9 | 1.3 | 4.7 |
| Have a spasm on the eyelids | 3.2 | 0.9 | 0.0 | 0.0 |
| Have a tremor in the limbs | 3.2 | 0.0 | 1.3 | 0.0 |
| Feel unwell | 0.0 | 0.9 | 0.7 | 1.6 |
| III : Projection of physical disintegration | 17.8 | 13.2 | 7.5 | 10.8 |
| T : Total | 13.8 | 11.9 | 7.7 | 17.6 |
| N | 93 | 106 | 149 | 64 |

Among part-timers, those working for six hours or more had more complaints of subjective fatigue feelings than had part-timers working for less than six hours. This result shows that subjective fatigue feelings are based on working hours (see Figure 9).

In the food department, fifty percent or more of part-timers working for six hours or more complained of subjective fatigue symptoms such as 'legs feel heavy', 'feel strained in the eyes', 'feel stiff in the shoulders' and 'suffer low back pain'. These results suggest that part-timers working six hours or more became more fatigued than part-timers working less than six hours and that the difference results from the difference in length of working hours.

Table 2. Percentage of workers complaining of the thirty subjective fatigue symptoms by job status (after work)

| | Section chief | Sub section chief | Full-timer | Part-timer (more than six hours) | Part-timer (less than six hours) |
|---|---|---|---|---|---|
| Feel heavy in the head | 34.1 | 14.3 | 8.3 | 13.2 | 8.8 |
| Get tired of the whole body | 31.8 | 28.6 | 20.8 | 12.4 | 4.1 |
| Legs feel heavy | 52.3 | 60.3 | 47.9 | 39.5 | 27.0 |
| Give a yawn | 11.4 | 12.7 | 18.8 | 9.3 | 8.8 |
| Feel the brain muddled | 36.4 | 17.5 | 18.8 | 14.0 | 5.4 |
| Become drowsy | 38.6 | 22.2 | 20.8 | 20.2 | 9.5 |
| Feel strained in the eyes | 56.8 | 47.6 | 25.0 | 43.4 | 48.0 |
| Become rigid or clumsy in motion | 6.8 | 12.7 | 8.3 | 7.0 | 3.4 |
| Feel unsteady in standing | 11.4 | 15.9 | 4.2 | 4.7 | 2.7 |
| Want to lie down | 15.9 | 22.2 | 14.6 | 10.9 | 2.7 |
| I : Drowsiness and dullness | 30.2 | 25.4 | 18.8 | 17.4 | 12.0 |
| Feel difficult in thinking | 29.5 | 14.3 | 10.4 | 2.3 | 2.0 |
| Become weary of talking | 4.5 | 12.7 | 8.3 | 7.8 | 1.4 |
| Become irritable | 9.0 | 12.7 | 4.2 | 5.4 | 2.0 |
| Unable to concentrate attention | 11.4 | 9.5 | 2.1 | 5.4 | 0.0 |
| Unable to have interest in things | 18.2 | 15.9 | 6.3 | 5.4 | 0.0 |
| Become apt to forget things | 9.0 | 6.3 | 4.2 | 6.2 | 9.5 |
| Apt to make mistakes | 4.5 | 6.3 | 2.1 | 2.3 | 2.7 |
| Feel uneasy about things | 18.2 | 7.9 | 6.3 | 7.8 | 2.0 |
| Unable to straighten up in a posture | 4.5 | 9.5 | 4.2 | 2.3 | 0.0 |
| No energy | 22.7 | 22.2 | 8.3 | 7.8 | 4.7 |
| II : Difficulty in concentration | 13.2 | 11.7 | 5.6 | 5.8 | 2.43 |
| Have a headache | 4.5 | 9.5 | 2.1 | 10.1 | 7.4 |
| Feel stiff in the shoulders | 43.2 | 25.4 | 27.1 | 51.9 | 31.1 |
| Suffer low back pain | 29.5 | 30.2 | 29.2 | 33.3 | 17.6 |
| Feel oppressed in breathing | 2.3 | 3.2 | 0.0 | 3.1 | 3.4 |
| Feel thirsty | 38.6 | 36.5 | 25.0 | 32.6 | 26.4 |
| Have a husky voice | 18.2 | 12.7 | 12.5 | 14.7 | 6.8 |
| Have a dizziness | 0.0 | 3.2 | 0.0 | 2.3 | 2.0 |
| Have a spasm on the eyelids | 4.5 | 0.0 | 0.0 | 3.1 | 0.0 |
| Have a tremor in the limbs | 2.3 | 0.0 | 0.0 | 2.3 | 0.0 |
| Feel unwell | 6.8 | 1.6 | 0.0 | 1.6 | 0.0 |
| III : Projection of physical disintegration | 15.0 | 12.2 | 9.6 | 15.5 | 9.5 |
| T : Total | 19.2 | 16.5 | 11.3 | 12.9 | 8.0 |
| N | 44 | 63 | 48 | 129 | 148 |

## Daily-life time structure

The results of the daily-life time structure survey showed that the off-duty time for a section chief was less than 11 hours.

For section chiefs, off-duty times of less than eleven hours appeared frequently. Seven of twenty eight cases had off-duty times of less than eleven hours. In these cases, it was difficult for section chiefs to obtain 6 hours of sleep. In fact, 30% of section chiefs slept for less than 6 hours.

Figure 10 shows that the off-duty time of part-timers was generally over 960 minutes, while section chiefs in the main had under 600 minutes of off-duty time.

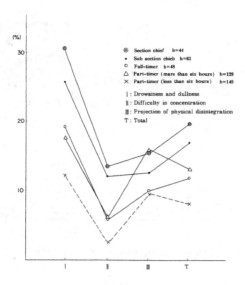

Figure 8. *Subjective feelings of fatigue by job status (after work)*

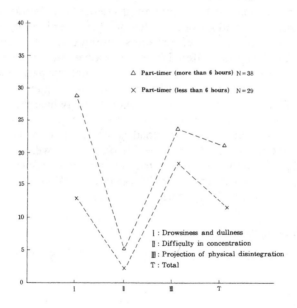

Figure 9. *Subjective feelings of fatigue of part-timers in the food department (after work)*

## Conclusion

Analyzing the work in the food department, most of the work involved stocking shelves. In the food department, the work of stocking shelves was the main task for part-timers. Part-timers were frequently in the squatting and bent over postures, and consequently, they frequently complained of stiff shoulders and low back pain. This

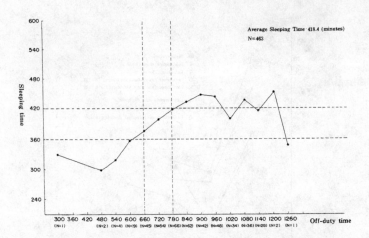

*Figure 10. The relationship between off-duty time and sleeping time*

indicates that improvements in working conditions in the food department are needed. The main improvement could be to reduce the frequency of uncomfortable postures by using adjustable carts and portable stepladders for stocking shelves.

These results suggest that, despite improvements in working conditions, the workload of cashiers is still greater than that for any other job category except section chiefs. However, on the whole, managers continue to feel more tired than other employees. Moreover, they suffered from cumulative fatigue caused by long working hours. In addition, I believe that increased reliance on part-timers places greater responsibility on managers and contributes to overall fatigue.

The workload of section chiefs was the highest. To reduce the workload of section chiefs, more effective use of part-timers is needed. The working hours for part-timers are comparatively short. Their workload is also lighter. Companies should train part-timers in the management of stores. If part-timers were promoted to assistant section chiefs, they would be able to take over some of those responsibilities. At present, section chiefs worked primarily as general workers, not as managers. A new management system should be created for the managers to reduce their workload.

## Acknowledgements

This reseach was supported by the Association of Employment Development for Senior Citizens.

## References

Itani, T., 1976, Health care for cash register operation in supermarkets. Journal of Science of Labour, 52, 585-597.

Kishida, K., Saito, M., and Hasegawa, T., 1985, Workload of managers and structure of their daily lives in two local factories. In: Toward the Factory of the Future (edited by H.-J. Bullinger and H.J. Warnecke) (Berlin: Springer-Verlag), pp.888-892.

Kogi, K., Saito, Y., and Mitsuhashi, T., 1970, Validity of three components of subjective fatigue feelings. Journal of Science of Labour, 46, 251-270.

Nakaseko, M., Nishiyama, K., and Hosokawa, M., 1975, Problems of reducing workloads in cash register operation. 1. Comparisons of workloads in usual cash register and electronic cash register operations. Japanese Journal of Industrial Health, 17, 168-169.

Nishiyama, K., Nakaseko, M., and Hosokawa, M., 1973, Cash register operators' work and its hygienic problems in supermarkets. Japanese Journal of Industrial Health, 15, 229-243.

Ohara, H., Mimura, K., Oze, Y., Itani, T., Ohta, T., and Aoyama, H., 1982a, Studies on cervicobrachial disorders among cash register operators. Part 1. Changes in characteristics of patients after improvement of working conditions. Japanese Journal of Industrial Health, 24, 55-64.

Ohara, H., Mimura, K., Oze, Y., Itani, T., Ohta, T., and Aoyama, H., 1982b, Studies on cervicobrachial disorders among cash register operators. Part 2. A review on clinical findings and working conditions of patients. Japanese Journal of Industrial Health, 24, 65-74.

Saito, Y., Kogi, K., and Kashiwagi, S., 1970, Factors underlying subjective feelings of fatigue. Journal of Science of Labour, 46, 205-224.

Soma, M., Watanabe, H., Kashiwagi, S., Nagata, Y., Tanaka, T., Teranodan, K., Osuga, T., and Kosugo, R., 1965, A factor analytic study of job structure. Bulletin of the Railway Labour Science Research Institute, 18, 121-130.

## 41. A COMPARISON BETWEEN NEW AND CURRENT METHODS OF DEBRANCHING TREES

T. Klen[*], E. Ahonen[*], Unto Kononen[*], Juhani Piirainen[*] and Juha Venalainen[**]

[*]Kuopio Regional Institute of Occupational Health
Section for Agriculture and Forestry, Finland

[**]Kuopio Research Institute of Exercise Medicine
Finland

### Abstract

A new method of cutting off branches was introduced into lumbering to raise productivity. So-called clean cutting has been replaced by a rougher sweep cutting method. In clean cutting, each bigger branch is removed separately, whereas in sweep cutting the chainsaw does long sweeping movements and cuts several branches with one movement. Loggers have found the sweeping method strenuous and they have not entirely accepted it in practice. The world champion chainsaw operator, Hannu Kilkki, has developed yet another cutting method which is a combination of the 6-phase clean cutting and the sweeping method. The main principle of this new method is to support the saw as much as possible against the trunk of the tree and leave out the long sweeping movements or apply them only to thin branches.

The aim of the study was to compare the method developed by Kilkki with the former sweeping method as regards the following factors: 1) work postures (measured with the OWAS method), 2) accident risks, 3) pulse rate, 4) oxygen consumption, 5) EMG, 6) productivity, 7) workers' subjective estimate of the debranching method.

The comparison was made using Norway spruce trees of the same age. The five subjects debranched trees alternately with both methods for 20-25 minutes at a time, after which they had a rest period. The rest period was sufficiently long to allow recovery before the next test period. Each subject had three cutting bouts with both methods. The subjects had been instructed in the new method and they had practised it for four months before the test. During this practice period, they also used the conventional method to maintain their skill in that method too. By the time of the test, they were familiar with both methods. Results indicated that, when working with the new method, the heart rate, oxygen consumption and EMG values (measured from the shoulder and back) of the loggers were slightly lower compared with the method used conventionally. Productivity was also a little lower (2%). The new method proved somewhat more advantageous as regards work safety, ergonomics aspects and work physiology than the current sweeping method. The new method is, therefore, recommended in spite of the slight loss in productivity.

**Keywords:** *physical workload; posture; risk; heart rate; oxygen consumption; electromyography; productivity; forestry*

## Introduction

Depending on the logging technique, the debranching of trees takes from 50 to 70% of the total effective working time of the chain saw operator. The productivity of the debranching thus has a decisive effect on that of the whole logging operation. Previously, the logger cut off each branch separately. This kind of work is obviously very slow from the point of view of work output. This so-called clean debranching (no stubs were allowed) was, therefore, replaced by a rougher sweep debranching method. In sweep cutting, the chain saw does long sweeping movements and cuts several branches with one movement. In this method the chain saw is mainly supported by the hands only. Loggers have found the sweeping method strenuous and that is why they have not entirely accepted it in practice. It takes from 3 to 5 sweeps to remove the branches from all sides of the trunk. When cutting the lowest branches the chain saw is very often invisible. This means a higher risk of kick back by the saw.

In order to combine the advantages of both described methods, the world champion chain saw operator, Hannu Kilkki, has developed a new method of cutting branches. The main principle of this new method is to support the saw as much as possible against the trunk of the tree and leave out the long sweeping movements or apply them only to thin branches. The logger moves only forwards and the lowest branches are cut after turning round the trunk. In this phase, when returning towards the stump, in addition to cutting off the lowest branches, the trunk will also be cross-cut into shorter logs. The large-sized trees are cross-cut into saw-logs when moving from stump to top. By using the intermediate forms of the working movements of the clean cutting and sweeping methods, the logger tries to cut several branches simultaneously, and at the same time he supports the saw against the trunk.

The aim of this study was to compare Kilkki's method with the sweeping method in regard to the following factors:
- work postures
- accident risk
- pulse rate
- oxygen consumption and ventilation
- productivity.

## Material and methods

### Subjects

The subjects were five healthy and experienced loggers. Their ages ranged between 34 and 49 years with a mean of 43 years. Before the field experiments the loggers underwent a medical examination. They also performed a series of tests in the laboratory which included the measurement of maximal oxygen uptake ($VO_2$max), ventilation, and pulse rate on a treadmill. According to this test the physical condition of the subjects was sufficient for this kind of work. The $VO_2$max varied from 3.54 to 4.68 l/min and from 41.7 to 61.5 ml/kg, respectively.

### The design and course of the study

The comparison measurements were made using Norway spruce trees of the same age. The five subjects debranced trees alternately with both methods for 20-25 minutes at a time, after which they had a rest period. The rest period was sufficiently long to allow recovery before the next test period. Each subject had three cutting bouts with both methods. The average length of the trees was about 25 meters. The subjects had been instructed in the new method and they had practised it for four months before the test. So, the working movements and the whole working chain and procedure, so-called 'kinetic melody' of work had become very familiar and automatic, releasing the higher cortical functions for other purposes. The subjects worked at their normal work pace using their own chain saws with new chains. The work clothes and personal protective equipment were specially designed for the loggers. The length of the guide bar was 13 inches. During the practice period of the new method the subjects also used the conventional method to maintain their skill in that method too. By the time of the test they were familiar with both methods.

**Methods of measurements**

The oxygen consumption and ventilation of the subjects were measured with a portable Morgan Oxylog device. The data were recorded on tape (Uher CR 210-stereo taperecorder). The weight of these two devices was 3 kg together. The data were later printed using a Bruel & Kjaer 2309 plotter.

Pulse rate was measured with a portable Marquette 8500 Holter taperecorder. EMG was measured with the EMG analyzer, ME 3000 microcomputer (size 16 x 7.5 x 2.5 cm) which was fastened to the back of the test person. This device works as an independent data collecting and recording unit and performs electromyography simultaneously in two muscles. This device was a prototype during the experiments. EMG was measured in the upper trapezius and erector spinae muscles.

Work postures were observed every ten seconds according to the OWAS method which is a classification system (Salonen & Heinsalmi, 1979). The method also gives a rough estimate of the injury risk, inconvenience of some posture combinations and information on how soon work postures should be improved. The observations were recorded in the memory of a Micronic data collection device, from where they were transferred to a Wang computer. Because all the measurements could not to be carried out during every measurement period, the summary of the measurements taken is shown below:

| I debranching | II debranching | III debranching |
|---|---|---|
| EKG | EKG | EKG |
| EMG | Oxygen consumption | EMG |
| OWAS | and ventilation | OWAS and videotaping |

The accident risk of working movements was observed visually without any systematic method. The number and size of branches were estimated by taking a sample of 10 trees debranched by each method. The size was measured as the diameter of the trunk. The duration of working periods was measured as minutes and the output as the number and size of branches removed and also as cubic meters of debranched trees. The volume was calculated with the aid of tree length and diameter at breast height (d 1.3 m). Productivity was calculated both as the sum of the area of removed branches per minute and as the volume of debranched trees per minute.

*Results*

**Productivity**

The productivity with the new method was about 3% lower than with the current sweeping method during this experiment, estimated on the basis of both branches removed and volumes of the trees. The productivity varied from 127 to 160 litres per minute, estimated according to the trunk volume.

**Pulse rate, oxygen consumption and ventilation**

The pulse rate was on average about 3% lower when the new method was used compared with the current method (Table 1). In this experiment, the pulse rate was 121 beats per minute with the new method and 124 with the current method.

*Table 1. Pulse rate (beats per min) during current and new work methods*

| SUBJECT | Measurement | | | | | | | |
|---|---|---|---|---|---|---|---|---|
| | I | | II | | III | | I-III | |
| | $\bar{x}$ | s | $\bar{x}$ | s | $\bar{x}$ | s | $\bar{x}$ | % |
| I | | | | | | | | |
| current | 84 | 6 | 140 | 5 | 134 | 6 | 119 | |
| new | 74 | 10 | 130 | 5 | 127 | 7 | 110 | -7.6 |
| II | | | | | | | | |
| current | 105 | 20 | 102 | 23 | 107 | 14 | 105 | |
| new | 106 | 18 | 97 | 17 | 107 | 11 | 103 | -1.9 |
| III | | | | | | | | |
| current | 126 | 23 | 151 | 8 | 142 | 3 | 140 | |
| new | 131 | 15 | 151 | 4 | 145 | 4 | 142 | +1.4 |
| IV | | | | | | | | |
| current | 112 | 3 | 112 | 3 | 121 | 3 | 115 | |
| new | 108 | 3 | 107 | 4 | 116 | 4 | 110 | -4.3 |
| V | | | | | | | | |
| current | 134 | 14 | 142 | 11 | 141 | 13 | 139 | |
| new | 135 | 14 | 145 | 13 | 135 | 12 | 138 | -0.7 |
| Change | | | | | | | | -2.6% |

The average decrease in oxygen consumption with the new method was 6% compared to the current one (Table 2). The ventilation decreased more than the oxygen consumption, i.e. 8% on average.

The relative load according to pulse rate was calculated by taking the pulse rate at rest into account, but as regards oxygen consumption, this correction was not made. The relative physical load on the basis of oxygen consumption rose to 52% with the new method and to 56% with the current method of $VO_2$max. According to the pulse rate, the respective values were 50 and 52% (Table 3).

**Electromyography (EMG)**

The electric activity (in micro volts, uV) in the back (erector spinae) was 7% lower, and in the shoulder (trapezius) 5% lower with the new method than with the current one (Table 4)

**Accident risk and work postures**

Based merely on visual observation by the researchers, the accident risk, especially that of kick-back of the saw, seems to be less for the new method. As for working postures, the difference between the methods can be seen in the postures of the back.

In the new method, awkward 'bent-twisted' postures were considerably fewer, but mere bent postures clearly more numerous than in the current method. As for the upper and lower limbs, the differences were of no importance.

Table 2.  Oxygen consumption (l/min) with the current and new work methods

| SUBJECT | $\bar{x}$ | % | s |
|---|---|---|---|
| I | | | |
| current | 1.64 | | 0.10 |
| new | 1.56 | -4.9 | 0.13 |
| II | | | |
| current | 2.14 | | 0.10 |
| new | 1.95 | -8.9 | 0.11 |
| III | | | |
| current | 2.33 | | 0.07 |
| new | 2.15 | -7.7 | 0.24 |
| IV | | | |
| current | 2.11 | | 0.17 |
| new | 1.93 | -8.5 | 0.20 |
| V | | | |
| current | 2.85 | | 0.12 |
| new | 2.83 | -0.7 | 0.19 |
| Average | | -6.1% | |

Table 3.  Relative load, %max with the current and new work methods

| Subject | | Oxygen consumption | Pulse rate |
|---|---|---|---|
| I | current | 42 % | 43 % |
| | new | 40 % | 37 % |
| II | current | 52 % | 43 % |
| | new | 47 % | 42 % |
| III | current | 64 % | 66 % |
| | new | 59 % | 67 % |
| IV | current | 60 % | 40 % |
| | new | 55 % | 35 % |
| V | current | 61 % | 68 % |
| | new | 60 % | 68 % |

Table 4. EMG (uV) with the current and new methods

| SUBJECT | Shoulder | | Back | |
|---|---|---|---|---|
| | uV | % | uV | % |
| I | | | | |
| current | 29.5 | | 57.5 | |
| new | 24 | -18.6 | 54.5 | -5.2 |
| II | | | | |
| current | 58 | | 66.5 | |
| new | 48.5 | -16.4 | 57.5 | -13.6 |
| III | | | | |
| current | 56.5 | | 86.5 | |
| new | 68 | +20.4 | 84 | -2.9 |
| Average | | -4.9% | | -7.2% |

## Discussion

Because of the small number of subjects, the measurement levels, especially those of cardiorespiratory strain, muscle tension and productivity, cannot be generalized. Some rough conclusions can nevertheless be drawn concerning the differences between the two methods. The results indicate that when working with the new method the heart rate, oxygen consumption and EMG values measured from the shoulder and back seem to be slightly lower compared with the method used conventionally. Also productivity was somewhat lower. On the other hand, an exact determination of productivity would necessitate estimation of output over a long period of normal work.

In spite of the limited possibility of generalizing the results of this study, the physical load was now roughly on the same level as in many earlier studies. For example Kukkonen-Harjula and Rauramaa (1984) obtained a pulse rate level of 125 beats per minute in debranching and cross-cutting, which is very near the corresponding results of this study. Also the relative physical load according to oxygen consumption was now near the result of Kukkonen-Harjula and Rauramaa, although it was a bit higher in our study. The relative load now rose over to 50% of $VO_2max$, thus exceeding the overloading limit often applied in forest work.

The interpretation of working postures was problematic. The bent postures were more numerous in the new method but the most awkward 'bent-twisted' postures were more numerous in the current method. Such postures need to be corrected immediately. These 'bent-twisted' postures in the current method are due to the long sweeping movements, and occur particularly when the logger is sweeping the lower side of the trunk.

As for accidents, from visual observations the new method seems to be less risky. This is because in the current method the saw chain and the sawing point are often not seen by the logger. If the top of the saw chain hits a branch when sweeping the lower side of the tree, it may sometimes cause a kick-back. What makes the situation even worse is that very often the working movement and kick-back are directed in the same direction. In such a case the kick-back reverts to the back of the leg where there is normally no safety padding. In addition, working movements are ballistic, which means that the reaction time of the human being is too slow to give the muscles the order to stop a movement which has already begun. Also the working movements of the current sweeping method are rather violent.

In conclusion, the new method proved slightly more advantageous than the current method as regards the variables measured in this study, but the deviations were small. Logging work with a chain saw is, however, so physically strenuous that all improvements, even small, must be taken into account. The new method is therefore recommended in spite of the slight loss in productivity. Many loggers have already adopted the new method.

## References

Kukkonen-Harjula, K. and Rauramaa, R. Oxygen consumption of lumberjacks in work with a power saw. Ergonomics 27 (1984): 1, 59-65.

Salonen, A. and Heinsalmi, P. OWAS-tyoasentojen havainnointijarjestelma. Sarja B No.50. Suomen itsenaisyyden juhlavuoden 1967 rahasto, Helsinki 1979, 150 s.

# Part VII
Environmental Ergonomics

# 42. AN ERGONOMICS EVALUATION OF CLEANROOM WORK

### Yeong-Guk Kwon

*Department of Industrial Engineering, Kwandong University
Kangreung, Kangwondo, (210-701), Korea*

**Abstract**

Cleanrooms are widely used in high technology industries. Currently within the microelectronics industry, there is an explosive growth in the number of cleanrooms. Therefore, special consideration of cleanroom workers is needed to understand the work induced stresses from contamination avoidance, clothing requirements, and confinement (Czaja, 1983). This paper evaluates a cleanroom environment from an ergonomics viewpoint. It is mainly concerned with the productivity of the cleanroom workers and the enhancement of the working environment. The first study (Part 1) is a survey of a cleanroom environment. The second study (Part 2) is an experiment designed to focus on the garment-related problems found in the first study using the environmental chamber.

**Keywords:** *environment; clothing; posture; discomfort; cleanrooms; workplace*

## On-site survey of cleanroom workers

### Design and methods

A questionnaire for evaluating significant ergonomics factors in the cleanroom environment was developed by the investigators after an on-site observation of a cleanroom facility and interviews with cleanroom workers and coordinators. Nine questions which addressed important ergonomics considerations in a cleanroom environment were developed. The nine questions dealt with the following topics:

1. Seating
2. Visual Fatigue
3. Illumination
4. Garment Heat Retention
5. Garment Vision or Breathing Interference
6. Garment Design
7. Work Place Layout
8. Physical Workload
9. Mental Workload.

A seven-point scale was used to evaluate the ergonomic design aspects related to each question in the survey: 1=Very Poor, 2=Quite Poor, 3=Poor, 4=OK, 5=Good, 6=Quite Good, and 7=Very Good. Comments were also encouraged.

Five factors were considered when analyzing responses to each question:
1. Work Area in Cleanroom
2. Sex of Respondent
3. Years of Cleanroom Experience
4. Weight of Respondent
5. Height of Respondent.

## Analysis

Statistical analyses were conducted to determine if work area, years of cleanroom experience, sex, weight, or height effects were statistically significant. Dependent variables were responses to the nine questions dealing with ergonomics considerations of the cleanroom environment.

Four work areas were defined for this study. The diffusion and wet process areas were classified as the HUMID area. The PHOTO area combined all photo-related areas. The PROBE area combined all probe-related areas. The ETC included all areas other than the HUMID, PHOTO, and PROBE areas.

If years of experience in the cleanroom were less than one, it was classified as LT1. If years of experience were between one and five, it was classified as G1L5.

Weight was classified to make approximately equal numbers in each category for the survey data: Subject weight was classified as either light, medium or heavy. Height was similarly classified as short, medium or tall.

## Results

The mean and standard deviation for responses to each of the nine questions are shown in Table 1 and Figure 1. Seating (Q1), garment heat retention (Q4), garment vision or breathing interference (Q5), and garment design (Q6) appear to be the greatest obstacles to improved productivity in the cleanroom.

The sex, area, and experience effects for each of the nine questions are also shown in Table 1 and Figure 1. The design of the cleanroom appears to favour male workers, although both males and females work in the cleanroom, and the percentage of females is often greater than that of males. Female workers in this survey reported more discomfort in seating (Q1), garment vision or breathing interference (Q5), and physical workload (Q8) than males. Seating was not well provided for or was unavailable for most workers. Even when it was provided, it was not designed for female workers, and chairs were generally not adjustable.

The work area effect was not significant except for the seating problem (Q1) in the PHOTO area. When examining the interactions between area and sex of the respondent, females had more difficulty than males with the physical workload (Q8) in the PHOTO and PROBE areas compared to the HUMID and ETC areas. The garment heat retention problem (Q4) appeared to be most significant in the HUMID area.

The category of years of cleanroom experience was not shown to be statistically significant in most cases. Seating (Q1) and garment vision or breathing interference (Q5) problems increased when subjects had more experience. This may indicate that the more experienced workers were more aware of those problems than were their less experienced counterparts. However, years of experience was not statistically significant.

The height and weight of workers were not found to be significant in the analysis of the survey questionnaire.

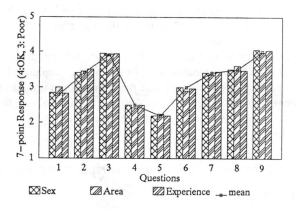

Figure 1. Graphical summary of the cleanroom survey

Table 1. Summary of responses to the cleanroom survey

|  | Survey Questions | | | | | | | | | |
|---|---|---|---|---|---|---|---|---|---|---|
|  | 1 | 2 | 3 | 4 | 5 | 6 | 7 | 8 | 9 | N |
| **Overall** | | | | | | | | | | |
| Mean | 2.8 | 3.4 | 3.9 | 2.5 | 2.2 | 3.0 | 3.4 | 3.5 | 4.0 | 54 |
| Std Dev | 1.3 | 1.1 | 1.1 | 1.3 | 1.1 | 1.4 | 1.3 | 1.3 | 1.2 | 54 |
| **Sex** | | | | | | | | | | |
| mean | 2.9 | 3.4 | 4.0 | 2.5 | 2.2 | 3.0 | 3.4 | 3.8 | 4.1 | |
| Male | 3.2 | 3.6 | 4.0 | 2.6 | 2.6 | 3.1 | 3.6 | 3.8 | 4.1 | 25 |
| Female | 2.5 | 3.2 | 3.9 | 2.4 | 1.8 | 2.9 | 3.2 | 3.2 | 4.0 | 29 |
| **Area** | | | | | | | | | | |
| mean | 3.0 | 3.5 | 3.9 | 2.4 | 2.3 | 2.9 | 3.4 | 3.6 | 4.0 | |
| ETC | 3.2 | 3.4 | 3.8 | 2.5 | 2.5 | 2.8 | 3.7 | 4.1 | 3.5 | 10 |
| PHOTO | 2.0 | 3.5 | 4.0 | 2.6 | 2.0 | 3.2 | 3.5 | 3.1 | 4.3 | 22 |
| PROBE | 3.5 | 3.2 | 3.7 | 2.7 | 2.2 | 2.9 | 3.4 | 3.4 | 3.8 | 13 |
| HUMID | 3.3 | 3.7 | 4.2 | 1.9 | 2.3 | 2.7 | 2.8 | 3.8 | 4.3 | 9 |
| **Experience** | | | | | | | | | | |
| mean | 2.8 | 3.5 | 3.9 | 2.5 | 2.2 | 3.0 | 3.4 | 3.5 | 4.0 | |
| LT1 | 3.1 | 3.9 | 4.2 | 2.6 | 2.4 | 3.4 | 3.9 | 3.8 | 4.3 | 14 |
| G1L5 | 2.8 | 3.3 | 3.6 | 2.2 | 2.2 | 2.5 | 3.1 | 2.9 | 3.8 | 16 |
| GT5 | 2.6 | 3.3 | 4.0 | 2.7 | 2.0 | 3.0 | 3.3 | 3.7 | 4.0 | 24 |

## Conclusions

Responses to the questionnaire reported by 54 currently employed and trained cleanroom workers provided some insight into the ergonomic design characteristics of their work environment.

Female cleanroom workers consistently reported more ergonomic design difficulties than their male co-workers. Apparently, cleanroom design criteria are more appropriate to the needs and anthropometric characteristics of the male than they are to the female cleanroom worker.

Ergonomic design features of the cleanroom were rated poorly with regard to the garment and its heat retention, interference with vision and breathing, and general design features.

Ergonomics considerations of the workplace, especially proper design of seating, have not had an impact in the cleanroom. Much of the emphasis in the cleanroom appears to be aimed at process control and the human operator has been somewhat neglected. However, as the process control problems are solved, the impact of the human operator and his or her ability to function efficiently in the cleanroom will become more critical.

## A *laboratory experiment to explore the restriction of cleanroom garments*

### Design and methods

After completion of a survey to determine potential ergonomics problems in the cleanroom, dominant problems appeared to be garment heat retention, garment vision or breathing interference, and garment design (Ramsey et al., 1988). An experiment was designed to focus on these problems and to use the Texas Tech environmental chamber to simulate a cleanroom environment. The temperature of the environmental chamber was set at $72^{\circ}F$ with relative humidity maintained at 60%.

Several different sizes of cleanroom garments were borrowed from a microelectronics manufacturing firm. In order to familiarize the subjects with the experimental tasks and procedures, they first performed the simulated tasks in their street clothing. Three sizes of garments were assigned to each subject; under, over, and exact size. The sizes were based on each subject's height and weight which were measured before the experiment. The garment sequence and task sequence were randomly selected for each subject. Subjects did not know the garment sequence or the task sequence. On the fifth trial, a randomly repeated size which was not the same as that of the fourth trial was used.

A series of three tasks required the subjects to use a wide range of body movements which are typically encountered in a cleanroom. These tasks were:

*1. Seated task*

The subject played a blackjack computer game for 2 minutes. The task simulated seated work such as writing or data entry and was expected to reveal garment restriction associated with seated tasks.

*2. Arm (upper activity) task*

The subject placed a cassette box on a shelf located at the subject's shoulder height. The subject was instructed to push the box two feet along the shelf with his or her right hand and then push it back to its original position with the left hand. The subject then removed the cassette box from the shelf using both hands and lowered it to the table. The subject was then instructed to open the cassette box and run over five cassettes, close the box and place it back on the shelf using both hands. This series of activities was repeated five times. The task simulated reaching and twisting tasks such as the loading and unloading of wafer trays.

*3. Whole body task*

The subject took the cassette box from the previous workstation, walked approximately ten feet to another workstation where he or she bent over a table and with both hands released the cassette box, allowing it to go down a slide to approximately two inches above the floor level. The subject then bent down to a squatting position, reached under the table with both hands, and retrieved the cassette box. The subject stood up and walked ten feet back to the original station, placed the cassette box on the table, opened it, turned over five cassettes, closed the box and repeated the entire sequence two more times. This task simulated walking, bending, and squatting tasks similar to those found in cleanroom processing.

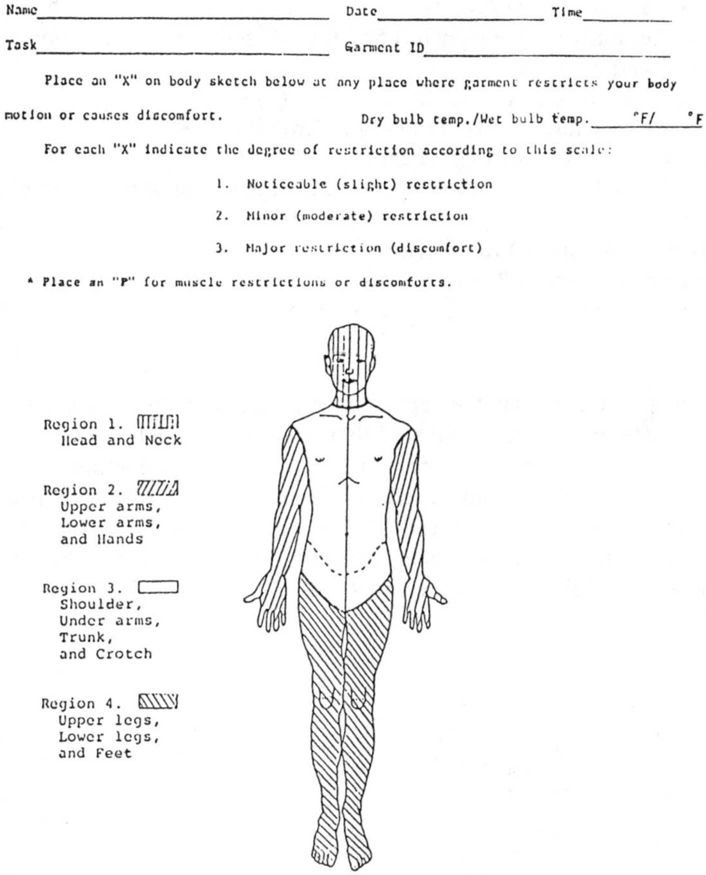
Name________ Date________ Time________

Task________ Garment ID________

Place an "X" on body sketch below at any place where garment restricts your body motion or causes discomfort. Dry bulb temp./Wet bulb temp.____°F/____°F

For each "X" indicate the degree of restriction according to this scale:

1. Noticeable (slight) restriction
2. Minor (moderate) restriction
3. Major restriction (discomfort)

* Place an "P" for muscle restrictions or discomforts.



*Figure 2. The response form for the four body regions*

## Analysis

The dependent variable frequency of restriction for the muscle and the garment represented a numerical count of the places where restriction was reported (FM and FG). The variable intensity of restriction for the muscle and the garment represented a summation of the degree of restriction values indicated by the three-point scale system previously described (IM and IG). The paired t-test was performed to determine differences between fractional replication and same size of garment data. The main independent variables were task, garment size, body region, and sex. The response form for the muscle and garment restrictions is shown in Figure 2.

The task variable (TASK) had three levels as follows:
1. Seated Computer Blackjack Game (1)
2. Upper Activity Task when Standing (2)
3. Whole Body Task of Bending, Squatting, and Walking (3).

The size variable (SIZE) had four different sizes as follows:
1. Street Clothing (Street)
2. Under Size Garment (Under)
3. Exact Size Garment (Exact)
4. Over Size Garment (Over).

The body region variable (REGN) had four different parts as follows:
1. Head and Neck (1)
2. Upper Arms, Lower Arms, and Hands (2)
3. Shoulder, Under Arms, Trunk, and Crotch (3)
4. Upper Legs, Lower Legs, Feet (4).

Subjects graded the restriction during the experiment using a three-point scale as follows:
1. Noticeable (slight) Restriction
2. Minor (moderate) Restriction
3. Major (considerable) Restriction (or Discomfort).

## Results

The paired t-test between the repeated size data and the original data showed no statistical difference. This indicated that a learning effect from repeating that task was not significant. Task, size, region, and sex effects are summarized in Table 2.

Figure 3 shows the summary chart of task effect. Garment restrictions were significantly different for each task. Task 3 showed the greatest garment restriction and Task 1 the least. The garment restriction data appeared to be a function of the level and range of the active movement.

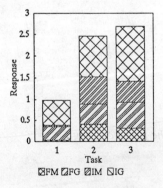

*Figure 3. Task effect*

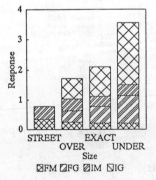

*Figure 4. Size effect*

Analysis of data with respect to the size variable (n = 240) is summarized in Figure 4. For the frequency of restriction of garment, all sizes were statistically different; the undersized garment being most restrictive, followed by the exact size and the oversize the least restrictive. Since there was no "garment" for the street clothes trial, those means were zero. The intensity of the restriction of garment showed a similar

Table 2. Frequency and intensity for muscle and garment, where FM = frequency of muscle, FG = frequency of garment, IM = intensity of muscle and IG = intensity of garment

| Task | FM | FG | IM | IG | N |
|---|---|---|---|---|---|
| 1 | 0.03 | 0.32 | 0.03 | 0.60 | 320 |
| 2 | 0.40 | 0.49 | 0.64 | 0.93 | 320 |
| 3 | 0.31 | 0.62 | 0.49 | 1.27 | 320 |
| Size | FM | FG | IM | IG | N |
| STREET | 0.33 | 0.00 | 0.45 | 0.00 | 240 |
| OVER | 0.24 | 0.38 | 0.41 | 0.69 | 240 |
| EXACT | 0.20 | 0.58 | 0.33 | 0.99 | 240 |
| UNDER | 0.21 | 0.93 | 0.37 | 2.04 | 240 |
| Region | FM | FG | IM | IG | N |
| 1 (HEAD) | 0.02 | 0.50 | 0.03 | 0.82 | 240 |
| 2 (ARMS) | 0.13 | 0.75 | 0.23 | 1.52 | 240 |
| 3 (TRUNK) | 0.45 | 0.57 | 0.73 | 1.23 | 240 |
| 4 (LEGS) | 0.38 | 0.08 | 0.57 | 0.15 | 240 |
| Sex | FM | FG | IM | IG | N |
| FEMALE | 0.39 | 0.67 | 0.63 | 1.38 | 240 |
| MALE | 0.20 | 0.41 | 0.31 | 0.78 | 720 |

pattern to the frequency of the restriction of garment. However, exact size and over size were not statistically different. This might be due to the non-uniformness of the human body and the wide range of the categorized sizes.

Analysis of data with respect to the body region variable (n = 240) is summarized in Figure 5. Analysis of the frequency of garment restriction indicated that region 2 (ARMS) had the highest frequency. This was apparently due to the large number of reports of discomfort from the gloves. Region 3 (TRUNK) and region 1 (HEAD) were next highest, but not different from one another. Garment restriction on region 3 (TRUNK) was often related to the crotch and underarms. A similar pattern was reported for the intensity of garment restriction. The mean values were consistently higher (approximately double) and the order was the same except for region 3 (TRUNK). For the intensity of garment restriction the relative restrictions on region 3 (TRUNK) were higher. However, region 3 (TRUNK) was not significantly different from region 2 (ARMS).

Figure 6 shows the summary chart of the sex effect. Analysis of data with respect to sex is represented in Table 2. There was a strong and consistently observable relationship for both frequency of restriction and intensity of restriction. Reports of both frequency and intensity for muscle and garment restrictions were much higher for the females than for males. Unisex garments, which are typically designed on male anthropometric considerations, probably contribute to this situation.

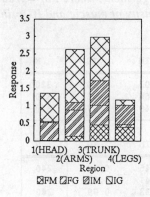
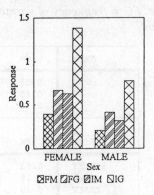

Figure 5. Region effect          Figure 6. Sex effect

**Conclusions**

As expected, the more active tasks resulted in more muscle discomfort and garment restriction compared to a seated task. However, the seated cleanroom tasks should not be ignored because sitting for a longer period of time may result in more garment and muscle complaints.

The mean values associated with body frequency and intensity of garment restriction were twice as high for the undersize garment compared to exact size garments. Oversize garments had fewer restrictions than exact size garments, which was expected.

Analysis of the regions of the body where garment interferences and discomfort were observed indicated problems in the trunk or central body region. High responses of discomfort associated with the head and arms were primarily due to the criticisms of the mask and gloves respectively.

*Conclusion*

The enhancement of the working condition and productivity for cleanroom workers can be easily accomplished if management were concerned about the actual working population. In addition, minor changes in the mask and gloves can enhance more productivity and comfort in the cleanroom setting.

*References*

Czaja, S. J., "The Role of Ergonomics on Cleanroom Environments," Panel Discussion of Human Factors in Work Environments, Human Factors Society - 27th Annual Meeting, 1983.

Ramsey, J. D., Smith, J. L., and Kwon, Y. G., "Ergonomics in the Clean Room", Proceedings of the 10th Congress of the International Ergonomics Association, Sydney, Australia, 1988.

# 43. HEAT TRANSFER CHARACTERISTICS OF INDUSTRIAL SAFETY HELMETS

John D. A. Abeysekera[*], Ingvar Holmer[**] and Christer Dupuis[**]

[*]Department of Human Work Sciences, Lulea University, Sweden

[**]Division of Climate Physiology
National Board of Occupational Safety and Health, Solna, Sweden

**Abstract**

The heat transfer characteristics of commercially available helmets (with and without modifications) were tested using a thermal manikin head, positioned in a climatic chamber controlled at +20°C (approx.) ambient temperature and an average relative humidity of 35% and a simulated solar heat load, viz. a radiant heat source lamp. The heat flux required to maintain a constant surface temperature on the manikin head was measured, programmed and recorded through a computer. The heat transfer coefficient was then calculated for different helmet shells with changes in environmental factors, e.g. air velocity and radiant heat, and human factors, e.g. hair. The results revealed that ventilated helmets (with holes on the shell) were much better for heat transfer than unventilated helmets. The helmet shell painted with a metallic (aluminium) surface had the highest heat transfer characteristics when compared to other coloured surfaces. There were also optimum advantages with helmet shells made out of PVC/ABS material. These findings can be used to improve comfort in the design of helmets for use in tropical or warm environments. The method used seems reliable and useful for testing heat transfer characteristics through head gear and other personal protective wear.

**Keywords:** *environment; heat stress; discomfort; methodology; manikin; safety helmets*

## Introduction

The popular method of protecting the head from impacts and falling objects adopted by people in both industrialized and developing countries is the use of safety helmets. In tropical climates and in hot industries the thermal discomfort caused by the use of safety helmets has been a major problem (Stroud and Rennie, 1982; Vayrynan, 1983). The thermal discomfort of helmets has been mainly responsible for the unpopularity and the non-use of helmets in hot environments (Abeysekera and Shahnavaz, 1988). Through design modifications heat discomfort in helmets can be reduced to a minimum.

The head is a major and an effective body area for heat removal (Proctor, 1982). Since a helmet covers the head, in order to reduce the effects of heat discomfort, it is important to minimise the effects of blockage in heat dissipation and maximise heat transfer through the helmet.

This paper describes a method of measuring dry heat transfer through different helmet shells and discusses the results of such measurements and possible design modifications to improve heat transfer in helmets that could be used in warm climates and in hot industries. The results of the heat transfer measurements would make it possible to find ways and means of developing a helmet shell that has the optimum heat dissipation properties.

## Method

A thermal manikin, a human model whose surface temperature was controlled at 34°C and which simulated the skin temperature of a human, was used to measure the dry heat transfer component through different helmets. Similar models for measuring heat loss from the foot and insulating properties of footwear (Elnas et al., 1985) and for measuring heat loss from the hand and insulating properties of winter mittens (Elnas and Holmer, 1980), have been used successfully. Though only the head of the manikin was of significance for the measurements intended, the whole bust consisting of 23 zones was used for the experiment. The manikin also wore a grey wig (artificial hair) on the head to simulate human hair. A controlling unit was programmed through a computer to record the heat flux required to control the surface temperature of each zone of the manikin to 34°C. For the purpose of this experiment a computer program was written to record the controlling heat flux at every minute. The manikin was positioned on a table and kept in a climatic chamber which was controlled to an average ambient temperature of 20°C and average humidity of 35%. Solar load or radiant heat was simulated by using an infrared 375 watts lamp hung about 68 cm directly over the head of the manikin. The air velocity was controlled in the climatic chamber. The controlling unit, computer terminal and the recorder were situated outside the climatic chamber.

The following procedure was followed:

The manikin was heated until its power consumption had attained steady state levels. The heat flux required for two zones (the head area covered by a helmet) to maintain the manikin surface temperature at 34°C was measured for 30 min (one measurement every minute) on every helmet used and every test condition. The average of the last 10 min was calculated and recorded on the computer printout. In the pilot testing programme it was shown that the measurements of heat flux were consistent during the last 10 min in the 30 min test run.

Two helmet types manufactured and commonly used in Sweden and helmets made of popular shell materials, viz. polycarbonate and polyvinyl chloride/acrylonnitrile butadiene styrene (ABS) copolymer, were used for most of the tests. Two other types with shell materials, viz. fibre glass and polyethylene were also used to compare the heat transfer effects of different shell materials.

The effects on heat transfer of shell material, ventilation holes, surface characteristics, colour as well as the environment, were tested. Factors such as the effect of new paint when compared with the original paint as well as the effect of hair on heat transfer were also tested.

When modified helmet shells were tested, the environmental factors, e.g. radiant temperature and air velocity, were kept constant. When environmental conditions were changed, the same type of helmet was used. For all tests the ambient temperature in the chamber was controlled at around 20°C dry temperature and 35% humidity.

The following tests were conducted:

(a) With and without helmet    -    ---

(b) Material effects:    - Polycarbonate, ABS, Fibre Glass and Polythene

(c) Colour effects:    - White, Blue, Green, Black, and Metallic (aluminium)

(d) Surface effects:    - Glossy and Matt

(e) Air speed effects:    - One type of helmet at air speeds (0.2 & 0.8 m/s)

(f) Hair effects:    - Grey hair wig used

(g) New and original paint:    - One type of helmet used

(h) Ventilation effects:    - Ventilation holes (circular & rectangular) either all over or along the crown.

Since solar radiation has a considerable impact on head heat exchange, all experiments were performed with a simulated solar heat load, as previously mentioned. This factor was measured by its effect on head heat loss; in other words the measured heat loss represents the net effect on convective and radiative heat losses and radiative heat gain. No attempt was made to evaluate separately the convective and radiative components. All results are expressed in terms of a combined heat transfer coefficient defined by the following equation:

$h = Q/(Ts-To)$ where

$h$ is the combined heat transfer coefficient by convection and radiation in $W/M^{2o}C$

$Q$ is the heat flux required to maintain surface temperature of the manikin in $W/M^2$

$T(s)$ is the surface temperature of the manikin in C, and

$T(o)$ is the operative temperature of the environment in C, where $T(o)$ was measured with globe temperature in the same position as the head, but with the manikin removed.

For reliability of measurements it was necessary to ascertain the extent of heat transfer measurement difference due to changes in the orientation angle of the helmet or any other unknown factor likely to show a difference. Nine trials were conducted in which the experimenter removed and replaced an unmodified helmet on the manikin which accounted for minor variations in the orientation angles. All other conditions were kept constant. The trials were conducted on 9 different days but not necessarily on consecutive days. From the results of the trial, the differences that can occur due to changes in orientation angles, electrical fluctuations, etc., have been worked out to +/-0.98 (2 x S.D. x $\sqrt{2}$ = 2 x 0.35 x $\sqrt{2}$, for 95% confidence limits, where S.D. is the standard deviation of the heat transfer coefficients in the 9 trials), assuming that the standard deviation is the same in all the situations. Any differences beyond +/-0.98 which can occur due to changes in orientation angles and/or electrical fluctuations can be considered significant. When interpreting results it was assumed that the same order of magnitude prevails from the minimum to the maximum heat transfer coefficients obtained.

## Table 1. Results of heat transfer tests showing the summary of averages of the heat transfer coefficient

Conditions in general: White, glossy, polycarbonate and unmodified safety helmet under environmental conditions of low air speed (0.2 m/s) and a source of radiant heat (375 watts) and manikin with grey hair wig, unless otherwise stated,

( h ) W/M² °K

| (a) With and no helmet | h | (e) Air speed effects | h |
|---|---|---|---|
| 1. With helmet and with grey hair | 3.69 | 1. Low 0.2 m/s, + helmet | 3.69 |
| 2. Without helmet and with grey hair | 0.48 | 2. High, 0.8 m/s, + helmet | 6.48 |
|  |  | 3. High, ventilated helmet | 9.91 |
| (b) Material effects | h | (f) Hair effects | h |
| 1. ABS | 7.10 | 1. Without hair and with helmet | 14.32 |
| 2. Polyethylene | 5.26 |  |  |
| 3. Fibre glass | 4.95 | 2. With hair (grey) and with helmet | 3.69 |
| 4. Polycarbonate | 3.69 |  |  |
| (c) Colour effects | h | (g) New and orig.paint | h |
| 1. Metallic (new paint) | 7.64 | 1. New paint (white) | 5.01 |
| 2. Black (new paint) | 4.53 |  |  |
| 3. White | 3.69 | 2. Original paint (white) | 3.69 |
| 4. Blue | 2.94 |  |  |
| 5. Green | 2.55 |  |  |
| (d) Surface effects | h |  |  |
| 1. Matt white (new paint) | 5.66 |  |  |
| 2. Matt blue (new paint) | 4.95 |  |  |
| 3. Glossy white | 3.69 |  |  |
| 4. Glossy blue | 2.69 |  |  |
| (i) Ventilation effects | | | h |
| * 1. 64 circular holes, all over, metallic new paint, ABS shell | | | 9.23 |
| 2. No holes, metallic new paint, ABS shell | | | 7.73 |
| * 3. 75 circular holes, all over, white, polycarbonate shell | | | 7.24 |
| 4. No holes, white, polycarbonate shell | | | 3.69 |
| # 5. 12 rectangular holes on crown, green polycarbonate shell | | | 4.02 |

\* Diameter of each hole = 6 mm (approx)
\# Diamensions of opening = 30 mm x 8 mm (approx)

## Results

The heat transfer coefficients in Table 1 (a to h) were calculated with the average measurement of heat flux in the two head zones obtained from the last 10 minute average during each 30 minute test run.

### (a) With and without helmets

Wearing a helmet increased the heat transfer (3.69) when compared to a no helmet situation (0.48). The possible cause may be the absorption of radiant heat into the hair or head surface when a helmet is not worn. Therefore in the presence of radiant or solar heat there seem to be advantages for heat transfer when a helmet is worn and vice versa in the absence of solar heat.

## (b) Material effects

ABS had the best heat transfer characteristics (7.10). This material is used in modern helmet shells. Polycarbonate which was previously a popular shell material had the lowest heat transfer properties (3.69).

## (c) Colour effects

Metallic colour far supersedes other colours (7.64) in heat transfer, followed by black (4.53). But it is noted that metallic and black were new paints applied to the helmet shell. Helmets with these colours are not usually commercially available. It is also noted that heat transfer was higher by 1.32 (see (g) 5.01 - 3.69) when a helmet with a new paint was compared with the same helmet with the original paint. But it is observed that even after deducting 1.32 from metallic and black coloured helmets, the metallic colour still has the best heat transfer characteristics (6.32), followed by the white shell (3.69) (the black shell becomes 3.21).

## (d) Surface effects

Since helmets with matt surfaces are not easily available, helmets painted with matt paint were used for these tests. If the allowance of 1.32 for the new paint is deducted from the heat transfer coefficient obtained for matt surface helmets, the difference in heat transfer properties between the matt and glossy surfaces becomes small (4.34 and 3.69 for white and 3.63 and 2.69 for blue respectively). Strangely the matt surface gave slightly higher heat transfer characteristics.

## (e) Air speed effects

At high air speed (0.8 m/s) there was a marked increase in the heat transfer viz. 6.48, when compared to low speed (0.2 m/s) when it was 3.69. It is also noted that a ventilated helmet (with ventilation holes) also largely increased the heat transfer (9.91) when compared to the unventilated helmet (6.48) at high air speed.

## (f) Hair effects

It is very clearly evident that the hair is a very good insulator of heat. The heat transfer without hair (14.32) is very much larger when compared to heat transfer with hair (3.69). The hair also absorbs radiant heat which is perhaps another cause for the low heat flux required with hair.

## (g) New and original paint

New paint had slightly higher heat transfer (5.01) when compared to old or original paint (3.69) of the same colour and surface characteristics. The bright surface shine of a new paint, which reflects radiant heat more than absorbing it, may be the cause of higher heat transfer values on newly painted helmets.

## (h) Ventilation effects

The objective in the test was to see whether ventilation holes improved heat transfer characteristics. The unventilated ABS shell (painted with metallic new paint) had a high heat transfer (7.73). It is believed that the high value was due to the metallic colour, new paint and ABS material. But the same helmet when ventilated had a still higher heat transfer value (9.23). In the case of a white polycarbonate shell with the original glossy shell surface, the ventilated helmet had a much higher heat transfer value (7.24) when compared to the unventilated helmet (3.69). Therefore, it can be concluded that ventilation holes significantly contribute to heat transfer when compared with unventilated helmets.

## Discussion

Investigation of the results of heat transfer tests reveal that a helmet shell made out of ABS material with a metallic surface, either glossy or matt, and provided with ventilation openings on the shell has optimum facilities for dry heat transfer from within the shell to outside. It seems there are many factors, viz. shell material, ventilation, etc., and environmental factors such as air velocity and radiant heat that can affect the rate of heat transfer through helmets. Any modification to improve heat transfer in helmets by the designer may be possible if either the shell material and/or ventilation openings can bring about a significant change in the heat transfer.

A study conducted in South Africa on temperature changes within hard hats as effected by ventilation holes has shown that the holes had no significant effect in the temperature of the wearer's scalp, or the air above the head (Van Graan and Strydom, 1968). But a pilot study conducted by Abeysekera and Shahnavaz (1988) has indicated that though there was no significant difference in scalp temperatures between ventilated (with ventilation holes) and unventilated helmets, the average increase in scalp temperatures was lower in ventilated helmets when compared to unventilated helmets (ventilated helmets were felt to be less hot by the wearers). The cooling advantages of ventilation holes in fire fighters' helmets have been also demonstrated in a study by Reischl (1986). In the present study the criterion tested was the hear transfer and not temperature changes within the helmet shell. The scalp temperature may not be a true indicator of thermal comfort or ventilation property of head gear. If there is an increase in heat transfer from the head, the body temperature regulation can be improved in hot environments, even though the scalp temperature has increased, decreased or not changed. Therefore in hot environments the overall feeling of thermal comfort can be achieved if the heat transfer through clothing or head gear can be improved. Since the head is believed to be a major area for heat removal (Proctor, 1982), it becomes important to use head gear that has the optimum heat transfer properties, in order to reduce heat discomfort.

The first principles of aerodynamics show that the passage of air through circular holes is smoother and greater than the passage of air through rectangular or square holes of the same area. It was not possible to test this principle on helmets. In the study of fire fighters' helmets by Reischl (1986) the truth of this principle was evident. In the current study with the green polycarbonate helmet shell where the holes were rectangular, the heat transfer was comparatively lower than other helmets which had circular holes (Table 1, h).

It is assumed that the crown of the helmet is the best area to site the holes in order to get optimum air circulation and to minimise the greenhouse effect. This facility is already used with advantage in sports helmets for ice hockey and cycling, etc, in Sweden. The air passage enters through the gap between the harness and shell, to escape from the crown area on the principle of the chimney effect. In the field study conducted by Abeysekera and Shahnavaz in Sri Lanka (1988), it was revealed that there were many complaints of sweating over the sweat band area of the helmet harness. Sweat band material of improved sweat absorbing properties can partly solve this problem. But it is presumed that provision of ventilation holes over the area of the sweat band, (i.e. in front, of the sweat band), may further alleviate the problem of sweating on the forehead where the sweat band fits.

Apart from the dry heat transfer, good ventilation can improve the evaporative heat transfer. Lastly, it has to be mentioned that the presence of holes on the helmet shell can reduce the impact resistance or safety performance and such a helmet will also fail to meet the electrical insulation test. Another disadvantage with holes is the entry of rain water or dust through the holes. But by appropriate design changes, e.g. a facility for opening and closing the holes, etc. these limitations can be overcome to a large extent.

## Conclusion

Provision of ventilation holes in helmets has a significantly positive effect on dry heat transfer and possibly evaporative heat transfer too. A helmet ventilated with holes can be designed with less bulk (as the bulk only serves to improve ventilation which can now be obtained with ventilation holes), and therefore less weight. The use of ABS material with a metallic outer finish are other features worth considering in the new design. Further work on how ventilated helmets can comply with impact and other requirements is needed.

## References

Abeysekera J.D.A., and Shahnavaz H., 1988, Ergonomics evaluation of modified industrial helmets for use in tropical environment. Ergonomics, 31(9), 1317-1329.

Elnas S., Hagberg D., and Holmer I., 1985, Electrically heated model for foot heat balance simulation. Arbete och Halsa, 1985:17, pp.1-24.

Elnas S., and Holmer I., 1980, Thermal evaluation of handwear using an electrically heated handmodel. Research Report 1980:38, National Board of Occupational Safety and Health, Solna, Sweden.

Proctor T.D., 1982, A review of research relating to industrial helmet design. Journal of Occupational Accidents, 3, 259-272.

Reischl U., 1986, Fire fighters' helmet ventilation analysis. American Industrial Hygiene Association Journal, 47(8), 546-551.

Stroud P.G., and Rennie A.M., 1982, Comfort and acceptability of safety helmets - A preliminary investigation. Institute for Consumer Ergonomics, University of Technology, Loughborough, UK.

Van Graan C.H., and Strydom N.B., 1968, Temperature changes within hard hats as affected by ventilation holes. Internationale Zeitschrift fur angewandte Physiologie einschliesslich der Arbeitsphysiologie, 26, 282-289.

Vayrynen S.T., 1983, Protection of the head and eyes in forestry work. Scandinavian Journal of Work, Environmental and Health, 9, 203-207.

# 44. THE LOWEST LIMIT OF ENVIRONMENTAL TEMPERATURE FOR OFFICE WORK

Tianlin Li, Zunyong Liu, Haichao Huang and Yongzhong Yu

*Department of Applied Physiology
Institute of Occupational Medicine, CAPM, China*

**Abstract**

Five healthy men and women were examined. The experiments were carried out in a climate controlled chamber at different temperatures, 25-20, 18, 15, 12, 10 and 7°C. Relative humidity was 40-50% and the wind speed was 0.05-0.08m/s. The results showed that the skin temperature of the finger decreased with the decrease in environmental temperature. There was a linear correlation between finger skin temperature and environmental temperature. The regression equation was: $Y = 1.029x + 5.4017$. Close correlations were also found among amplitude, wave shape, rheolimbgram, environmental temperature and finger skin temperature. The results showed that 15°C is the lowest limit for temperature to ensure normal activities of the hands and 18°C is the lowest ergonomic limit to protect the body from feeling cold.

**Keywords:** *environment; environmental temperature; skin temperature; rheolimbgram; discomfort; offices*

## Introduction

In most parts of China the temperature in winter is rather low and so heating systems are necessary for work places. To maintain suitable working conditions and economical utilization of energy sources, reasonable heating periods and room temperatures must be arranged. The aim of this work was to obtain scientific bases for setting the lowest suitable limits with regard to ergonomics and hygiene, for environmental temperatures in offices.

## Methods

### Subjects

Five healthy men and women, aged 25-45, were selected. Experiments were carried out in a climate chamber. Temperatures were adjusted at 25-20°C, 18°C, 15°C, 12°C, 10°C, and 7°C. Relative humidity was 40-50% and wind velocity was 0.05-0.08m/s.

All subjects wore the same kinds of warm clothes to keep skin temperature stable at 32.5-33.5°C. Experimental parameters were determined at 20-25°C before starting the experiment. During the experiment they were tested every half hour. Duplicate data were obtained for each test. Experiments lasted for 2 hours.

### Experimental parameters

The skin temperatures of four covered points on the body (breast, back, upper arm and upper leg) and the middle part of the 4th finger of the left hand were recorded using a Hilfsstromregler potentiometer. The rheolimbgram of this finger was also studied. Comfort was graded as comfortable, cold, slightly cold and very cold.

## Results and discussion

### Skin temperature

In Table 1, average skin temperatures of subjects at different temperatures were shown to stay at 32.5-33.5°C. This indicated that the clothes can keep subjects warm enough during the experimental period, although the skin temperatures of the fingers were lowered in accordance with the lowering of the environmental temperature. The changes were significant between the different temperatures. A regression equation was obtained:

$Y = 1.029x + 5.4017$

where Y represents skin temperature of the finger
and x represents environmental temperature

*Table 1. Skin temperatures of the finger and body at different environmental temperatures*

| Environmental Temperature (°C) | Skin Temperature of Finger (°C) | Average Skin Temperature of Body (°C) |
|---|---|---|
| 20-25 | 31.2 | – |
| 18 | 28.7** | 33.5 |
| 15 | 19.9** | 33.2 |
| 12 | 17.6* | 33.1 |
| 10 | 14.2** | 32.5 |
| 7 | 12.6 | 33.5 |

\*\* $p<0.01$
\* $p<0.05$

### Rheolimbgram of the finger

Changes in value of a, h, l, h/a obtained from the rheolimbgram corresponded well with the change in environmental temperature (see Table 2).

From the results, it is shown that the values of a, h, h/a all decreased with the lowering of environmental temperature. When subjects rested quietly at 15°C for 2 hours, the blood flow in the finger is less than half the normal flow, therefore 15°C was the most significant temperature limit. At 7°C, some unusual phenomena may occur due to cold stimulation.

### Frequency of appearance of abnormal waves

Abnormal waves, such as flat plateau, three peaks, low level etc. appeared in the rheolimbgrams. The frequency of appearance of abnormal waves also correlated with environmental temperature (Table 3). Between 25-20 and 15°C, it was 3.5-5.6%; at 12-10°C, it was 20-27%; at 7°C, it was 56.7%. When environmental temperature was lower than 12°C, the chances of the appearance of an abnormal rheolimbgram wave increased significantly.

Table 2. Changes of parameters in the rheolimbgram with environmental temperature

| Environmental Temperature (°C) | a (sec) | h (ohm) | h/a | L (cm) |
|---|---|---|---|---|
| 20-25 | 0.116 | 0.433 | 3.925 | 0.588 |
| 18 | 0.095* | 0.364 | 3.986 | 0.409 |
| 15 | 0.062* | 0.185 | 3.936 | 0.365 |
| 12 | 0.077 | 0.156 | 2.296 | 0.398 |
| 10 | 0.070 | 0.108 | 1.558 | 0.346 |
| 7 | 0.100 | 0.145 | 1.764 | 0.159 |

\* $p < 0.05$
"a" indicates the maximum velocity of blood flow in expanded artery.
"h" indicate height of peak
"h/a" index of inflow volume of blood
"L" is the height of rebound wave which may indicate the elasticity of the blood vessel.

Table 3. Frequency of appearance of abnormal waves in the rheolimbgram

| Environmental Temperature (°C) | Total Case | Abnormal Waves | |
|---|---|---|---|
| | | case | % |
| 20-25 | 29 | 1 | 3.5 |
| 18 | 18 | 1 | 5.6 |
| 15 | 20 | 1 | 5.0 |
| 12 | 20 | 4 | 20.0 |
| 10 | 20 | 5 | 25.0 |
| 7 | 30 | 17 | 56.7 |

Table 4. Subjective feelings at different environmental temperatures

| Environmental Temperature (°C) | Subjective Feeling | Per cent of Subjects (%) |
|---|---|---|
| 18 | Comfortable | 94.4 |
| 15 | Slightly cold | 25.0 |
| 12 | Cold | 70.0 |
| 7 | Very cold | 100.0 |

**Comfort**

According to the subjective feelings reported by subjects, the relation between comfort and temperature is as shown in Table 4.

The above results showed that the subjects had enough clothing to keep warm under the experimental temperature range, because the skin temperature of the covered parts of the body remained stable at above 32°C, but at 15°C or below for 2 hours, 25-100% of subjects complained of coldness. The cold feeling originated from the uncovered part of the body and dispersed to the whole body. To put on more clothes without raising the environmental temperature was useless in keeping subjects warm enough. Some design for warming the bare parts of the body was needed to avoid the feeling of coldness.

At an environmental temperature of below 7°C, all subjects felt cold. The volume of blood flow was 65.9% less than the normal blood flow. Loss of elasticity and the influence on normal moving functions were noticed.

In conclusion, the results showed that with an environmental temperature of 15°C, people can maintain normal active motion and a finger skin temperature not lower than 20°C. So, this temperature could be considered the lowest hygienic limit of environmental temperature in the office; on the other hand, with an environment temperature of 18°C, there is no feeling of coldness, while the skin temperature of the fingers was 28.7-29.0°C and normal blood flow as well as normal finger function was maintained. Therefore, 18°C would be the lowest critical ergonomic temperature in offices.

# 45. CABIN ATTENDANTS' WORKING ENVIRONMENT

Rebecca Orring[*] and Lena Erneling

Occupational Health and Safety Department
Scandinavian Airlines, Sweden

[*]Now at Swedish Telecom Networks

### Abstract

The working environment of cabin attendants working for Scandinavian Airlines System (SAS) in Sweden has been studied in two surveys, in 1979 and 1988. The purpose of the surveys was to collect information about problems in the working environment and effects on cabin attendants' health, to suggest improvements and to identify areas where further study is needed. Although measures have been taken to improve the working environment for cabin attendants during the period between the two surveys, the results show that the main problems are unchanged. The major sources of discomfort in the physical working environment are cabin air quality, particularly tobacco smoke, noise and physical workload. Factors contributing most to job dissatisfaction are shortage of time, the physical working environment and early morning flights.

**Keywords:** *cabin attendants; environment; health and safety; fatigue; discomfort; workplace; stress; cosmic radiation; job satisfaction; musculoskeletal disorders; working hours; aircraft*

## Introduction

In 1979-1980 a comprehensive survey of the cabin attendants' (C/As) working environment in the Scandinavian Airlines System (SAS) was carried out. The survey was undertaken by the Department of Human Work Sciences at the University of Lulea on behalf of SAS and the Scandinavian Cabin Crew Association (SCCA), Stockholm. The purpose of the survey was to collect information about the C/As' working environment, its problem areas and effects on C/As' health.

In 1987, SAS and SCCA initiated a new programme to find practical solutions to some of the problems associated with the C/As' working environment. The purpose of the first stage of the programme was to evaluate the outcome of the 1979 study, and to collect upto date information on working environment issues.

## Survey of working environment 1979

The 1979 survey consisted of a study of various physical factors in the cabin environment, a laboratory study of workload in the handling of service equipment, a questionnaire study covering various aspects of C/As' work and an extensive literature study.

In a report summarizing the results of the project (Ostberg and Orring, 1980), suggestions were made as to where efforts to improve the working environment should be concentrated. These recommendations are discussed later in the paper. In the study of physical factors in the cabin environment (Sundback and Tingvall, 1980), measurements were made of levels of noise, infrasound, vibration, concentration of charged particles and the thermal environment. High noise levels were found, while infrasound and vibration levels were low. The concentration of charged particles in the cabin air was in general low, although unexplained peaks were recorded. The workload study (Winkel, Ekblom and Tillberg, 1980) consisted of observational studies and laboratory measurements of workload in beverage service and cart handling. Results of the study of beverage service showed that the average workload for the neck and underarm muscles was too high, and unacceptably high static loads on muscles in the neck, underarm, overarm and possibly shoulder were also identified. It was concluded that the extreme loads on the arm would not be eliminated by simply redesigning the beverage pot, but that modification of the whole work routine was required to achieve an improvement. In the study of cart handling it was concluded that handling of wide-body carts required forces exceeding existing recommendations.

The questionnaire study (Orring and Ostberg, 1980) covered various aspects of C/As' working environment and working conditions. The main purpose of the study was to obtain some quantitative information about C/As' attitudes towards their working environment and the effects on health. Interviews with 26 C/As were carried out to collect information which was then used to draft a preliminary version of the questionnaire, which was revised a number of times as a result of opinions and suggestions for change from other researchers, C/As and other representatives of the company. The final version of the questionnaire consisted of 86 questions covering a wide range of topics relating to the working environment. The questionnaire was sent to 961 C/As normally based in Stockholm and Malmo, including those temporarily stationed abroad. The total response rate was 83%. Some of the results from this study will be presented later in the paper and compared with results from the 1988 study.

## Survey of working environment 1988

The main purpose of the second survey was to
- give a broad overview of the C/As' working environment and its major problem areas in 1988
- identify high priority areas for future improvements
- follow up the outcome of the 1979 survey.

The survey was carried out by the Occupational Safety and Health Department within SAS on behalf of the Cabin Crew Safety Committee in Stockholm.

The primary source of information was a questionnaire sent to all SAS C/As in Sweden. Additional information was obtained from interviews with C/As, administrative personnel in the Operations Division and staff from the Occupational Safety and Health Department.

Approximately half of the questions were taken either directly or with slight modification from the previous questionnaire. While the emphasis in the 1979 survey was on psycho-social aspects of the working environment and working conditions, in 1988 more emphasis on the physical working environment was required. A number of questions relating to incidence of musculoskeletal disorders was also included.

The final questionnaire consisted of 65 questions, covering the topics shown in Table 1.

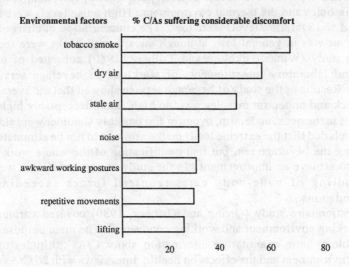

Figure 1. *Principal causes of discomfort in the physical environment*

Table 1. *1988 questionnaire topics*

| | |
|---|---|
| Background information | Job content and work organization |
| Physical working environment | Occupational perspective |
| Health problems and sick leave | Information and communication |
| Fatigue and stress | Fear of flying |
| Sleep and time zone shifts | Job satisfaction and dissatisfaction |
| Meal breaks | Occupational safety and health |

To facilitate analysis, fixed response categories were provided for all questions except a final request for comments.

The questionnaire was tested on a small sample of C/As, and after modification was sent to all 1132 C/As employed by SAS Sweden. A reminder was sent to those who had not responded after approximately three weeks. The total response rate was 81%.

## Questionnaire results

A detailed presentation of the results of the survey can be found in Erneling et al. (1988). A brief summary of results is given here.

**Physical working environment**

The results of this study show that several types of galley and inflight service equipment needed to be improved. Sinks and standard removable storage units achieved particularly poor ratings. The study did not, however, identify why C/As consider certain types of equipment to be bad. A committee was therefore formed to identify problems and possible solutions.

The factors in the physical working environment that caused most discomfort were cabin air quality (dry and stale air, tobacco smoke), noise and physical workload (see Figure 1). A comparison with the previous study shows that the environmental factors causing most discomfort are identical. The only environmental factor that is reported as causing more discomfort in 1988 than in 1979 is dust. This suggests that cleaning of aircraft needs to be improved.

**Health problems**

The most frequently experienced health problems reported in the questionnaire were swelling of stomach and legs, stomach pains and digestive disorders, colds, blocked nose, coughs and ear and eye complaints (see Figure 2). Some of these problems, particularly swelling of stomach and legs, are well known effects of flying. C/As should be given information concerning how these effects can be minimized.

The questions concerning musculoskeletal disorders reveal that the highest workloads are imposed on the wrists, the hands and the shoulders, according to C/As' assessment, whereas the percentage of C/As reporting musculoskeletal disorders was highest for shoulders and lower back. Compared to other occupational groups the incidence of musculoskeletal disorders appears to be high but further studies are needed in order to reach any firm conclusions. The results show the need for more information for C/As about lifting techniques and other aspects of manual handling than is provided today.

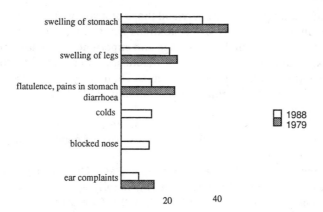

Figure 2. *% C/As reporting often/always experiencing various health problems*

**Social and psychological issues**

Tiredness after flights is a common problem; C/As in the short-haul group attribute this tiredness to physical exhaustion, while others reported lack of sleep as the primary cause. Sleeping problems are widespread, and are caused chiefly by early morning flights in the short-haul group, and by time zone shifts in the long-haul group. Efforts should be made to investigate ways of minimizing the problems, since they disrupt social life and have a negative effect on well-being.

Time available for meal breaks, timing of meal breaks and opportunity to relax during meal breaks were reported as inadequate by many C/As. However, the quality of crew meals is generally considered to be acceptable.

Fear of flying is reported less frequently by C/As in this study than in the 1979 study; 27% of C/As report frequent or occasional fear of flying in 1988 as compared with 39% in 1979.

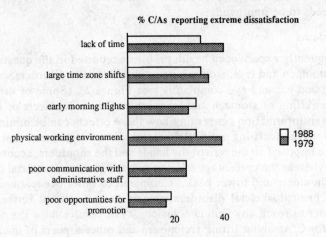

Figure 3. Principal causes of job dissatisfaction

## Organizational issues

A comparison of the results of this study and those of a study of the attitudes of other SAS employees in a wide variety of occupations shows that C/As are less satisfied with opportunities for using their knowledge and experience in their work than other SAS employees. Training for the job is for the most part considered to be adequate.

Stress due to shortage of time is the factor most commonly reported as contributing to job dissatisfaction for C/As both in the 1988 and 1979 studies. Efforts must be made to find ways to alleviate this.

The majority of C/As have not considered changing jobs during the past two years, and are more certain now than in 1979 of their ability to cope with the demands of the job over a longer period of time.

Communications between C/As and flight deck, and C/As and ground staff, are considered to function well. However, it was generally agreed that communication between C/As and administrative staff within the Operations Division function neither well nor badly. The results show a slight improvement when compared with those of the previous study, but attempts should be made to improve this situation.

## Job satisfaction and dissatisfaction

The two factors contributing most to C/As' job satisfaction are colleagues and opportunities for contact with other people. In general, ratings of job satisfaction are higher in this study than in 1979.

The factors contributing most to C/As' job dissatisfaction are shortage of time, the physical working environment, early morning flights, poor communication with administrative staff, and poor opportunities for promotion (see Figure 3). In general, ratings for job dissatisfaction are lower than in the 1979 study, with the exception of change of colleagues, which contributes more to job dissatisfaction now than in 1979.

## Changes in working environment

While most of the problems identified in the 1979 survey have been subject to further study, few have been completely solved. A brief discussion of some of the most important issues follows.

*Tobacco smoke* is still reported by C/As to be a major cause of discomfort. Some improvements have been made since 1979, for example smoking of pipes and cigars is no longer permitted on board, and air is no longer recirculated in the ventilation systems of aircraft used for long distance flights. No-smoking signs have been made larger to draw passengers attention to the fact that smoking is not permitted. A major improvement was made just after the completion of the 1988 survey, when smoking on domestic flights in Sweden was banned.

Considerable efforts have been made during the past eight years to reduce *noise* levels on board SAS aircraft. The new aircraft purchased by SAS are less noisy than their predecessors. In addition, several technical and operational modifications have been carried out to reduce noise on board existing aircraft. Despite these efforts noise levels still need further investigation.

By using more carts and trolleys and fewer standard units it has been possible to reduce the *physical workload* for C/As. A new coffee pot, introduced in 1987, was designed to reduce the strain on the arm, shoulder and the back when serving coffee. Ergonomics aspects have been important design criteria in planning the galley areas for the new long-haul aircraft. C/As' work is still physically demanding and further improvements are needed to develop better and more ergonomic work procedures and equipment.

The problem of *cosmic radiation* is still a cause of concern for many C/As. The Occupational Health and Safety Department has investigated radioactive doses received by C/As. The study is not yet finished but the results so far show no reason for concern, or give rise to any further measures.

Since the previous study *ozone* and its health effects have been subject to further investigation. Since 1986 SAS has installed ozone filters on all long-haul aircraft as it was suspected that ozone could have been the main factor causing discomfort to C/As on certain flights.

*Stress due to shortage of time* is still a major problem for C/As, particularly on short-haul routes. Changes in service, such as the introduction of business class, have exaggerated rather than reduced the problem. There has been no systematic attempt to follow up this problem since the 1979 study.

No further studies have been carried out concerning *charged particles*. The exact effect of charged particles on the human being is still not known but there is no conclusive scientific evidence as to their effects.

Effects of *time zone shifts* have not been studied further by SAS since the 1979 study.

Problems of *fatigue and depression* have not been subject to further investigation within SAS since the previous study. However, changes within the organization have made it possible to reach C/As with these problems at an earlier stage, and the counselling function at the Occupational Health and Safety Department has been extended to provide help.

C/As experiencing extreme *fear of flying* can obtain individual help from counsellors either within the Occupational Safety and Health Department, or elsewhere. Courses with the primary aim of helping C/As to deal with passengers' fear of flying have been made available on a voluntary basis. It has been hoped that these courses will also help C/As to come to terms with their own fears and anxiety.

In the early 1980's, much effort was directed to improving the *occupational identity* of C/As within the company. These efforts were directed both towards the C/A group (for example, specially designed service courses) and towards changing attitudes within the company concerning C/As' role in production.

*Disturbance of body fluid balance and swelling of the stomach* are health problems which, although not incapacitating, are a cause of discomfort for many C/As. Studies of C/A health carried out by other airlines and research institutions have also identified these problems as typical for C/As.

## Discussion and conclusions

The questionnaire studies reported in this paper were carried out in order to identify working environment issues of particular concern to C/As. A comparison between the results of the two studies shows that while problems in the physical working environment are quite similar, attitudes to many aspects of work and work organization are more positive. The surveys serve the purpose of documenting this progress, and were thus useful as evaluation tools.

As tools for planning improvements in the working environment, the surveys were less successful. For a number of reasons, no attempt was made to use the results of the first study for this purpose. Although the results of the second study were discussed when plans for further studies were made, several other factors determined the priority given to the various problem areas selected on the grounds of the survey results.

As a means of drawing attention to working environment issues, both surveys were quite successful. The first survey was carried out at a time when interest for these issues was high, and in many respects a source of conflict between the company and the union. When the second study was initiated, working environment issues were overshadowed by other aspects of working conditions, such as retirement benefits for C/As. In both cases the surveys succeeded in focussing attention on the working environment, both among C/As and other SAS personnel.

The following major problem areas have been identified in this survey:
- sleep and time zone shifts
- physical workload
- cabin air quality
- stress due to shortage of time
- information/communication between cabin crew and administrative personnel
- stomach complaints.

Of these areas, sleep and time zone shifts were given highest priority by company and union representatives. A project was initiated to study these problems in detail. A literature study has been completed (Akerstedt, 1989), and plans have been made for a study of sleep/wake disturbances, using objective measurement techniques as well as subjective assessments.

## References

Akerstedt, T (1989) A review of sleep/wake disturbances in connection with displaced work hours in flight operations. National Institute for Psychosocial Factors and Health, Karolinska Institute, Stockholm, Stress Research Report No 218.

Erneling, L, Orring, R and Joachimsson, A (1988) Cabin attendants' working environment study. SAS Occupational Health and Safety Department, Stockholm.

Orring, R and Ostberg, O (1980) Cabin attendants' working environment - a questionnaire study. University of Lulea Technical Report 1980:74T.

Ostberg, O and Orring, R (1980) Cabin attendants' working environment. (In Swedish). University of Lulea Technical Report 1980:73T.

Sundback, U and Tingvall, B (1980) Study of the physical working environment for SAS cabin attendants. (In Swedish). University of Lulea Technical Report 1980:57T.

Winkel, J, Ekblom, B and Tillberg, B (1980) Physical load of cabin attendants' work in civil aviation. (In Swedish). University of Lulea Research Report TULEA 1980:23T.

## 46. A STUDY ON THE VISUAL PROBLEMS EXPERIENCED BY PAINT INSPECTION WORKERS USING MEASURES OF VISUAL ACCOMMODATION

Naofumi Hirose[*], Shinobu Akiya[*], Susumu Saito[**],
Kimiko Koshi[**] and Sasitorn Taptagaporn[***]

[*]Department of Ophthalmology, School of Medicine
University of Occupational and Environmental Health, Japan

[**]National Institute of Industrial Health
Ministry of Labour, Japan

[***]Department of Public Health and Environmental Science
Faculty of Medicine, Tokyo Medical and Dental University, Japan

### Abstract

Research was carried out into the visual ergonomics problems of paint inspection workers in an automobile factory where the lighting of the workplace was found to be the problem of most concern. A study of the problem was performed in three phases: (1) a survey of visual complaints from paint inspection workers and control groups, (2) an analysis of physiological changes in pupil size at work, and (3) an experiment showing how this type of lighting environment affects the human eye, especially visual accommodation. The questionnaire survey indicated that the paint inspection workers more often reported visual complaints than other workers. Ergonomics evaluations of the visual environment also found glare to be a problem at the workplace.

**Keywords:** *visual fatigue; discomfort; glare; visual accommodation; pupil size; inspection*

### Introduction

It is increasingly recognized that inspection tasks in modern industries such as monitoring displays, or bottle inspection frequently involve high visual demands.

In this company, workers engaged in paint inspection often complained of severe eye strain. According to the field study, the workplace was illuminated with many rows of fluorescent lights placed both on the ceiling and walls at intervals of about 30cm. Because there were no filter screens or shields being used, obvious reflections of luminaires were superimposed on the plain surface of an automobile. The illuminating intensity of the surface ranged from 450 to 2860 Lx, luminance from 5.2 to 200 ft.L. Accordingly, diffuse reflections are unevenly distributed across the surface. To detect small scratches or patches accidentally made during the painting process, the observer must watch the surface of the automobile closely while at the same time viewing the reflected images of the glare source.

Usually, reflection glare has been considered hazardous, but it can also be useful in some ways. For example, reflecting light towards the observer leaves a dark defect on a bright background because the scratches or patches on the plain surface change the direction of the specular reflection. However, in this workplace, there can be little doubt that the inappropriate lighting condition produced a number of adverse effects on the physical characteristics of the human eye.

There are many studies available concerning inspection, but the traditional studies seldom acknowledge the physiological aspects with regard to the visual environment. Therefore, this study used a questionnaire survey, pupillary response at work, and a simulated experiment of the visual environment.

## Methods

### Questionnaire survey

A questionnaire was used to gather information about several measures of subjective symptoms among paint inspectors and control groups. The relationship between the frequency and composition of the subjective symptoms of eye fatigue among workers was investigated on the basis of accumulated data from the field study. Enquiries into subjective symptoms consisted of 35 items classified into 7 categories. The following topics were examined by the questionnaire:

(A) General eye fatigue
(B) Eye pain
(C) Irritation of the eye
(D) Difficulty in focusing
(E) Blurred vision
(F) Headache
(G) General condition.

To evaluate subjective symptoms, the following indices were taken into account;
1) For general eye fatigue (A), the frequency of eye fatigue experienced by the workers was recorded using a three point scale ranging from good to extremely tired.
2) The frequency of affirmative answers to all items (T) and to items B-F was recorded. The frequency of complaints was calculated on the following formula (Yoshitake, 1971):

$$\frac{\text{gross number of complaints in a group}}{\text{number of items} \times \text{number of members of each group}} \times 100\ (\%)$$

Two control groups were studied consisting of six employees from the testing department (group I), 30 workers from the assembly sections in the same factory (group II), and eight inspectors (group III), a total of 44 subjects. The average age of all the subjects was 33.4 years. The average age of group III was 31.4 years.

### Pupil response at work

The pupil responses of the paint inspectors were monitored by means of the Iriscorder (Hamamatsu Photonics) while performing their tasks. Using a semiconductor TV camera, pupillary response was monitored continuously while at work without disturbing the performance of the subjects. The pupil size of three subjects from the paint inspection section was measured using the video image (Taptagaporn and Saito, 1990).

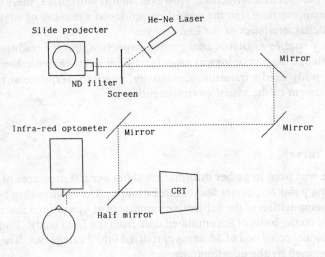

Figure 1. Schematic diagram of the simulation study

While performing this task, pupillary response was examined under three lighting conditions which were changed experimentally in three levels, namely real situation (bright), medium and lower conditions. The light source power was decreased to 70% in the medium condition and a filter screen was used in the lower lighting condition.

After the experiments, the workers were interviewed about visual comfort which was classified into five levels.

### Effects of glare on visual function through the simulation study

Recent studies on the response of the human lens accommodation suggest a new approach to visual problems associated with tasks requiring high visual demands. The infrared optometer allows precise measurements of visual accommodation, which is the changes in the curvature of the human crystalline lens. In addition, continuous monitoring enables the acquisition of data on the dynamic response of the human eye in respect of latency time, and velocity and amplitude of accommodation (Saito et al., 1989).

In an effort to examine the responses of the human eye, we have developed a laboratory simulation of this visual environment. This experiment was divided into two phases.

In the first phase, subjects had to add up three numbers that appeared one by one on the CRT display (NEC model PC-KD 852) which were presented at a distance of 50cm from the subject's eye. Five metres away from the eye was the rear screen on which a dot pattern was projected as noise. The dot pattern was plotted evenly at regular intervals. With the use of a half mirror, the system was arranged so that the near target (characters on the CRT) and the noise (dot pattern on the screen) were both in the same visual axis. By putting neutral density optical filters in the pathway, the illumination levels of the noise were varied logarithmically in five steps in the range 18.4 to 211 $cd/m^2$. The laser spot on the screen was presented as a far target. A motor driven shutter switched the near target (character on CRT with noise) and far target (laser spot on the screen) at 16 seconds duration. Subjects were requested to gaze at the two targets alternatively. Preceding the experiment, which was carried out in a complete dark room, all subjects were prepared to undergo retinal dark adaptation for

10 minutes. A modified NIDEK AR-1100 infrared optometer was used to measure lens accommodation and pupil size of the subjects' right eyes (Figure 1). A chin rest was used to maintain a stable head position.

The second phase was the control study on lens accommodation without glare among the same five subjects. The near target was the character displayed on the CRT screen and the far target was a "+" projected on the rear screen. Due to these distances, the dioptric change between the two targets was the same (1.8 dioptres (D)) as in the first phase of the experiment. An electronic timer limited exposure duration for each test target to 8 seconds.

For the data analysis, the lens accommodation signal was converted into analogue voltage and then fed to a pen recorder and a magnetic tape recorder. The automatically selected 8 seconds of data were averaged by the computerized data processor. Three aspects of lens accommodation were analysed in this study. They were latency time (sec), velocity (D/sec), and amplitude (D) of accommodation. Figure 2 shows the normal response of the human lens accommodation.

The pupil image was recorded continuously on the video tape recorder and fed to the frame analysis. While focusing on its stationary portion, an analysis of the pupil size was performed in this study, exempting its transient portion which came out immediately after shifting the target. The ages of the five subjects who underwent this experiment ranged from 24 to 34 years (mean 28.4). All cases of refractory errors were corrected by appropriate lenses, except emmetropia.

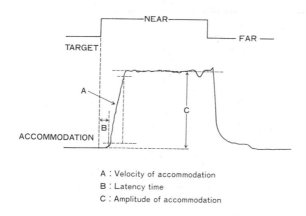

A : Velocity of accommodation
B : Latency time
C : Amplitude of accommodation

*Figure 2. An example of the accommodation response*

## Results and discussion

### Questionnaire findings

The results showed that paint inspectors complained more about their eye symptoms on almost every factor than did other workers from the same factory.

The incidence of workers who experienced some eye fatigue was 78%, which was a little higher than that of group III (75%). On the other hand, accumulated data of general eye fatigue (Figure 3) showed a remarkably high score among group III.

Table 1 demonstrates the frequency of symptoms corresponding to each component for the three groups. Note that in almost every factor, group III workers reported many more complaints than the other two groups.

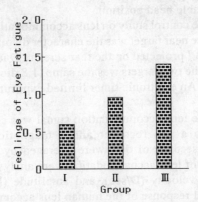

*Figure 3. Accumulated data for general eye fatigue among workers in the 3 groups of workers*

*Table 1. Percentages of each pattern of symptoms (B to G, T represents the total symptoms) among the 3 groups of workers.*

|   | I | II | III |
|---|---|---|---|
| B | 11.1 | 7.6 | 25.0 |
| C | 16.7 | 11.7 | 22.9 |
| D | 16.7 | 12.2 | 12.5 |
| E | 8.4 | 7.2 | 12.5 |
| F | 4.2 | 5.8 | 12.5 |
| G | 11.7 | 12.3 | 16.3 |
| T | 11.5 | 10.1 | 19.9 |

## Pupil responses of the paint inspection workers

The averaged values of pupil diameter, its variation, and subjective evaluation while the subjects were inspecting paint are shown in Figure 4.

The pupil size varied as a function of the intensity of illumination, with the state of accommodation, and with the interest and value of the stimulus and task context. With regard to accommodation, the larger the pupil diameter the less was the depth of focus and the more was the effect of the lack of optical conjunction of the retina and the target. In other words, under high illumination levels the pupil constricts, the depth of field increases, and the targets within that field are seen clearly whether the retina and target are conjugate or not.

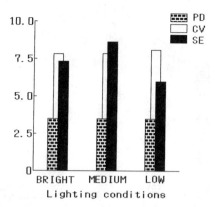

*Figure 4. Pupil diameter and visual comfort under 3 lighting conditions during visual work (mean of 3 subjects), where PD = pupil diameter (mm), CV = coefficient of variation of pupil size (%) and SE = subjective evaluation of visual comfort (five point scale)*

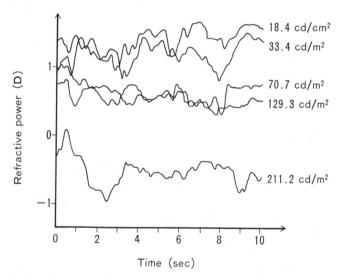

*Figure 5. Averaged traces of accommodation while viewing near targets with glare (an example from one subject)*

Under the three lighting conditions, the pupils of the paint inspectors were so constricted because of the too highly illuminated environment that there were almost no differences in averaged pupil diameter. This implies that working in this lighting environment required the maximum retinal adaptation which results in small pupil size. It can also be assumed that pupillary response would have little contribution in regulating the luminance in all the conditions and can be one of the causes of visual fatigue. Thus the lens accommodation, the other important factor in accommodation, came into consideration in this study.

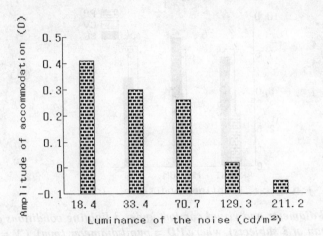

*Figure 6. The relationship between amplitude of lens accommodation and luminance of noise (r = -0.967). The averaged value under the control study was 0.72 D*

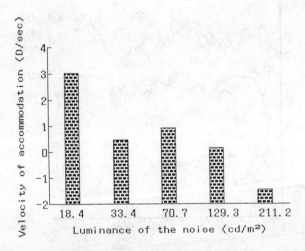

*Figure 7. Relationship between velocity of accommodation and luminance of the noise (r = -0.889)*

### Results of the experimental study

Glare produced significant effects on both accommodation and pupil size. Physiological responses of the human eye were distributed as the noise level increased. Figure 5 is an example from one subject. Shown in this figure are averaged traces of accommodation while viewing near targets with glare. It was apparent that the amplitude of accommodation decreased with the increase in the noise luminance. It was also noticed that the potential of accommodation was drawn further toward the side of the noise than the near target.

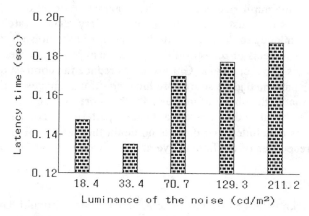

*Figure 8. Relationship between latency time and luminance of noise (r = 0.905). Analysis of t-test, there was a significant difference (t =0.003)*

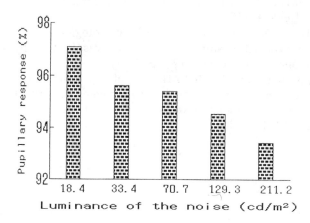

*Figure 9. Relationship between pupil size and the luminance of the noise (r = 0.905), pupillary change rate was calculated on the following formula:*

$$\frac{\text{pupil diameter while viewing near target}}{\text{pupil diameter while viewing far target}} \times 100\ (\%)$$

The relationship between the amount of noise (in log scale) and the amplitude of accommodation is illustrated in Figure 6. The higher luminance of the noise, the lower was the amplitude of accommodation. Compared with the control study, analysis of the t-test revealed that there was a significant difference (p = 0.013). Furthermore, on average with all five subjects tested, velocity (Figure 7) as well as amplitude of accommodation decreased as the amount of noise increased and latency time (Figure 8) was likely to increase. Saito et al. (1989) investigated the physiological characteristics of lens accommodation. They also examined the relationship between several indices of lens accommodation and subjective evaluation of visual comfort. They concluded that the velocity of accommodation was highly correlated with subjective evaluation, which is also the case in this study. It was remarkable in the experiment that all three selected indices of lens accommodation showed clear correlation with the noise luminance.

Figure 9 shows the pupil size change plotted against the level of noise luminance. Pupil diameter was demonstrated in terms of pupillary change rate. The pupillary response had a tendency to decrease with the increase in the noise luminance.

These disturbing effects of reflection images are also found in other situations like VDT workstations (Boyce, 1987). Glare sources reduce the contrast of the image of interest. Also, a patterned glare source produces additional images at different optical distances from that of the object of interest. Furthermore, focused glare sources cause visual difficulties because they possess a textured pattern that competes with the task information. It is concluded that there is no doubt that reflection glare distracts the physiological responses of the human eye and may cause visual fatigue.

## Acknowledgments

We gratefully acknowledge Mr. Yoshinao Asako and Mr. Hiroyuki Kojima of Tokyo Institute of Polytechnics for their technical assistance throughout the study.

## References

Boyce, P. R. 1987, Lighting the display or displaying the lighting, Work with Display Units 86 (Elsevier), 340-349.

Saito, S., Ishikawa, K. and Hatada, T. 1989, Physiological evidences of superiority of positive type CRT among information displays, Advances in Human Factors/Ergonomics (Elsevier), 12A, 536-541.

Taptagaporn, S. and Saito, S. 1990, How display polarity and lighting conditions affect the pupil size of VDT operators, Ergonomics, 33, 201-208.

Yoshitake, H. 1971, Relations between the symptoms and the feelings of fatigue, Ergonomics, 14, 175-186.

# Part VIII
# VDT Ergonomics

# 47. THE EFFECT OF VDT DATA ENTRY WORK ON OPERATORS

Chuansi Gao[*], Rongtai Cai[**], Lei Yang[**], Guogao Zhang[**], Demao Lu[*] and Qiyuan She[*].

[*]*Department of Ergonomics*
*Safety & Environmental Protection Research Institute*
*Ministry of Metallurgical Industry, Renjia Rd, Qingshan Wuhan*
*430081 P.R. China*

[**]*Department of Occupational Health*
*School of Public Health Tongji Medical University*
*Hangkong Rd, Wuhan 430030, P.R. China*

## Abstract

Some physiological and biochemical measures before and after task performance were used to delineate the ergonomic effects of different VDT data entry work on operators. Twenty-nine healthy Chinese students were chosen and divided randomly into the simple and complicated data entry groups. The subjects were instructed to work as quickly and accurately as possible for 150 minutes. The results showed that performance fluctuated over time, decreased significantly after 50 minutes of work, followed by an improvement in performance and finally, at the end of the work period, performance again improved, reflecting a motivational phenomenon. The changes in physiological parameters revealed that operators' fatigue occurred after data entry work. The adrenaline excretion in urine showed a tendency to increase after simple data entry work. The noradrenaline excretion showed a tendency to decrease after complicated data entry work.

**Keywords:** *visual display terminals; arousal; stress; catecholamines; critical fusion frequency; productivity; data entry*

## *Introduction*

The widespread use of visual display terminals (VDTs) has improved the efficiency of message transmission between man and machine and reduced the physical workload of the operators, but has brought about new occupational health and ergonomics problems.

VDT work can be divided into three types, data entry, data retrieval and dialogue (Benz et al., 1983). Data entry operators read numbers, symbols and/or words from a source document; remember them, then enter them into the computer via a keyboard; finally they look at the screen to check whether the information is correct or not. The eye and hand movements form a simple cycle. So VDT data entry work is monotonous, repetitive, and more harmful to health than other types of VDT work (Grandjean, 1984; Bergqvist, 1984). Recently several studies have been reported on the effect of data entry work, but not much attention has yet been paid to data complexity and work performance. VDT use in China is just beginning, but with the development of the economy, VDTs will be used increasingly. In particular, it should be established whether or not many VDTs imported from industrialized countries are suitable to the physiological and psychological characteristics of Chinese operators. It is therefore necessary to carry out a comprehensive ergonomic study on VDTs and their operators. In this paper the before/after study of physiological and biochemical parameters was used to delineate the effects of simulated realistic simple and complicated data entry work on health and performance, providing a scientific basis for rational work organization and regulation.

## Methods

Using the same work environment and equipment, the subjects entered data into the computer for 150 min without a break. Work performance was recorded automatically once every 10 min. The before/after complaints, physiological and biochemical parameters, and neurobehaviour were also studied.

### Subjects

Twenty-nine healthy Chinese students (23 male, 6 female) were chosen as subjects, and divided at random into 2 groups, the simple data group (11 male, 3 female) and complicated data group (12 male, 3 female). The average age of the subjects was 26.9 years (1 SD = 3.9).

### The experimental task

An APPLE II microcomputer was used, displaying green letters. The illuminance and contrast remained unchanged during the task. The working environment was air-conditioned.

The two groups of subjects entered two types of data respectively, the simple data (any integer from 10 to 99, e.g. 38, 99 etc) and the complicated data (decimal, e.g. 3.7486, 0.5692 etc). The source data were produced and printed at random by computer.

The subjects were instructed to familiarise themselves with the experimental procedure and environment before the experiment. Two subjects were studied each day. Firstly, their complaints were elicited by questionnaire, and all the physiological parameters were tested. Then the subjects entered data for 150 min continuously without any break. After data entry, the parameters were tested once again, and the subjective complaints were also again recorded.

### The measures recorded

The following methods were used and measures recorded;
- Critical flicker frequency (CFF), the tester was made in China (Type:ZPJ).
- Simple reaction time (SRT), the tester was made in China (Type:JFM-821).
- Aiming test, this was carried out according to the Neurobehaviour Core Test Battery (recommended by WHO) (Liang, 1987).
- Diastolic blood pressure in standing position (DBPSP), DBPSP of subjects' right arms was measured with a standing sphygmomanometer.
- Collection of urine and the fluorometric analysis of catecholamines (Andersson et al., 1974; Huang et al., 1987). The adrenaline(A) and noradrenaline (NA) in urine were tested with a Hitachi F-3000 fluorospectrometer. The A and NA concentrations were averaged by creatinine (C) in the same urine sample.

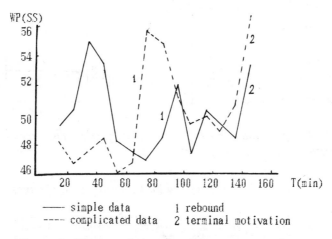

*Figure 1. Work performance of data entry work*

- Data entry work. After eliciting subjective complaints, measurement of parameters and collection of urine, the subjects were instructed to work as quickly and correctly as possible according to the "data entry work programme" designed by the authors. The correct, incorrect and total entry were recorded automatically by the computer once every 10 min.

These measures were recorded both before and after work and in some cases during work.

## *Results*

### Work performance (WP)

The first 10 min work performance (correct entry) was removed because of large inter-individual differences exhibited during this period. The raw scores of performance were transformed into standard scores (SS). Figure 1 shows the curve of data entry work performance values arranged by 10 min intervals. The maximum performance occurred in the 30-40 min period in the simple data group, then decreased significantly after 50 min of work. There was a statistically significant difference in average performance scores between the 60-70 min and 30-40 min periods (t=2.2440, p.<0.05), followed by a rebound, and increase again in the last 10 min.

The average performance scores were relatively low in the first hours in the complicated group, followed by a significant rebound in the 70-80 min period. There was a statistically significant difference compared to the 60-70 min performance (t=2.3415, p.<0.05). During the last 10 min, performance increased markedly. There was a highly significant difference compared to the 50-60 min performance (t=2.7826, p.<0.01).

Between the two groups, there was a significant difference only in the 30-40 min period (t=2.2918, p.<0.05). The average work performance fluctuated over time. Rebound and terminal motivation phenomena were observed from both groups.

### Health effects

The before/after measures related to health are shown in Table 1. SRT and DBPSP were statistically different between the two groups.

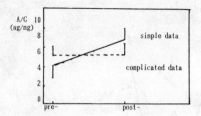

Figure 2. The pre- and post- A/C excretion with data entry work

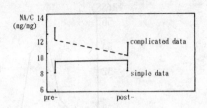

Figure 3. The pre- and post- NA/C excretion with data entry

Table 1. The pre- and post-physiological changes

| parameters | groups | pre (X+SD) | post (X+SD) | differences |
|---|---|---|---|---|
| CFF (Hz) | simple | 22.4+2.2 | 21.5+2.4* | -0.9 |
|  | complicated | 21.9+1.8 | 21.1+1.5* | -0.8 |
| SRT (ms) | simple | 244.8+41.6 | 287.0+51.7** | 42.2 |
|  | complicated | 280.2+63.7 | 292.3+71.3 | 12.1 |
| Aiming (stan- | simple | 52.6+8.1 | 50.3+9.0 | -2.3 |
| dard scores) | complicated | 50.4+10.4 | 46.8+11.9* | -3.6 |
| DBPSP (mmHg) | simple | 75.0+8.6 | 76.6+10.8 | 1.6 |
|  | complicated | 73.3+7.1 | 77.9+9.0* | 4.6 |

* $p<0.05$  ** $p<0.01$

## Catecholamine excretion in urine

Because of the obvious inter-individual difference in A and NA excretion in urine, the data appeared abnormally distributed. We therefore transformed the data into normal distribution using the formula [X'=log(x+1)], and then conducted a paired t-test. The adrenaline excretion in urine showed a tendency to increase after simple data entry work, but was not significantly different to that before work (p.<0.05). The noradrenaline excretion showed a tendency to decrease after complicated data entry work, but was not significantly different to that before work either. The before/after A and NA excretion showed no significant difference between the two groups as shown in Figures 2 and 3.

## Discussion

### Data entry work performance

It is assumed that VDT work with an air-conditioned environment and light physical workload is a comfortable occupation. But in this experiment, many subjects complained about the work being monotonous and tedious (82.8%). There were also visual and musculoskeletal complaints. The performance fluctuated over time, and there were rebound and terminal motivation phenomena at the end. This result is consistent with those of Floru et al. (1985), which were associated with the auto-arousal and cerebral compensatory effort during monotonous and repetitive mental work. The average performance scores were relatively low in the first hours in the complicated data group, which may be because subjects had not adapted to the task or were not operating the VDT efficiently. Based on the results that showed that performance decreased markedly in the 50-60 min period, there should be a rest break given after 50 min data entry work to guarantee high work efficiency. The ideal and concrete parameters indicating mental work capacity have not so far been established.

## Physiological changes after work

CFF may reflect fatigue and shortage of oxygen in the central nervous system (Saito and Tanaka, 1981; He et al., 1983). CFF decreased significantly after work in both groups. The results are consistent with Saito and Tanaka's (1981), indicating that the visual resolving ability decreased, and that central nervous system fatigue occurred after work. This is consistent with the subjects' complaints.

SRT is a measure of "Attention/Reaction Speed". When the operators are tired, SRT is increased with the development of protective inhibition and a decrease in the excitability of the central nervous system. The results showed that SRT was more significantly increased after work than before work in the simple data entry group, a result consistent with Floru et al.'s (1985). The increase in SRT was significantly different between the two groups after work, indicating that the simple data entry task had a more marked effect on "Attention/Reaction Speed" and eye-hand coordination. Aiming is a neurobehaviour test indicating "Psychomotor Stability". The results showed that the aiming scores after work were lower than those before complicated data entry work, indicating that the complicated data entry work has led to a deterioration in "Psychomotor Stability".

The results showed that the DBPSP increased 4.6 mmHg after the complicated data entry work. According to Van Ameringen et al.'s (1988) studies, the DBPSP can reflect intrinsic job stress, the stronger the stress, the higher the DBPSP. Thus the results confirm that there was a considerable amount of stress accompanying the complicated data entry work.

## Catecholamine excretion

Catecholamines include adrenaline(A), noradrenaline (NA) and dopamine (DA). A and NA levels are high in blood and urine. NA is excreted mainly from the sympathetic nerves and A from the adrenal glands. The excreted A and NA reach the blood, and then are excreted to the urine from the kidneys. Many factors (diseases or stress, for example) can affect the A and NA excretion in urine. The amounts of A and NA in the blood and urine can be used as parameters of VDT workload (Tanaka et al., 1988). Patkai's study showed that the A and NA excretion increased in mental work requiring high concentration (Patkai, 1971).

The levels of A and NA excretion in urine shown in Figures 2 and 3, are consistent with Tanaka et al.'s (1988) data. The excreted A and NA did not change significantly after work, which may be associated with light workload and low sympathetic nerve excitability. The differences in A and NA between the two groups after work indicate that the simple and the complicated data entry work have different effects on A and NA excretion. Some studies have reported that the maximum catecholamine excretion occurs 3 hours later in the urine than in the blood (Akerstedt and Levi, 1978). We collected urine immediately after VDT work only. The maximum catecholamine excretion may not yet have occurred. Therefore, in future studies, data complexity, work difficulty and mental stress should be included in order to study the dynamic state of A and NA excretion including excretion after breaks, and the relationships with work performance. If possible, the more stable and sensitive Radioactive Enzyme Method should be used to examine the A and NA in the urine.

## Conclusions

The simulated VDT data entry work can realistically reflect the practical job. Correct data entry can feasibly indicate performance levels, and the curve of work performance can reflect the dynamic state of mental work capacity. Fluctuating performance was associated with "auto-arousal and cerebral compensatory effort". Because performance decreased significantly after 50 min of work, short breaks should be introduced after 50 min of continuous work in the case of the real job. The changes in physiological parameters revealed that operator fatigue occurred after data entry work. The adrenaline excretion in urine showed a tendency to increase after simple data entry work. The noradrenaline excretion showed a tendency to decrease after complicated data entry work. The differences in performance, DBPSP and neurobehaviour between the two groups indicated that there was much stress accompanying the complicated data entry work.

## Acknowledgments

The authors are greatly indebted to Mr Tang Cibing, Miss Tao Youci, Mr. Yu Ri'an, Feng Guoming, Zu Zongsui and Zhang Kaiye for their assistance.

## References

Akerstedt, A. and Levi, L. 1978, Circadian rhythms in the secretion of cortisol, adrenaline and noradrenaline. European Journal of Clinical Investigation, 8, 57-58.

Anderson, B. et al. 1974, Analyzer fluorescence method. Clinica Chimica Acta, 51, 13-28.

Benz, C. et al. 1983, Designing VDU Workplaces, (Verlag TUV Rheinland, Koln), 21-110.

Bergqvist, V.O. 1984, Video display terminals and health. Scandinavian Journal of Work, Environment and Health, 10(suppl 2).

Floru, R. et al. 1985, Psychophysiological changes during a VDU repetitive task. Ergonomics, 28, 1455-1468.

Grandjean, E. 1984, Ergonomics and Health in Modern Offices, (Taylor & Francis Ltd., London), 240-247.

He, B. et al. 1983, Experimental Psychology, (Beijing University Publishing House, Beijing), 526-528.

Huang, M. et al. 1987, Test of catecholamines in healthy adults' blood and urine. Journal of Shanghai Medical Test, 2, 133-135.

Liang, Y. 1987, Introduction to the Neurobehavior Core Test Battery recommended by WHO. Industrial Hygiene and Occupational Diseases, 13, 331-339.

Patkai, P. 1971, Interindividual differences in diurnal variations in alertness, performance and adrenaline excretion. Acta Physiologica Scandinavica, 81, 35-46.

Saito, M. and Tanaka, T. 1981, Eyestrain in inspection and clerical workers. Ergonomics, 24, 161-173.

Tanaka, T. et al. 1988, The effects of VDT work on urinary excretion of catecholamines. Ergonomics, 31, 1753-1763.

Van Ameringen, M. R. et al. 1988, Intrinsic job stress and diastolic blood pressure among female hospital workers. Journal of Occupational Medicine, 30, 93-97.

# 48. TECHNOLOGICAL CHANGE AND WORK RELATED MUSCULOSKELETAL DISORDERS: A STUDY OF VDU OPERATORS

Choon-Nam Ong, J. Jeyaratnam and W. C. Kee

*Department of Community, Occupational and Family Medicine*
*National University of Singapore*
*Kent Ridge, Singapore 0511*

## Abstract

A study was conducted to evaluate the prevalence of musculoskeletal complaints among 161 word processing operators, 43 data entry operators, 49 clerical officers, 206 VDU operators, 270 LPD operators and 383 school teachers. The results show that the prevalence increased significantly with VDU working hours. Job activity also affected the prevalence; data entry operators were noted to have a higher prevalence than the other groups while teachers had the lowest. Although there was no direct association with age, the prevalence was more pronounced among working mothers aged 26 to 35. It was also noted that Malays and Indians reported a higher prevalence than their Chinese counterparts. These findings suggest that job activities and duration of work could have some influence on the observation. Social and psychological factors would also need to be taken into consideration.

**Keywords:** *psychosocial factors; stress; musculoskeletal disorders; visual display terminals; discomfort; ageing; liquid plasma displays; word processing; data entry; teachers; ethnic groups*

## Introduction

Musculoskeletal symptoms of the upper extremities have probably been around since men undertook repetitive tasks. Early examples of the problem include the various craft palsies or cramps, such as threader's wrist and brewer's arm. It is interesting to note that despite increased mechanisation in recent decades, musculoskeletal problems have increased dramatically. It is only relatively recently, however, that measurement of the prevalence and incidence of these problems have been studied more systematically and the detailed analysis of associated factors and their causations have been investigated.

Occupational disorders of the musculoskeletal system of office workers are also not new. Ramazzini (1713) highlighted problems experienced by scribes and notaries: "increasing driving of the pen over paper causes intense fatigue of the hand and the whole arm because of the continuous and almost tonic strain on the muscle and tendons, which in course of time results in failure of power in the right hand".

Today, the incidence of such conditions appears to be increasing in various parts of the world, in particular Australia (Sharrod, 1985). Ferguson (1984) reported that musculoskeletal problems of office workers are serious and repetition strain injuries are reaching epidemic proportions. It is however, important to mention that the terminologies used to define musculoskeletal symptoms related to occupational activities are confused. This is partly because symptoms reported were of various types and they often occur in different parts of the body. The problems can range from numbness, stiffness, fatigue, cramps to tremors. The body parts affected are usually neck, shoulders, lower back and wrist.

Such terminology confusion could hamper the compiling of statistics relating to the occurrence of the symptoms (Ferguson, 1984). Furthermore, there are no specific pathological or clinical features to confirm such symptoms. On many occasions, studies were carried out based on analysis of semi-objective findings such as local tenderness or pain.

Recent studies have indicated that psychosocial factors do play a part in the development of the disorder. Such invocations should be treated with great circumspection. Clearly, occupational and environmental factors play an important part in the pathogenesis of the resulting syndrome. A study examining these factors together with the circumspections of the workplace is an essential prerequisite for a more valid study.

The present study attempts to examine some to the factors which we believe would influence the prevalence of musculoskeletal symptoms of computer terminal operators, in particular full-time visual display unit (VDU) operators.

## Methods

383 school teachers from 14 primary schools and 715 VDU operators from five large statutory boards in Singapore took part in this study. After a general briefing about the purpose of the investigation each subject completed a questionnaire at the workplace. The survey was conducted in small groups, office by office, under the supervision of one of the investigators (WCK). The questionnaire was designed in such a way that it can be readily answered by the respondents. The language used was simple and can be self-administered. Prior to the survey proper, several pilot studies were conducted to correct for ambiguity in the questions. Results of the pilot study were not included in the final data analysis.

A total of 1098 questionnaires were returned fully completed. The rest were incomplete and the respondents could not be traced back for rectification. The overall response rate was 96.8%.

The questionnaire consisted of 4 sections. Section 1 was on biodata, past and present occupations, domestic duties, games, sports and work environment/attitudes towards work environment and working conditions. Section 2 was on musculoskeletal problems of the past and present. This section was modified from a Nordic questionnaire (Kuorinka et al., 1987).

## Results

Table 1 shows the demographic profiles of the 1112 subjects studied. 459 operators who used computer terminals for their work participated in this study. Among them 270 were from a telecommunication centre using liquid plasma displays, while the other 459 used visual display units (VDU). The reference group was made up of 383 teachers. They rarely used computers for their work. Those who used VDUs for more than two hours per week were not included in the final data analysis. The mean age for the whole study population was 34 years old. Average working experience with a VDU was about 12 years. Clerical officers appeared to have less experience with VDUs, with an average experience of 5.1 years.

Table 1. Mean age and working experience of the study population

| Category | n | M ± SD | range (years) | experience |
|---|---|---|---|---|
| School teachers | 383 | 38.4 ± 9.0 | 19 - 58 | - |
| Word processing operators | 161 | 34.9 ± 7.4 | 23 - 54 | 13.2 |
| Data entry operators | 43 | 37.2 ± 6.2 | 22 - 47 | 12.4 |
| Clerical Officers | 49 | 31.4 ± 6.0 | 20 - 41 | 5.12 |
| Tele. operators (VDU) | 206 | 33.1 ± 6.1 | 24 - 51 | 12.7 |
| Tele. operators (LPD) | 270 | 31.9 ± 4.9 | 21 - 53 | 12.1 |
|  | 1112 | 33.9 ± 6.8 | 19 - 58 | 11.2 |

Table 2. Prevalence of musculoskeletal complaints with different categories of operations

| Category | Neck | Shoulder | Lower back | Arm | Hand/Wrist |
|---|---|---|---|---|---|
| Teachers | 45.5 | 25.3 | 37.1 | 4.1 | 26.0 |
| Word processing operators | 49.7 | 37.2 | 49.0 | 9.3 | 17.4 |
| Data entry operators | 61.5 | 53.8 | 58.7 | 15.4 | 15.4 |
| Clerical Officers | 55.0 | 46.1 | 45.0 | 9.7 | 20.0 |
| Tele operators (VDU) | 62.1 | 39.2 | 51.7 | 11.6 | 19.9 |
| Tele. operators (LPD) | 64.8 | 51.4 | 53.3 | 11.8 | 17.4 |

Table 2 shows the prevalence of musculoskeletal complaints in relation to job activities. Data entry operators were noted to have a higher prevalence than other groups. School teachers had the lowest prevalences on all body parts.

The neck was the body part most complained about by all groups of workers. The elbows and arms were the least. Table 3 shows the prevalence of musculoskeletal discomfort in relation to the duration of daily work with VDUs. Significant differences in neck, shoulder and low back pain were found between teachers and VDU operators. Of neck pain, the prevalence increased from 19.7% for 2-4 hours of work to 46% for more than six hours VDU work. Similarly, of low back pain, prevalence rose from 34% to well over 67%. It is, however, interesting to note that those who do not use VDUs (viz school teachers) had a high prevalence of hand and wrist complaints.

*Table 3. Prevalence of musculoskeletal complaints with number of hours working with VDUs*

| No of hours Working with terminals | n | Neck | Shoulder | Lower Back | Arm | hand/wrist |
|---|---|---|---|---|---|---|
| 0 | 383 | 44 | 25.0 | 36.9 | 3.9 | 25.9 |
| 2-4 | 68 | 38 | 19.7 | 34.3 | 10.3 | 11.9 |
| 5-6 | 66 | 59 | 47.0 | 53.0 | 7.5 | 22.7 |
| >6 | 543 | 62 | 46.2 | 67.4 | 11.6 | 18.4 |
| p: | 1060 | <0.001 | <0.001 | <0.001 | <0.001 | NS |

*Table 4. Prevalence of musculoskeletal complaints by years working with VDUs*

| Years working with terminals | n | Neck | Shoulder | Lower back | Hand/wrist |
|---|---|---|---|---|---|
| 0 | 383 | 44.0 | 25.0 | 37.0 | 23.0 |
| <1 | 15 | 66.4 | 10.0 | 44.0 | 4.8 |
| 1-2 | 9 | 64.8 | 29.1 | 16.6 | 8.0 |
| 2-5 | 42 | 60.5 | 36.0 | 15.7 | 19.5 |
| >5 | 637 | 66.0 | 38.3 | 49.7 | 16.7 |
| p: | 1086 | NS | 0.05 | 0.01 | 0.05 |

*Table 5. Prevalence of musculoskeletal complaints among school teachers*

| Age group | n | Neck | Shoulders | Low Back | Hand/wrist |
|---|---|---|---|---|---|
| 21 - 25 | 42 | 45.0 | 17.0 | 37.5 | 2.3 |
| 26 - 30 | 71 | 45.0 | <u>31.0</u> | 38.8 | 11.2 |
| 31 - 35 | 29 | <u>52.0</u> | 27.6 | <u>48.3</u> | <u>14.0</u> |
| 36 - 40 | 45 | 31.0 | 30.0 | 33.0 | 8.9 |
| 41 - 45 | 102 | 46.0 | 21.0 | 36.0 | 19.6 |
| >45 | 92 | 47.8 | 26.0 | 33.0 | 27.2 |

Table 4 shows the prevalence of musculoskeletal problems and working experience with VDUs. The results suggest that those who have more than five years experience had higher prevalence than those who had less. It is however, also noted that those newcomers who have had less than one year experience also tend to have a high prevalence.

There was no direct association of musculoskeletal complaints with increase in age (Tables 5 and 6). One important feature noted was that operators less than 40 years old reported more complaints than those who are 41 and above. Prevalences for both

VDU users and non-users appeared to have the same trend. The data in Table 5 also reveal that the highest prevalence was usually reported by the 26 to 35 age group. This holds true for both VDU operators and school teachers.

Table 7 shows the prevalence of musculoskeletal complaints among the three ethnic groups. It was noted that Malays and Indians reported a higher prevalence than the Chinese VDU operators.

Table 6. Prevalence of musculoskeletal complaints among VDU operators (including LPD operators)

| Age group | n | Neck | Shoulders | Low Back | Hand/wrist |
|---|---|---|---|---|---|
| 21 - 25 | 67 | 61.2 | 43.0 | 55.2 | 7.3 |
| 26 - 30 | 210 | 58.0 | 41.7 | 56.1 | 6.6 |
| 31 - 35 | 247 | 66.8 | 44.9 | 55.7 | 6.8 |
| 36 - 40 | 104 | 56.7 | 42.3 | 57.0 | 10.0 |
| 41 - 45 | 46 | 54.3 | 26.0 | 41.0 | 6.0 |
| > 45 | 37 | 40.5 | 37.8 | 27.0 | 16.0 |

Table 7. Prevalence of musculoskeletal complaints by ethnicity

| Ethnic Group | n | Neck | Shoulders | Low Back |
|---|---|---|---|---|
| Chinese | 292 | 149(51.0%) | 127(43.5%) | 129(44.2%) |
| Indian | 130 | 82(63.1%) | 61(46.9%) | 75(57.7%) |
| Malay | 247 | 172(69.6%) | 168(68.0%) | 159(64.4%) |

## Discussion

### Work related factors

Questionnaire surveys on physical discomfort in VDU work usually report a relatively high frequency of musculoskeletal complaints, often considered to be job related. However, such subjective complaints depend on a wide variety of interacting factors. Ergonomics, bio-demographical and psychosocial factors may have strong implications on the findings. Therefore, it is not unusual to find in a given situation that some subjects have more physical complaints, while others report less.

In this study a group of non-VDU users was studied. The results here clearly demonstrated that musculoskeletal complaints reported by the non-VDU users were significantly lower than the VDU operators. The prevalences for neck, shoulder and lower back problems were much more common among VDU operators. Nevertheless, the school teachers appeared to have a higher prevalence of hand and wrist complaints. This suggests that the nature of work activity may contribute towards the differences.

Both task design and workstation design play important roles in the development and prevention of musculoskeletal strain. In an earlier study, we have shown that stiffness and pain in the hand and wrist were relatively common amongst VDU operators (Ong et al., 1981). In the present study, the results also showed that data entry operators also tend to have a higher prevalence when compared to other VDU operators. This was due to the repetitive movements of keying combined with long hours of uninterrupted sitting at the VDU workstations. The results from the present study further showed that the prevalence for musculoskeletal pain was higher among LPD operators when compared to VDU operators, this is probably due to the badly designed workstations used by the operators which are totally unadjustable for fitting operators of different anthropometric dimensions.

## Bio-demographical factors

Several earlier epidemiological studies have suggested that musculoskeletal strain is usually related to ergonomics factors at the workplace (Ong, 1984; Starr et al., 1985). Few studies have taken into consideration bio-demographic or psychosocial factors (Jeyaratnam et al., 1989). Table 7 shows a significant difference in complaints ($p<0.05$) among three ethnic groups working at the same workplace. This suggests that social-cultural background may play an important role in the perception of pain. There are, however, other factors that need to be considered. These include health and nutritional status, family support, family size as well as activities outside working hours.

The results here also suggest that musculoskeletal complaints were more pronounced among females aged 26 to 35. Further analysis indicates that this age group is made up mainly of working mothers. In addition to their occupation they have domestic duties and other maternal activities to attend to. Fatigue from work constraint may be similar for all age groups but domestic chores and child-bearing may aggravate the symptoms (Ong and Phoon, 1987). Prenatal and perinatal stress have an effect on their health, too.

The results on age and musculoskeletal complaints are consistent with the findings of Sauter (1984), i.e. older workers did not appear to have more complaints than their younger counterparts. One possible explanation is the selection or survival phenomenon. Alternatively, the more active lifestyle of younger workers may aggravate the symptoms.

Nevertheless, it is interesting to note that the prevalence of operators using LPDs and VDUs are quite different (Table 2). Those who used LPDs were reported to have a higher prevalence than those who used VDUs. The difference in workplace layouts may have contributed to this observation.

In summary, the introduction of new technology needs to take into consideration some of these social-cultural factors in addition to the ergonomics factors.

## References

Ferguson D. 1984. The new industrial epidemic, Medical Journal of Australia, 140: 318-319.

Jeyaratnam J, Ong C N, Kee W C, Lee J and Koh D. 1989. Musculoskeletal symptoms among VDU operators. In: M J Smith and G Salvendy (Eds). Work with Computers, Elsevier, Amsterdam, pp 334-337.

Kuorinka I, Jonsson B, Kilbom A, Vinterberg H, Biering-Sorensen F, Andersson G and Jorgensen K. 1987. Standardised Nordic questionnaires for the analysis of musculoskeletal symptoms, Applied Ergonomics, 18: 233-237.

Ong C N. 1984. VDT work place design and physical fatigue. In: E Grandjean (Ed). Ergonomics and Health in Modern Offices, Taylor & Francis, London, pp 477-483.

Ong C N, Hoong B T and Phoon W O. 1981. Visual and muscular fatigue in operators using visual display terminals. Journal of Human Ergology, 10: 161-171.

Ong C N and Phoon W O. 1987. Influence of age on performance and health of VDU workers. In: B. Knave and P-G. Wideback (Eds), Work with Display Units 86, Elsevier, Amsterdam, pp 211-214.

Ramazzini B. 1713. De Morbis Artificum - Diseases of Workers. (English edition, Hafner Publishing, New York, 1964).

Sauter S L. 1984. Predictions of strain in VDU users and traditional office workers. In: E Grandjean (Ed). Ergonomics and Health in Modern Offices, Taylor & Francis, London, pp 129-134.

Sharrod F. 1985. 'Kangaroo paw' or what? Medical Journal of Australia, 142: 376.

Starr S J, Shute S J and Thomson C R. 1985. Relating posture to discomfort in VDU use. Journal of Occupational Medicine, 27: 269-271.

# 49. THE EFFECTS OF THE VISUAL CONDITIONS OF VDT VIEWING ON PUPIL SIZE

## Sasitorn Taptagaporn[*] and Susumu Saito[**]

[*]Department of Public Health and Environmental Science
Faculty of Medicine, Tokyo Medical and Dental University
1-5-45, Yushima, Bunkyo-ku, Tokyo 113, Japan

[**]Department of Industrial Physiology
National Institute of Industrial Health, Ministry of Labour
6-21-1, Nagao, Tama-ku, Kawasaki 214, Japan

### Abstract

Based on physiological and psychological indices, positive CRT display polarity (dark characters on bright background) was ascertained from the study to be more appropriate for VDT operators than the negative one. Differences in pupil size and subjective visual comfort while undertaking visual tasks were examined for both display polarities under three different illumination levels. Pupil size was not affected greatly by different illumination levels or luminance contrasts of visual objects while working with a positive display. The majority of the subjects preferred working with the positive display under normal office illumination levels.

**Keywords:** *visual display terminals; CRT displays; display polarity; environment; illumination levels; pupil size; visual fatigue; discomfort*

## Introduction

To ascertain which type of CRT display polarity, positive (dark characters on a bright background) or negative (bright characters on a dark background), is more comfortable for the eyes of VDT operators, some experimental studies on VDTs have been carried out for the last decade. For instance, positive display types have been considered to be better when lens accommodation and pupils were examined (Saito et al., 1989) or when other factors such as visual acuity and illuminance contrast were considered (Bergqvist, 1984). Up to now, there have been very few studies regarding the pupil as one of the most important physiological indices for VDT work. This paper developed on the basis of a study by the authors (Taptagaporn et al., 1990). The reason why consideration of the pupil came into the study is that clear vision will partially be acquired by pupillary contraction in the case of those people who lack the necessary lens accommodation, especially among the aged group (Charness, 1985). Not only is accommodation necessary for visual acuity, but also pupillary contraction which reduces the refractive errors of the lens (Grandjean, 1988). The preliminary study carried out by the authors on visual fatigue from VDT work ascertained only the decrease in accommodation, but no effect on pupil size was found. Furthermore, illumination levels, luminance, and luminance contrast were found to be the most important conditions in the VDT workplace (Gobba et al., 1988). Laubli et al. (1981) also revealed that there was a significant relationship between lighting conditions and the incidence of eye impairments. Hence, the pupil and lighting conditions were taken into account in this study. The objectives of the study were to find the effects of environmental lighting conditions and display polarity on pupil size; and to find the relationship between pupil diameter and subjective visual comfort while undertaking the assigned visual tasks.

## Methods

### Experimental procedures

The VDT workstation was simulated in the study. The ambient illumination was set up for three levels: dark (20 lx), medium (500 lx), and bright (1,200 lx) lighting conditions. There were two visual tasks for each subject. Firstly, the visual task was to count all the numeral '1's among a set of different numerals which were generated and presented on both positive and negative CRT display polarity. Secondly, the subjects had to view the CRT, script, and keyboard consecutively for each duration of 8 sec for seven trials, on both positive and negative display polarity. In order to maintain a distance of 50 cm between their eyes and viewing all objects (CRT, script, and keyboard), the subjects were supported at their chins in front of the CRT display as shown in Figure 1.

*Figure 1. The experimental equipment for the second visual task showing the CRT, the script and the keyboard, the Iriscorder, and the chin-support*

After accomplishing both tasks, the subjects were interviewed with regard to visual comfort while undergoing the tasks on both display polarities under three lighting conditions.

### Subjects

So groups were not different, the subjects were ten students: one female and nine males; with an average age of 24.1 years; range 21-27 years. Typing skills were not required.

### Apparatus and analysis

A personal computer (NEC model PC-9801) was used to create the visual tasks. The CRT display (NEC model N5913) was adjusted to obtain a screen contrast of 10:1 (the luminance of the bright area was 41.2 $cd/m^2$). The subject had to put on an

Iriscorder (Hamamatsu Photonics) connected on-line to a VTR to record the pupil image. Only the right eye was recorded and the pupil diameter was measured by a Percept Scope (Hamamatsu Photonics model C 3160). The pupils of all subjects recorded by VTR were converted into analogue voltage and fed to a Pen Oscillograph. Evaluated pupil diameters were obtained in terms of mean value and standard deviation.

## Results

Figure 2 shows the differences in pupil diameter measured when the subjects were doing the first visual task on each positive and negative display polarity under three lighting conditions. Pupil diameter was smaller when working with a positive display under all three lighting conditions. The variation in pupil diameter among the three lighting conditions was greater when the visual task was on the negative display. The individual differences of ten subjects were also greater with the negative display and became smaller when the environmental illumination level increased. The variation in pupil diameter while undergoing the task, as well as the mean pupil diameter size, was smaller when the illumination level increased.

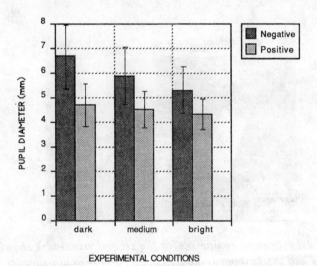

Figure 2. The pupil diameter of ten subjects while undertaking visual tasks on negative and positive CRT display polarities under three lighting conditions

Analysis of variance was carried out to confirm that all factors including the subject, display polarity and lighting condition had influenced pupil diameter ($p.<0.01$). The t-test analysis also showed that there was no significant difference in pupil diameter when working with a positive display polarity under medium and bright lighting conditions. While working with a negative display polarity, however, there were significant differences in pupil diameter among all lighting conditions ($p.<0.01$).

The subjective evaluation obtained by interviewing subjects on how they felt about visual comfort while undertaking the visual tasks under each experimental condition revealed that the condition evaluated as most comfortable was working with a positive display under the medium lighting condition, while the worst was that with a positive display under the bright light condition (Figure 3). The relationship between pupil

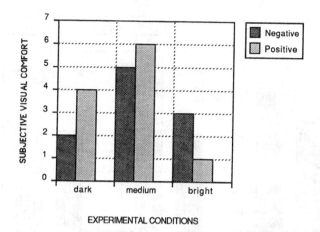

*Figure 3. Subjective visual comfort of ten subjects while undertaking the tasks on negative and positive CRT display polarities under three lighting conditions*

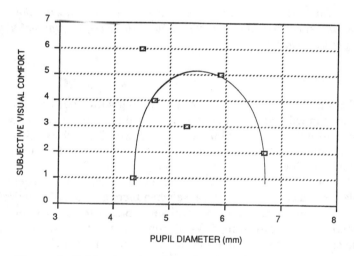

*Figure 4. Subjective visual comfort in relation to pupil diameter*

diameter and the subjective visual comfort was not found in the study by the analysis of simple linear regression ($r = 0.13$). However, the conditions evaluated most comfortable fell in the rank of medium pupil size, while the two least comfortable fell in the largest and smallest pupil size in the study (Figure 4).

The results from the second visual task are shown in Figure 5. The differences in pupil diameter among three viewing targets when working with a positive display were smaller than with a negative one, under medium and bright lighting conditions. The least differences in pupil diameter were obtained from working with the positive display in the medium lighting condition, while the most were obtained working with the positive display in the dark room. The differences in pupil size while viewing CRT, script and keyboard were found to be smaller when the illumination levels increased.

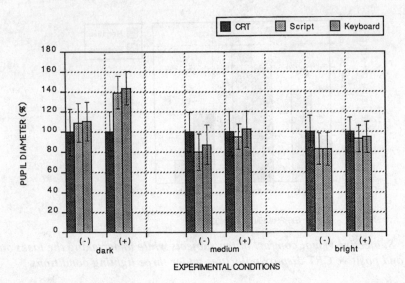

*Figure 5. Pupil diameter of ten subjects in percentages while viewing the CRT display, script, and keyboard on negative (-) and positive (+) display polarities under three lighting conditions*

## Discussion

Since pupil size was found to be smaller when working with a positive display from the study, it was implied that more depth of field (and of focus) was obtained by the VDT operator. This is the advantage for the operators working with a positive display, whose accommodative power decreases when working with VDTs (Grandjean, 1980), especially among the elderly. When considered together with lighting conditions, the positive display polarity caused smaller differences in pupil size. Therefore, the positive display is considered better because when there are unstable environmental lighting conditions as in an actual workplace, small effects on pupil size are considered negligible. Moreover, most VDT workstations utilize sunlight as one of the light sources so it is not possible that the illumination level all over a room is consistent.

The relationship between pupil diameter and the subjective visual comfort was not found in the study by the analysis of simple linear regression. However, the result of the subjective evaluation indicated that the condition most appreciated was working with the positive display in the medium lighting condition (500 lx), which is more or less the same as the usual office illumination level.

When the VDT workstation was simulated in the study (visual task 2), the positive display was again considered better because the differences in pupil diameter among the three viewing objects when working with the positive display were less than those of the negative one. Working with the negative display required more frequent retinal dark and/or light adaptation which is probably a cause of eye fatigue. When pupil size has stabilized or the pupil is in constant motion, it is said to be in a resting position. Therefore, the negative display induced more eye fatigue which added to general fatigue as a consequence (Grandjean, 1988). Moreover, the positive display type has a similar characteristic as a source document (paper), this might help to reduce luminance contrast among visual targets. In cases where working in a dark room is unavoidable, the negative display should be more applicable according to the results of the study. However, the positive display is recommended provided that local illumination is prepared for source documents and keyboards in order to minimize the differences in surface luminances. High luminance contrasts between screen, source document, and the surroundings were found to be associated with an increase in eye troubles (Laubli et al., 1981).

## Conclusion

While there is no consistent evidence for improved performance or comfort with either polarity, the ANSI/HFS standard does note that flicker is less likely with a negative polarity while reflected glare of patterned objects is less noticeable with a positive one (Snyder, 1989). However, the physiological and psychological evidence from this study ascertained the superiority of a positive CRT display to a negative one especially in the usual office illumination levels (approximately 500 lx). It is recommended from the study that all visual objects in a VDT workstation should be adjusted to have the least luminance contrasts, at the appropriate illumination levels, in order to avoid eye fatigue caused by frequent retinal adaptation or unstable pupil size.

## Acknowledgments

The authors are very grateful to Prof. T. Takano of the Tokyo Medical and Dental University and Dr. S. Koshi of the National Institute of Industrial Health for their kind advice and encouragement throughout the course of the study. We also thank the subjects involved.

## References

Bergqvist, U. 1984, Video display terminals and health, Scandinavian Journal of Work, Environment and Health, 10, suppl.2, 31-33.

Charness, N. 1985, Aging and Human Performance, (John Wiley & Son), 3-9.

Gobba, F. M., Broglia, A., Sarti, R., Luberto, F., and Cavalleri, A. 1988, Visual fatigue in video display terminal operators: objective measure and relation to environmental conditions, International Archives of Occupational and Environmental Health, 60, 81-87.

Grandjean, E. 1980, Ergonomic Aspects of Visual Display Terminals, (Taylor & Francis, London), 1-12.

Grandjean, E. 1988, Fitting the Task to the Man, (Taylor & Francis, London), 231-239.

Laubli, Th., Hunting., W. and Grandjean, E. 1981, Postural and visual loads at VDT workplaces II. Lighting conditions and visual impairments, Ergonomics, 24, 933-944.

Saito, S., Ishikawa, K. and Hatada, T. 1989, Physiological evidences of superiority of positive type CRT among information displays, Advances in Human Factors/Ergonomics (Elsevier), 12A, 536-541.

Snyder, L. H. 1989, The ANSI/HFS standard for visual display terminals, Information Display, 5, 20-23.

Taptagaporn, S. and Saito, S. 1990, How display polarity and lighting conditions affect the pupil size of VDT operators, Ergonomics, 33, 201-208.

# 50. THE EFFECTS OF VDT POLARITY AND TARGET SIZE ON PUPIL AREA

### Masaru Miyao and Sin'ya Ishihara

*Department of Public Health*
*Nagoya University School of Medicine*
*Nagoya 466, Japan*

### Abstract

Negative and positive CRT display conditions were compared. Twelve healthy female volunteers were subjects. A white CRT in negative and positive polarity was used to view the target. Two targets, 10 mm and 31.6 mm in diameter, were used. The pupil area was measured for every 10-s load using infrared videopupillography. The results showed that the pupil area with the negative displays was significantly larger than with the positive displays. With negative polarity, the larger the target, the more the pupil constricted. When the CRT was used in positive polarity, the pupil dilated in proportion to the target size. The results indicated that screen polarity had a marked influence on pupil area and that a difference in pupil area resulted from a difference in the target size.

**Keywords:** *visual display terminals; CRT displays; pupil size; display polarity*

## Introduction

Eye strain, neck and upper limb syndrome, along with mental stress, constitute the most frequently voiced complaints of VDT operators (Bennet et al., 1984; Grandjean, 1984; Scalet, 1987). It is clear that eye strain is affected by the need for constant eye focus and by the features of the characters or figures displayed on a CRT display. Thus, the size, colour, flicker, resolution and other factors involved with the characters have been studied from many different standpoints (ANSI, 1988; Miyao et al., 1989). CRT polarity has also been much discussed. The much used negative displays, for example, have a low mean luminance which results in a large luminance contrast between the bright characters and the dark display background. The resulting greater visual load has been pointed out in the literature (Bauer and Cavonius, 1980).

There has yet to be any published study on the effects on pupil size area of factors such as polarity or target size. In the present study, we compared positive and negative CRT display conditions, varying the size of the target in the centre of the screen, for any effect on the pupil area.

## Materials and method

The subjects were 12 healthy female volunteers aged 20-21 years, with overall correcting visual acuity of more than 20/20 and no past history of ophthalmological diseases. The laboratory was windowless with flourescent lighting arranged to provide constant illuminance of 170 lx on the horizontal. The testing period was from 11:00 am to 4:00 pm. An NEC Model PC8853n colour display (14") was used as the CRT. Subjects used a chin rest secured at a point from which their eyes would be 50 cm away from the display. The eyes of subjects were exposed to infrared rays originating from an angle poise lamp positioned at an angle in front and passing through an infrared filter. The subjects' right eyes were photographed from the front with an infrared camera.

For the test loading, for both positive and negative CRT displays, subjects gazed at the target, which was a circle presented for 10 s each in the centre of the display with two different sizes. Target sizes of 10 mm and 31.6 mm in diameter were used. In the case of negative polarity, a white target was made to appear suddenly on the dark background. A dark target was presented on the white background with positive polarity.

As shown in Figure 1, the pupil area of the right eye of each subject was measured for every 10-second load using infrared videopupillography (Hamamatsu Photonics Inc). The right eye was shown on a monitor TV set, with the pupil displayed in white; the high contrast conditions made it stand out against all other portions, which were darkened. The entire pupil was observed thus by the monitor display, and the size (pupil area) could be measured by the number of active phosphor elements. The measured area was sampled 40 times per second and inputed by a personal computer as a digital signal and recorded on a floppy disk. To calibrate pupil area, a black circle with a diameter of 4.62 mm was measured, and this was used to calculate the actual pupil area. In order to evaluate the change in pupil, average values of pupil areas were calculated for five seconds from 5 to 10 sec.

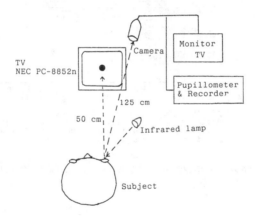

Figure 1. Experimental layout.

## Results

Figure 2 gives the mean plots of the data for 10 seconds in the 12 subjects for the respective conditions. Average values of initial pupil areas were approximately 32 $mm^2$ (6.4 mm in diameter) for the negative polarity display, and approximately 10 $mm^2$ (3.6 mm in diameter) for the positive polarity display. The pupil areas at the start points of the respective loads show rapid changes because the target was displayed suddenly. The pupil areas for the negative display dilated obviously because of a low mean luminance. In the case of the positive polarity, pupil areas showed less changes or weak constrictions.

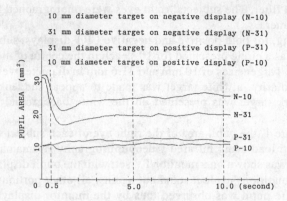

Figure 2. Changes in mean values of pupil area in 12 subjects obtained under different conditions

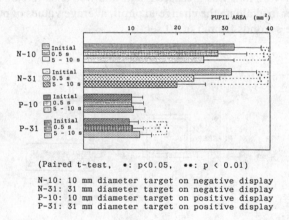

Figure 3. Changes in mean values of pupil area in 12 subjects under various conditions (mean +/-1 S.D.)

In Figure 3, a statistical test was carried out using the Wilcoxon matched pair signed rank test. On the positive screen with the 31.6 mm in diameter target and the negative display, the pupil size was significantly dilated. But a 10 mm diameter target on the positive display hardly changed the pupil area.

Analysis of variance for absolute differences between initial pupil size and backward values is calculated in Tables 1 and 2. The 0.5 s values and the mean values from 5 s to 10 s are used with the initial values in Tables 1 and 2, respectively.

The pupil area in the negative displays was significantly larger than in the positive displays. With negative polarity, the larger the target the more remarkably the pupil constricted. The largest pupil was observed with the 10 mm diameter target on the dark background, while the smallest pupil appeared when the 10 mm diameter target was displayed against the white background.

Table 1. *Analysis of variance for differences between initial and 0.5 s values*

| Sources of variation | SS | DF | MS | F | Significance |
|---|---|---|---|---|---|
| Polarity | 337.13 | 1 | 337.13 | 95.07 | 0.0001 |
| Target size | 74.83 | 1 | 74.83 | 21.10 | 0.0001 |
| Interaction | 53.28 | 1 | 53.28 | 15.02 | 0.001 |
| Residual | 156.04 | 44 | 3.55 | | |
| Total | 621.27 | 47 | 13.22 | | |

Table 2. *Analysis of variance for differences between initial values and mean values from 5 s to 10 s*

| Sources of variation | SS | DF | MS | F | Significance |
|---|---|---|---|---|---|
| Polarity | 657.79 | 1 | 657.79 | 48.02 | 0.0001 |
| Target size | 116.28 | 1 | 116.28 | 8.49 | 0.01 |
| Residual* | 616.37 | 45 | 13.70 | | |
| Total | 1390.44 | 47 | 29.58 | | |

\* Interaction is pooled in residual.

## Discussion

The results indicate that the factor with the greatest effects on pupil size proved to be the target luminance. The relationship between this target luminance and pupil diameter was analyzed by Reeves (1918) and Crawford (1937). According to Reeves, when light intensity changed at around $10^9$, the pupil diameter was found to change from 8 mm to 2 mm. Crawford reported that when target luminance reached changes of approximately $10^6$, the pupil diameter changed from about 5.7 mm to about 2 mm. Campbell (1960), using a synthetic pupil, varied the target luminance in order to obtain the pupil diameter allowing the maximum visual acuity. His results virtually agreed with the natural pupil diameter measurements of Reeves (1918) and Crawford (1937).

In a pupil reflex there is also a near reflex in addition to the aforementioned light reflex. When one looks at a target near at hand, the pupil contracts together with the convergence movement of both eyes and the ensuing accommodation. In the present experiment, the visual distance was constant at 50 cm. In this kind of load distance, the convergence reflex is not taken to be definite. There is a diurnal variation in pupil area. Our present experiment was run between the hours of 11:00 am to 4:00 pm, so this time zone would be fairly stable.

There are few studies on the relationship between VDT and pulpillary area. Zwahlen (1984) had subjects focus 96 consecutive times on 4 different locations (screen, keyboard, document, and wall) of a VDT workstation, in order to measure successive changes in pupil diameter. As a result, in young persons (aged 19-22), the pupil diameter became slightly larger, depending on the target luminance, whereas in older persons (61-66 years of age) the pupil diameter obviously grew smaller, with fewer changes on record.

The results of our study indicate that screen polarity had a marked influence on pupil area. A difference in pupil area results from a difference in target size. With a positive display, the pupil invariably contracted. The white area (background) luminance was 71.8 cd/m$^2$ against a dark area (target in the centre of the display) luminance of 1.2 cd/m$^2$, so it was not surprising that a light reflex occurs when focusing on a positive VDT display. But it is important that, irrespective of target luminance, a difference in pupil area results from a total luminance of the display.

Moses (1975) performed a geometric estimate of depth of field. According to him, when actual accommodation is at 1 m (1 D), if the pupil diameter is 2 mm the target should be distinct even at 0.94-1.06 m; on the other hand, when pupil diameter is 4 mm, unless the target is at 0.97-1.03 m, it may not be seen clearly. In other words, as the pupil expands the depth of field becomes more shallow, and accommodation must in fact be more rigorous. In the light of the findings of Moses, where pupil area increases in terms of a negative display, the depth of focus obviously becomes more shallow. Hence, unless accommodation is exact, reading would presumably become impossible. Conversely, when pupil area contracts with a positive display, the depth of focus becomes deeper, making the load on accommodation presumably that much less. This should be tested empirically.

## Conclusions

We studied the effects of VDT polarity and target size on pupil size. The results showed that the pupil area in the negative displays was significantly larger than in the positive displays. With negative polarity, the larger target caused the pupil to constrict more remarkably. A difference in pupil area results from a difference in polarity and target size.

## Acknowledgments

The authors wish to thank Mr. Hirokazu Iguchi, Toyota Central Research and Development Laboratories, Inc., for his technical assistance. Thanks are also due to Professor Shin'ya Yamada, Nagoya University School of Medicine, for his valuable remarks.

## References

American National Standards Institute (ANSI) 1988, ANS for Human Factors Engineering of VDT Workstations, (The Human Factors Society, Santa Monica, CA).

Bauer, D., and Cavonius, C. R. 1980, Improving the legibility of visual display units through contrast reversal. In Ergonomic aspects of visual display terminals, (Edited by E. Grandjean, and E. Vigliani)(Taylor & Francis, London), 137-142.

Bennet, J., et al. (eds) 1984, Visual Display Terminals (Prentice-Hall, Englewood Cliffs NJ).

Campbell, F. W., and Gregory, A. H. 1960, Effects of size of pupil on visual acuity, Nature, 187 (No.4743): 1121-1123.

Crawford, B. H. 1937, The dependence of pupil size upon external light stimulus under static and variable conditions. Proceedings of the Royal Society, 121 B: 376-395.

Gould, J. D., et al. 1987, Reading from CRT displays can be as fast as reading from paper. Human Factors, 29: 497-517.

Grandjean, E. (ed) 1984, Ergonomics and Health in Modern Offices (Taylor & Francis, London).

Miyao, M., Hacizalihzade, S. S., Allen, J. S., and Stark, L. W. 1989, Effects of VDT resolution on visual fatigue and readability: an eye movement approach. Ergonomics, 32: 603-614.

Moses, R. A. 1975, Accommodation. In: Adler's Physiology of the eye, (Edited by Moses, R. A.) (The C.V. Mosby Company, Saint Louis), 298-319.

Qioyuan, J., et al. 1985, The pupil and stimulus probability, Psychophysiology, 22: 530-534.

Reeves, P. 1918, Psychological Review, 25: 399.

Scalet, E. 1987, VDT Health and Safety (Ergosyst Assoc., Lawrence, KA).

Zwahlen, H. T. 1984, Pupillary responses when viewing designated locations in a VDT workstation. In: Ergonomics and Health in Modern Offices, (Edited by E. Grandjean)(Taylor & Francis, London), 339-345.

# 51. THE SIGNIFICANCE OF CHANGES IN CFF VALUES DURING PERFORMANCE ON A VDT-BASED VISUAL TASK

### Tsuneto Iwasaki and Shinobu Akiya

*Department of Ophthalmology*
*University of Occupational and Environmental Health*
*1-1 Iseigaoka, Yahatanishi-ku*
*Kitakyushu, 807, Japan*

**Abstract**

Changes in critical fusion frequency (CFF) values in both eyes were recorded at intervals of 10 minutes during 30 minutes sustained performance on a VDU-based visual task. A simple mathematical addition task was presented to one eye (the loaded eye) while the other eye acted as the control. The relative decrease in CFF values for the loaded eye over the values for the control eye were proportional to the time required to perform the visual task. These results suggested that the decrease in the CFF value did not reflect a decline in the activity of central visual functions, but in the activity of the retina or the optic nerve.

**Keywords:** *visual display terminals; mental workload; visual fatigue; critical fusion frequency*

## Introduction

The significance of the decrease in critical fusion frequency (CFF) values to physiological factors is not yet clearly understood. In the field of fatigue research, it is said that the decrease in the CFF value is correlated with the onset of mental fatigue and the deterioration in arousal and consciousness levels of the brain centre (Simonson and Enzer, 1941; Baschera and Grandjean, 1979; Weber et al., 1980). In the field of ocular physiology, the CFF test is used to examine diseases of the retina and the visual pathway. This test is claimed to be very helpful in recording the change in the signal transmission function among the visual sub-systems (Hecht and Shlaer, 1936; Titcombe and Willson, 1961; Salmi, 1985). Thus the decrease in CFF value cannot only be physiologically taken to mean a decrease in the arousal and activity levels of the brain centre or the onset of mental fatigue, but can also be considered as reflecting retinal adaptation and impulse conduction in the entire visual system including the optic nerve and optic tract.

The spread of visual information processing tasks, including visual display terminal (VDT) based tasks, has increased the reported frequency of visual fatigue and eyestrain and increased the variety of factors responsible for visual fatigue and eyestrain. Measurement of CFF values is frequently employed to clarify complicated loads imposed on humans, including the problem of fatigue in particular, because of its ease of operation and its sensitivity. Clarification of the physiological meaning of the CFF value is an important issue when the effect of the working environment, which is likely to increase in complexity, on the visual function is investigated by means of flicker tests.

In this context, the change over time in CFF values of subjects who were loaded with a visual task was measured and the effect of the work load on the visual function of the subjects was experimentally studied together with clarification of the physiological phenomenon of flicker fusion.

## Method

### Subjects

Fifteen healthy (mean age of 19.5 years), emmetropic females without any ophthalmologic diseases, were subjects in the experimental work.

### Visual loading method

The visual task chosen for this study involved subjects judging whether a series of visually presented addition equations were correct or not (a modification of the Kraepelin test that involves mental work). Each equation was displayed in white characters of 30 cd m$^{-2}$ luminance on a CRT screen for a duration of 2 sec. The subjects were instructed to press the "1" key when the equation was correct (e.g., 1+2 = 3) and the "2" key when the equation was incorrect (e.g., 1+2 = 4). The subjects performed this judgement task for 30 min. The illuminance of the CRT screen was set at 50 lux. The task for 30 min was presented to only one eye, although both eyes were open.

A blackboard was located in front of the CRT display screen placed 50 cm from the subjects' eyes. There were two holes of 3 cm diameter in this board to fix the interpupil distance. The image of the addition equation (visual angle of about 1.7 degree) was given to one eye (the loaded eye) and to the other eye (the control eye) the background image of the CRT display was projected. The two images were fused by two prisms (see Figures 1 and 2). For the test period, the correct answer ratio was calculated and the work efficiency of subjects measured.

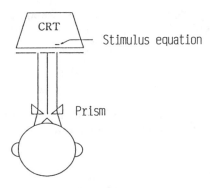

*Figure 1. The experimental layout*

### Measurement of CFF values

The CFF values of the subjects were measured with a flicker photometer using light-emitting diodes (Handy Flicker HF, Neitz). The colour of the light-emitting diodes was red (660 nm). The target, 8.7 mm in diameter, was presented to the subject and was turned on and off by a rectangular wave generator with a pulse duty ratio of

50% (visual angle of about 1.5 degrees at a visual distance of 33 cm). The subjects were asked to fixate on this target with one eye and to report the time when the target ceased to flicker by the ascending method. Three measurements were averaged to obtain the CFF value of the subject. The red CFF values of the subjects were measured before the start of the visual task and at 10, 20 and 30 min after the start of the task. Also in accordance with the method described by Ohshima (1959), the CFF dependence was calculated against the value obtained before the task.

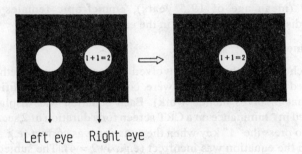

Figure 2. The displayed image

### Graded evaluation of subjective symptoms of eyestrain

Subjective measures of eyestrain were obtained by using a self-rating method: "1" represents the most favourable condition where subjects feel no eyestrain while "9" represents the worst condition. The subjective symptoms were scored before and after the task on two occasions; one when the visual task was presented to the right eye and the other when it was presented to the left eye.

### Statistical analysis

The differences in the CFF values at the corresponding points of measurement between the control eye and the loaded eye were tested with the t-test for significance.

## Results

### Work efficiency

The correct answer ratios at 10, 20 and 30 min after the start of the visual task were 97.8%, 96.9%, and 96.7%. The ratios tended to decrease but these differences were not significant.

### Changes in CFF values

The changes in CFF values of the loaded eye and the control eye are shown in Figure 3. At 0 min, the mean CFF values were not significantly different between the control eye and the loaded eye, but a significant difference was found between the eyes after 10 min. The decline in the CFF value for the loaded eye continued to the end of the visual task, but for the control eye, CFF values exhibited no decrease over the 30 min. In proportion to the duration of the visual task, the decrease in CFF values in the loaded eye were higher than those in the control eye, and there were significant differences between the CFF values at the respective points of measurement (p.< 0.01).

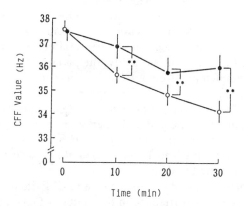

*Figure 3. Changes in CFF values for the control and loaded eyes, mean +/- 1 S.E. where ● - ● = control, O - O = loaded and \*\* = significant difference between the two eyes, p.< 0.01*

Table 1. Changes in CFF dependence in the two eyes

|  | 10 min | 20 min | 30 min |
|---|---|---|---|
| Control eye | 2.0% | 4.8% | 4.3% |
| Loaded eye | 5.0% | 7.5% | 9.3% |

The CFF dependence showed a similar pattern of changes in CFF value (Table 1). The dependent rate in the loaded eye was 5.0% at 10 min after the loading, 7.5% at 20 min, and 9.3% at 30 min. These values exceeded the critical point of 5% which corresponds to the limit set for performing mental work. For the control eye, the CFF dependence did not exceed 5% at the respective points of measurement.

**Complaints of eyestrain**

The mean score for eyestrain was 4.58 before the task commenced and 30 min after initiation of the task it was 6.20.

## Discussion

From these results, the decrease in CFF values and the increase in subjective symptoms, the experimental visual task displayed on the CRT display screen obviously imposed a load upon the vision of the subjects. The pattern of change differed with each eye, that is, the decrease in CFF value was larger in the loaded eye than in the control eye in the same visual environment. These results raise questions about the conventional physiological concept such as a drop in CFF values implying a deterioration of brain centre activity.

Many experimental studies have commented on the physiological meaning of CFF values. Walker et al. (1943) recorded the response pulses of the optic nerve, lateral geniculate body and cerebral cortex of monkeys to repetitive photic stimuli. The researchers found that the response stopped at photic stimuli of 62, 59 and 34 Hz for the optic nerve, lateral geniculate body and cerebral cortex, respectively, and concluded that flicker fusion mainly occurs in the cerebral cortex of monkeys. However, if flicker fusion mainly occurs in the cerebral cortex, decrease in CFF in both eyes should be the same. But, in our study, the pattern of CFF values recorded in the control eye was completely different from that recorded in the loaded eye CFF. This can be understood more readily as the appearance of a uniform drop in the physiological function of a given part of the visual system rather than as a difference in the effect on functionally different parts. The part with the most obvious functional differentiation in connection with the pathway of vision is the prechiasmal visual pathway such as the optic nerve and retina, rather than the brain centre, that is according to the results reported to date.

Iwasaki et al. (1989) have reported that the change in CFF did not reflect a scale of cerebral function because of the difference in the pattern of change in the CFF with the type of colour used. Osaka (1985) loaded subjects with a calculating task on a CRT screen for 30 min and measured CFF values with respect to various colours before and after the visual task. He reported that the reduction in the green and yellow CFF values was smaller than that in the red and blue CFF values and that eyestrain was lower with the green and yellow light sources than with the red and blue light sources. He discussed these differences in relation to the sensitivity peaks of the respective colours on the sensitivity curve for the retina. Kelly (1978) reported that CFF thresholds mainly depended on the action of visual and bipolar cells in the retina.

In the present study, although CFF dependence in the loaded eye exceeded the critical 5% limit (Ohshima, 1959), the dependence in the control eye was below 5% (Table 1). This means that performance on a mental task, such as one involving arithmetic calculations, will be more likely to be correct when it is presented to one eye than to another. This contradicts the results that would be predicted if decreases in CFF reflect deterioration of cerebral function.

From the results of the present experimental work it cannot be conclusively said that the decrease in CFF values reflects only the activity or consciousness level of the cerebrum. The results of this study suggest that CFF values obtained with a short task duration of 30 min constitute an index that indicates not only mental fatigue or the activity and consciousness levels of the cerebrum but also the deterioration of other physiological functions.

## References

Baschera, P. and Grandjean, E. 1979, Effects of repetitive tasks with different degrees of difficulty on critical fusion frequency (CFF) and subjective state, Ergonomics, 22, 377-385.

Hecht, S. and Shlaer, S. 1936, Intermittent stimulation by light v. the relationship between intensity and critical frequency for different parts of the spectrum, Journal of General Physiology, 19, 965-979.

Iwasaki, T. et al. 1989, The changes in colour critical flicker fusion (CFF) values and accommodation times during experimental repetitive tasks with CRT display screens, Ergonomics, 32, 293-305.

Kelly, D. H. 1978, Human flicker sensitivity to stages of retinal diffusion, Science, 202, 896-899.

Ohshima, M. 1959, The method of determination in the results of flicker test (1), Rodo Kagaku (Labor Science), 35, 423-426.

Osaka, N. 1985, Effects of VDT colour eccentricity, and adaptation upon visual fatigue, Japanese Journal of Ergonomics, 21, 89-95. (In Japanese)

Salmi, T. 1985, Critical flicker frequencies in MS patients with normal or abnormal pattern VEP, Acta Neurologica Scandinavica, 71, 354-358.

Simonson, E. and Enzer, N. 1941, Measurement of fusion frequency of flicker as a test for fatigue of the central nervous system, Journal of International Hygiene and Toxicology, 23, 83-89.

Titcombe, A. F. and Willson, R. G. 1961, Flicker fusion in multiple sclerosis, Journal of Neurology, Neurosurgery and Psychiatry, 24, 260-265.

Walker, A. E. et al. 1943, Mechanism of temporal fusion effect of photic stimulation on electrical activity of visual structures, Journal of Neurophysiology, 6, 213-219.

Weber, A. et al. 1980, Psychophysiological effects of repetitive tasks, Ergonomics, 23, 1033-1046.

# 52. THE COMPLEXITY OF VDT WORK CONTENT AND ITS RELATION TO VISUAL AND POSTURAL LOAD: WHAT AN ORGANISATIONAL ANALYSIS CAN TELL

Gunnela Westlander and E. Aberg

*The Division of Social and Occupational Psychology*
*National Board of Occupational Safety and Health*
*S-17184 Solna, Sweden*

**Keywords:** *visual display terminals; job characteristics; job variety; offices*

## Introduction

In this article an analytic method is proposed to answer the following question: How common is purely specialized VDT work relative to more varied computer work? The analysis is not conducted, as is common in the epidemiological tradition, in order to compare occupational groups specialized in data entry, information retrieval, interactive communication, word processing (Dainoff, 1981; Evans, 1985; Johansson & Aronsson, 1984; Knave et al., 1985; Smith et al., 1981; Smith, 1984; Westlander & Magnusson, 1988). Instead it is conducted at an organizational level with the aim of offering an approach that can provide a starting point for local changes. The method involves:
- characterizing VDT work as a whole with respect to its "content profile" within an organizationally-delimited unit;
- characterizing the content of VDT work by employee with respect to its degree of specialization (uniformity - variety);
- establishing to which VDT tasks specialization may apply;
- identifying any risk groups, i.e. those containing individuals who have extensive amounts of specialized VDT work (and who, it might be feared, suffer from physical wear and tear as a result).

## Measuring the content of VDT work

The proposed analytic model has been applied to data collected in 1987 and 1988. The study covered a number of workplaces (two social insurance offices, two street departments, one educational authority, one public child care department, one supplies department and one private food distributing company) where a variety of types of administrative work was undertaken; for this reason, the alternatives contained in the question 'Which type(s) of job do you have at the computer terminal?' were word and text processing, data entry, data acquisition, interactive communication.

**VDT work content profiles (organizational level)**

A starting point was to take the number of people who reported that they worked with the specific VDT tasks (word/text processing, data entry, data acquisition and interactive communication) either to a large or moderate (some) extent. This provided the basis for a quantitative assessment of how large a proportion of VDT workers was concerned with each category of VDT work, and also permitted a "content profile" for VDT work as a whole to be obtained for the department in question.

### Degree of specialization (variety-uniformity) at VDT work (individual and organizational levels)

As the study was conducted on the assumption that specialization in a particular type of VDT work would not necessarily apply to each individual, an index was constructed that was based on the number of types of tasks that the VDT user stated he or she undertook either to a large or moderate extent. In this way, it was possible to obtain a picture both of the variation in VDT work for the individual and the distribution in terms of uniformity-variety for the personnel as a whole. How many employees were allocated just one type and how many several types of jobs? In other words, was there a tendency for VDT work to be specialized, or did it tend to be composed of several types of tasks?

### Types of VDT tasks to which specialization may apply

Each specific question on VDT tasks is related (by cross-tabulation) to the index "Degree of specialization at VDT work".

### The identification of risk (people with both specialized and large amounts of VDT work)

Scores on the degree of specialization index can be cross-tabulated with the amount of VDT work as measured by time spent (Westlander, 1990). The number of people with both a high score on the specialization index and a large amount of VDT work (time spent) represents an indicator of the incidence of risk. The type of VDT task to which the risk applies can then be specified.

## Results

### Content profiles

Of the eight workplaces studied, five had similar content profiles: a predominance of data entry and data acquisition, and very little interactive communication and word or text processing. The two street department offices showed a more even distribution, while work at the private food distributing company was dominated by data entry.

### Degree of specialization (variety-uniformity) at VDT work

Four workplaces were clearly dominated by two different types of tasks per employee, while VDT work consisting of more than two risk types was more uncommon (Table 1). Work at the two supplies departments was predominantly specialized; more than half of the VDT workers were occupied with just one type of task. The index, therefore, provides a guide to the extent to which terminal-based work has become specialized or, alternatively, the extent to which there is a tendency for it to be more multifaceted. The percentage share of VDT workers with only one type of VDT task varies strongly from workplace to workplace. The two supplies departments have the highest degree of specialization while work in the traffic division of the street department is most varied; all VDT workers at this workplace have more than one VDT task.

Table 1. *Degree of specialization in VDT work: Number of VDT users with one or several types of VDT work (number of users by workplace with just one type of VDT work expressed as a percentage)*

| Workplace | Number of VDT users doing one or more types of VDT job to a *large* or *moderate* extent. | | | | | |
|---|---|---|---|---|---|---|
| | one type | | two types | three types | four types | |
| | n | % | n | n | n | |
| Public child care div. n=21 | 4 | 19 | 8 | 4 | | |
| Street dep't (acc'ts.) n=9 | 5 | 56 | 1 | 3 | | |
| Street dep't (traffic) n=3 | 0 | | 1 | 2 | | |
| Supplies dep't n=20 | 13 | 85 | 2 | 2 | | |
| Educ. authority n=30 | 8 | 30 | 14 | 4 | 1 | |
| Soc. ins. office I. n=51 | 13 | 25 | 36 | | | |
| Soc. ins. office II. n=40 | 6 | 17 | 25 | 7 | | |
| Food dist. company n=21 | 8 | 43 | 8 | 1 | 1 | |

## Types of VDT tasks to which specialization applies

The types of VDT tasks to which specialization applies are described in Table 2. From a total of 195 VDT users, 57 were engaged in "specialized" work. While in itself this is a large number, the proportion such people represented of VDT workers varies from workplace to workplace, as too does the pattern of tasks to which specialized work applied (the content profile).

## The identification of risk

It may be feared that those with specialized VDT work, i.e. just one type and a large amount of computer work, may find themselves in a position of risk from a work-related point of view. The study contains 28 people in this position, i.e. half of those VDT workers who have just one type of VDT task. Table 2 provides a breakdown of all 57 specialized workers by workplace and task type; figures for the 28 daily users are given in brackets.

It also emerges that only a few people are affected at each workplace, but - it should be noted - at all the workplaces studied. Are these people at risk?

Merging the data from all the workplaces permits the creation of two user categories of sufficient size to make a statistical test: the 12 data entry workers who use their computer terminal daily and the 13 working with data acquisition who also use the terminal daily. These two categories can be compared in order to investigate the relationship between the content of specialized VDT work and job strain. The measures of strain used in the study are self-reported vision and postural complaints. The results confirm the tendency detected in earlier studies based on larger quantities of material. Data entry workers tend to report more postural complaints (from neck, shoulders, shoulder blades, base of the spine, arms) than operators with data acquisition tasks ($\chi^2$ test p.<0.01 - p.<0.03). However, no difference between the groups in terms of vision difficulties emerges.

Table 2. Types of VDT tasks to which specialization applies (number of daily users in parentheses)

| | Workplace and number of VDT users with one type of task (to a large or moderate extent) occupied solely with: | | | |
|---|---|---|---|---|
| | data entry n | data acquisition n | interactive comm. n | word/text processing n |
| Public child care div. n=21 | 1 (1) | 3 (1) | | |
| Street dep't (acc'ts.) n=9 | | 2 (1) | | 3 (1) |
| Street dep't (trafic). n=3 | | | | |
| Supplies dep't n=20 | 5 (4) | 6 (2) | 1 (0) | 1 (1) |
| Educ. authority n=30 | 2 (1) | 6 (0) | | |
| Soc. ins. office I. n=51 | 3 (2) | 10 (6) | | |
| Soc. ins. office II. n=40 | 2 (2) | 3 (3) | 1 (0) | |
| Food dist. company n=21 | 6 (2) | | | 2 (1) |

It should be added that the degree of specialization (according to the specialization index) has no clear relation in itself to self-reported vision and postural complaints, not even if we restrict the analysis to the daily user category alone. This suggests that a combination of uniformity and type of task is a necessary starting point for the identification of the strains arising from VDT work.

The analyses presented here can be used to describe each individual organizational unit and thereby provide a characterization of the VDT work undertaken. This involves the simultaneous consideration of:

- the proportion of employees working with VDTs;
- how many of these have varied/uniform work tasks - the degree of specialization (and, from among specialized VDT workers, those for whom the task is undertaken daily for more than four hours a day);
- the particular tasks to which specialization applies.

This will provide a basis for discussion on changes to the work organization that may be desirable.

A comparison of the workplaces under study shows that the proportion of employees working at VDT terminals with uniform computer tasks varies from workplace to workplace. It is high in the supplies department (85%) and relatively low in the social insurance offices (25 and 17%). At these three places there are 7, 8 and 5 people respectively with working conditions characterized by the risk of ergonomic strain (4 hours a day or more, with just one type of computer task). All the workplaces except one (the street office's traffic department) have one or more employees in a situation of risk as it is defined here. The number of people concerned, however, is not such that more than individual interest and support can be obtained with respect to improvements in their situation at work. That the job tasks concerned vary from workplace to workplace also emerges from the results. The identification of these tasks should facilitate the implementation of adequate measures for the prevention of ill-health.

## Discussion

In the present analysis we have defined and characterized the organization of VDT work by considering its content and relating this to the amount of work (time spent). The purpose of the study has been to investigate what information on the situation of employees at a number of office units can be obtained in this way.

The proposal contained in this paper does not, of course, provide for exhaustive analysis. Its practical advantage is that the measurement/survey base covers relatively few aspects of work, with the result that data collection is not particularly time-consuming. The point with a study of this sort is to collect and analyze information to provide the best possible characterization of a workplace or workplaces; this can then provide a starting point for continued investigation and prevention (if necessary) at a local level. The questions posed which are factual by nature permit the key issues to be addressed directly.

Above all, the analysis provides a description of the situation that prevails at a particular point in time. It does not guarantee a lasting characterization over time as conditions at the workplace may change, sometimes quickly, as could be seen from a follow up study carried out approximately 18 months later. Two of the eight workplaces, the street office (accounts and supplies) and one of the social insurance offices were included in this follow up. New data using the same survey instrument (somewhat modified in certain respects) were collected. There had been changes in personnel at both workplaces; at the social insurance office there had been a very high rate of personnel turnover (approximately 50%) and a certain reduction in staffing levels. The possible existence of a trend towards more varied work could, therefore, only be investigated at a staff rather than individual level (see the introduction to this paper). The number of VDT users with only one type of computer task had fallen at both workplaces while, at one of them, the number of employees using computer terminals had risen.

The taxonomy that has been employed (data entry, data acquisition, interactive communication, word and text processing) provides an opportunity for carrying out a quick survey of the type of work undertaken at a particular office unit, and thereafter for assessing the degree of specialization, and the types of work to which specialization applies. As some types of ergonomics problems have shown a tendency to be related to different types of VDT tasks - something we have learned from the findings of epidemiological studies on VDT work and health - the organizational level-assessment described in this article provides an indication of the strains that may arise in offices.

## References

Dainoff, M. J. & Happ, A. (1981). Visual fatigue and occupational stress in VDT operators. Human Factors, 23(4), 421-438.

Evans, J. (1985). VDU operators display health problems. Health & Safety at Work, November, 33-37.

Johansson, G. & Aronsson, G. (1984). Stress reactions in computerized administrative work. Journal of Occupational Behaviour, 5, 159-181.

Knave, B., et al. (1985). Work at video display terminals. An epidemiological health investigation of office employees. I. Subjective symptoms and discomfort. Scandinavian Journal of Work, Environment and Health, 11, 457-466.

Smith, M. J., Cohen, B. G. F. & Stammerjohn, L. W. (1981). An investigation of health complaints and job stress in video display operations. Human Factors, 23(4), 387-400.

Smith, M. (1984). Health effects of VDTs: Ergonomic aspects of health problems in VDT operators. In Office Hazards: Awareness and Control, North West Center for Occupational Health and Safety, University of Washington, Seattle.

Westlander, G. & Magnusson, B. (1988). Swedish women and new technology. In G. Westlander & J. Stellman (Eds). Government Policy and Women's Health Care. The Swedish Alternative. Women & Health, vol 13 no 3/4. New York: The Haworth Press.

Westlander, G. (1990). Use and non-use of VDTs - organization of terminal work. Research findings from Swedish cross-site studies in the field of office automation. International Journal of Human-Computer Interaction, 2(2), 137-151.

# 53. ERGONOMICS AT SWIFTT

### F. W. Darby

*Department of Labour*
*New Zealand*

### Abstract

A government department used ergonomics methods in the evaluation and selection of VDU terminals and workstations, to be used nationwide by up to 4,000 staff. Tenderers' specifications were compared with a standard drawn up by management and unions for both terminals and workstations. Difficulty was experienced finding equipment which complied with all the requirements of the standard. Eventually, seven terminals which 'passed' the standard were identified, and a user trial was conducted to select the best one from the ergonomics viewpoint. The trial involved 58 subjects and gave a clear cut result which was shown to be both consistent and 'robust'. For the chairs and desks, however, the user trial gave a result which needed interpretation. This arose because of the short time that each subject sat on each chair or at each desk. In the ten minute trial subjects tended to equate softness with comfort. The results, when analysed carefully, however, were of value in making the selection. The department recognised that good equipment alone would not provide a problem-free workplace. Therefore, an environmental survey was commissioned. After a weeks' instruction, six SWIFTT staff conducted walkthrough surveys of lighting and thermal comfort in the ninety-six offices of the department. Much empirical and anecdotal information was gained, and awareness of these aspects was increased. Last, a safety training programme was developed for all staff due to use the computer equipment. This included the production of a video, an Occupational Health and Safety booklet for all staff ("GOSH" - Guide to Occupational Health and Safety) and training for trainers in health and safety issues. Because of the widespread uncertainty about computers that existed among staff in the department, it is likely the SWIFTT project would have encountered major difficulties if the ergonomic component had been neglected.

**Keywords:** *visual display terminals; workplace; environment; furniture; surveys*

## Introduction

The acronym SWIFTT stands for "Social Welfare Information for Tomorrow Today". It refers to a computer project of the Department of Social Welfare (DSW) in New Zealand, which was set up to install an on-line computer system for use by up to 4,000 staff. Unlike many government departments, very few of the DSW staff were familiar with computers, since previous data entry work had been done in a batch mode by a central agency. Ergonomics was therefore given a high priority at the outset, through both management and union initiatives.

Management wanted to provide the best possible equipment for their staff. The union was concerned that the well known problems associated with introducing VDUs were avoided. Neither could management afford the kind of interruptions that were possible if staff were not happy with the equipment, or the way in which it was introduced, or felt their long-term health was being compromised.

This article describes how visual display units and workstations for the project were selected, and mentions other ergonomics aspects of the project to date.

## Method and results - evaluation and selection of VDU terminals

Management and unions agreed on ergonomics standards for equipment at the outset. Tenderers were asked to state that their terminals complied with the relevant standards. From the terminals that actually did comply (very few) a short list was drawn up for inclusion in a user trial. The short list comprised two colour and seven monochrome screens (four white and three amber). For the trial, 58 staff were chosen at random from the department's offices in the lower half of the North Island. They attended the SWIFTT project office, where the trial took place, in groups of six. The trial ran over a period of one week, and care was taken to treat each group of subjects the same. The evaluation questionnaires were discussed with each group before they sat at the terminals, and they were shown how to adjust the furniture for optimum comfort. Subjects were asked not to speak about the trial among themselves, or to others in the department, until it was over.

Each subject was at a terminal for about 25 minutes, and did artificial data entry and data retrieval tasks for 20 minutes. They then answered brief, separate questionnaires about the screen and the keyboard. The responses were mostly a forced choice from among five answers ranging from 'excellent' to 'unacceptable'. Each terminal was the subject of between 32 and 43 trials.

The questions are reproduced in the Appendix.

The screen questions covered ease of reading the text and overall opinions about the terminal while the keyboard questions covered how the key 'felt' to the touch, key spacing, size, placement, and keyboard slope. There was space on the questionnaire for free comments, and some valuable ones were made. For example, one keyboard had a noisy spacebar, which several subjects noted would have been almost impossible to work with in a large office.

Table 1. Weightings used in the analysis of screen responses

|  | Original Weights | Variation A | Varn. B | Varn. C | Varn. D |
| --- | --- | --- | --- | --- | --- |
| Ease of reading the screen | 1.0 | 1.0 | 1.0 | 0.4 | 0.5 |
| Overall display appearance | 0.4 | 0.8 | 0.8 | 0.3 | 0.3 |
| Glare in screen | 0.2 | 0.4 | 0.8 | 0.1 | 0.1 |
| Colour of unit | 0.2 | 0.4 | 0.2 | 0.1 | 0.05 |
| Overall appearance of unit | 0.2 | 0.4 | 0.2 | 0.1 | 0.05 |

Results were analysed separately for the screens and keyboards. For each terminal, the responses from a selection of the questions were pooled to give a final score for each question. Each question was assigned a weight, and the weighted scores were added to give a final score. This enabled an order of preference to be assigned to each screen and keyboard. To test the robustness of the method the weightings assigned to each question were varied as shown in Table 1.

*Table 2. Ranks of screens under different weightings*

| RANK OF SCREEN | | Original Weights | Variation A | Varn. B | Varn. C | Varn. D |
|---|---|---|---|---|---|---|
| A | | 7 | 7 | 7 | 7 | 7 |
| D | | 4 | 5 | 6 | 4 | 4 |
| G | (white screens) | 2 | 1 | 1 | 1= | 3 |
| H | | 3 | 3 | 3 | 1= | 1 |
| B | | 6 | 6 | 4 | 6 | 6 |
| E | (amber screens) | 4 | 4 | 5 | 4 | 4 |
| F | | 1 | 2 | 2 | 1= | 2 |

With these different weightings, the screens were ranked as shown in Table 2.

It is evident that the ranks differ very little according to the weightings used, and that the method could therefore be regarded as 'robust'.

A further encouraging point was that the screens that scored similarly (pairs of terminals A and B; D and E; G and H), were exactly the same brand and model, differing only in display colour. This result is reflected in their rankings.

Results of the keyboard questions were treated the same way. Here the ranking of the keyboards was identical no matter what weighting was used.

The keyboard and screen results were combined (weight: screen 2/3, keyboard 1/3) to give an overall result.

## Method and results - selection and evaluation of chairs and desks

A similar process was used in the selection of the chairs and desks to be used by SWIFTT staff. Difficulty was again found in finding equipment that met the standards that management and union representatives of the department had drawn up.

In the end, a shortlist of eight chairs and twelve desks was chosen. In this trial 12 subjects used the chairs and desks over a period of two days.

### Chairs

Each subject sat on each chair for ten minutes (at least) and answered a brief questionnaire about it. In addition, each subject sat on each possible chair pair, and expressed a preference for one of the pair. In this way two sorts of evaluation of the chairs were possible.

The questions asked about the chair included what the subjects thought of the comfort of the backrest, seat pan, and chair overall; the ease of operation of the chair adjustments, and the location of the controls. The Appendix shows the questions asked.

Analysis of the results here posed problems when it was realised that a ten minute trial could not predict long-term comfort. Accordingly, a variety of weightings was investigated (see Table 3). Note that in variation B, the effect of comfort is neglected entirely.

Table 3. Weightings used in the analysis of chair responses

|  |  | Original Weights | Variation A | Variation B |
|---|---|---|---|---|
| Overall comfort of chair |  | 1.0 | 1 | 0 |
| Seat height | ease of adjustment | 0.2 | 1 | 1 |
| Backrest height |  | 0.2 | 1 | 1 |
| Backrest angle |  | 0.2 | 1 | 1 |
| Comfort of seat pan |  | 0.3 | 1 | 0 |
| Comfort of backrest |  | 0.3 | 1 | 0 |
| Seat height | ease of location | 0.2 | 1 | 1 |
| Backrest height |  | 0.2 | 1 | 1 |
| Backrest angle |  | 0.2 | 1 | 1 |
| Aesthetic quality |  | 0.2 | 1 | 1 |

Table 4. Ranks of chairs under different weightings

|  | Original Weights | Variation A | Variation B | Pair Comparisons |
|---|---|---|---|---|
| A | 5 | 4 | 5 | 4 |
| B | 6 | 7 | 7 | 5 |
| C | 3 | 5 | 3 | 3 |
| D | 8 | 8 | 8 | 8 |
| E | 7 | 6 | 6 | 7 |
| F | 1 | 1 | 2 | 2 |
| G | 4 | 2 | 1 | 66 |
| H | 2 | 3 | 4 | 1 |

With the different weightings, the chairs ranked as shown in Table 4. Here the results from the pair comparisons are also included.

From these results it is evident that the rankings are consistent, with the exception of chair 'G', which ranged from rank 6 to rank 1 when comfort questions were disregarded.

Chair 'G' was very hard in the padding compared to the others, which were all of a comparable softness. In a ten minute trial we should expect softness to be awarded high ratings for comfort. Over an eight hour working day, however, comfort may not equate to softness, and a harder padding may be preferred. This is the rationale behind the construction of chair 'G' (and indeed, of course, of some hard car seats).

After including questions of cost and local manufacture, it was found possible to recommend three chairs as suitable for use. The chairs had a variety of features, with two having a tilting seat pan. This is beneficial especially for clerical work where the line of sight is more vertical.

Thus, it will be possible for each staff member to have some choice in what they will sit on for the next five years or so. To assist in their choice of chair, a notice was circulated to staff about the three models, advising the differences between them, and points to look for when choosing.

**Desks**

In the trial of desks (including screen support arms) a similar method to that of the chair trial was used. Clear cut results were obtained, largely because of the very different quality in the desks offered. It was also quickly apparent that the mechanisms for supporting the screen (the swivel arms) also differed widely in quality, ease of use and stability. In the end, the simplest model of screen support arm was chosen.

A disappointing aspect of the desk manufacturers' tenders was that in spite of an invitation to discuss the project with the SWIFTT project staff, very few manufacturers did so. The specification called for a 'solution to a problem' rather than adherence to numbers, but all save two manufacturers offered a standard model. It was apparent, however that the maker of the desk which was preferred in the user trial had gone to a lot of trouble to provide a quality article.

## Environmental survey

The environmental survey was carried out by six SWIFTT staff after a week of instruction. They measured illuminances, temperatures and humidity in each departmental office; assessed glare from windows and luminaires, assessed the window covering and decor, and gave self-administered questionnaires about lighting and thermal comfort to a selection of staff. The survey took place over a week in October - spring in New Zealand.

The survey could not be termed definitive in that conditions were measured on one occasion during one day. To get an accurate picture, measurements over a week in winter and summer in each office would be a minimum.

Interesting results of the survey were: 18% of the illuminances were above 1000 lux; the temperatures in the office were very uniform over the whole country (despite its length of 1600 km); about 15% of questionnaire respondents regarded glare from windows and/or luminaires as a severe problem, and a third thought the window coverings to be inadequate.

Two unexpected results were: (i) the region with the least percentage reporting feeling 'warm' or 'hot' in summer was the northernmost, (ii) the region with the greatest percentage reporting feeling 'hot' in winter was the southernmost! These results were put down to the way people dressed.

Cross-correlations of illuminance and temperature measurements with questionnaire results showed that, for the population, these quantities did not affect the way people voted about their environment.

The experience gained by the six SWIFTT staff in carrying out the survey was of value. It enabled them to identify problem areas, and the accurate information they were able to give staff was in itself of value in affirming that the SWIFTT project was in earnest regarding staff welfare.

## Occupational health and safety training

The final aspect of ergonomics in the project was the development of a training package. A video was produced professionally which covered how to set up the chair, desk, and screen for optimum posture, how to avoid occupational overuse syndrome ("RSI"), possible radiation emissions and how to position the VDU for the best lighting.

A 30 page booklet repeated these points, and included the "Pocket Ergonomist" (Brown, 1986), a self-diagnostic guide for curing and preventing muscle aches and pains.

Trainers were instructed in the essential points of these items.

## Conclusions

Ergonomics has played an essential role in the SWIFTT project.

Apart from the initiatives and contributions it made in drawing up the ergonomics standards, the union displayed a keen interest in the conduct of the user trials, the environmental survey, and the content of the training programme. The potential health problems of large scale computer use have been very real to the DSW staff at all levels, and it is doubtful that the project would have been implemented without major industrial problems, had the ergonomics issues not been addressed as they were.

Some of the actions taken were perhaps more thorough than strictly necessary. However, the perception by DSW staff that the SWIFTT office was attempting to do the best for future computer users helped smooth the path for their introduction.

In several instances since the main ergonomics involvement, access to a scientific ergonomics opinion has meant that apprehension about potential work related health problems (photocopiers, laser printers, and occupational overuse syndrome for example) has been laid to rest quickly. Here it seemed that the goodwill and trust established during earlier phases of the project paid off.

There is another result that the Department of Social Welfare can now look to. This is that inevitable ergonomics problems that will crop up as the SWIFTT project is implemented can be dealt with from a base of understanding, rather than from scratch.

## Acknowledgments

The user trials were a team effort. Without the computer skills of Geoff Galbraith of the SWIFTT project office, the insistence on quality by John Good of PA Consultants, and the organizational talents of Glen Smith of the SWIFTT project office, the results would never have been obtained as smoothly as they were. The six Regional Implementation Officers who carried out the environmental survey are to be congratulated for their grace under pressure. I would like to thank Mr David Brown of the Group Occupational Health Centre, Sydney, for his advice during the trial of the chairs.

## Reference

Brown, D. A., 1986. The Pocket Ergonomist, Group Occupational Health Centre, Sydney.

## Appendix

Questionnaires

### A1 Visual Display Unit - Screen
1. What did you think of the ease of reading this screen?
2. Did you detect any flicker on this screen?
3. What did you think of the overall appearance of the display?
4. Did you notice any glare from the screen?
5. What did you think of the colour of the material of the terminal?
6. What did you think of the whole appearance of this VDU?
7. Could you adjust the contrast and brightness satisfactorily?
8. Were any letter pairs easy to confuse?

### A2 Visual Display Unit - Keyboard
1. What did you think of the way the keys felt on this keyboard?
2. What did you think of the spacing of the keys?
3. What did you think of the size of the keys?
4. What did you think of the colour of the keyboard?
5. What did you think of the placement of the function keys?
6. What did you think of the placement of the cursor keys?
7. What did you think of the overall appearance of this keyboard?
8. Were any pairs of keys placed awkwardly?
9. Did you notice any glare on the keys?
10. What did you think of the slope of the keyboard?

### B1 Chair
1. What did you think of the overall comfort of this chair?
2. What did you think of the ease of adjustment of the seat height?
3. What did you think of the ease of adjustment of the backrest height?
4. What did you think of the ease of adjustment of the backrest angle?
5. What did you think of the comfort of the seat pan?
6. What did you think of the comfort of the backrest?
7. What did you think of the location of the seat height control?
8. What did you think of the location of the backrest height control?
9. What did you think of the location of the backrest angle control?
10. What did you think of the aesthetic qualities of this chair?

### B2 Desk
1. What was your overall impression of this desk?
2. What did you think of the ease of adjustment of the keyboard height?
3. What did you think of the ease of adjustment of the VDU screen swivel?
4. What did you think of the ease of adjustment of the VDU screen height?
5. What did you think of the ease of adjustment of the eye-screen distance?
6. What did you think of the fore-aft knee room under the desk?
7. What did you think of the sideways knee room under the desk?
8. What did you think of the fore-aft knee room under the keyboard?
9. What did you think of the sideways knee room under the keyboard?
10. What did you think of the ease of adjustment of the desk height?
11. What did you think of the ease of stowing the keyboard?

# Part IX
Process Industries

# 54. ERGONOMICS CONSIDERATIONS FOR THE DESIGN OF A CRT-BASED PROCESS CONTROL SYSTEM

### Min Keun Chung, Jae H. Choi and Eui S. Jung

*Department of Industrial Engineering*
*Pohang Institute of Science and Technology*
*Pohang, Korea*

**Abstract**

With a general trend towards larger and more complex systems with centralized control, the operator's role in supervisory decision making in the control room becomes more important. Identification of potential man-machine interface (MMI) problems with CRT-based process control systems was attempted using a structured questionnaire survey. Based on the survey results, a laboratory experiment was carried out to investigate the effect of two display design parameters (grouping and colour consistency) upon operator performance in alarm detection with a process control CRT display. User performance was measured in terms of reaction time from the onset of an alarm sound until the completion of predefined response procedures. The results revealed that for highly complex tasks, mean reaction time (RT) with a colour consistent display was 10.5% faster than that with the original display, and RT with a grouped display was reduced by 8.2% in comparison with the colour consistent display.

**Keywords:** *display design; display format; colour coding; human-computer interface; process control; questionnaires*

## *Introduction*

With rapid technological development, human factors specialists have become concerned with the man-machine interfaces (MMI) of modern systems. One major concern is that, in larger and more complex systems with centralized control, an operator's role in supervisory decision making in a control room becomes more important. In centralized and automated systems, quite normal and to-be-expected human error may cause drastic losses. For instance, recent accidents at the nuclear power plant at Three Mile Island (TMI) in the U.S. and at the chemical company in Bhopal, India, revealed that operators in a control room are playing a more important role as the size and the complexity of systems increase. Another concern is the allocation of tasks to man and to automatic systems. Due to the introduction of computers, which are inexpensive but powerful in information processing, the work of decision makers and system operators has changed in several different ways. Automation does not remove humans from the system, but the role of man becomes high-level supervision and long-term maintenance and planning, instead of immediate control of system operations. Therefore, changes in instruction methods for operators and in system design need to be made (Rasmussen, 1986).

Process control systems are normally designed to display and control the varying status of process variables such as temperature, pressure and amount of oxygen, etc., during the production process. Recently process control systems have changed from panel-type, analogue control to computer-based CRT-type, digital control systems. The newer digital control systems involve sequential supervisory control, complicated operations, and the processing of more information. Such systems benefit us a great deal by facilitating the expansion and modification of the system, promoting energy saving, and curtailing the number of control room operators. Despite such advantages, there appear to exist some MMI problems, due to lack of understanding of the newer highly advanced systems and to unergonomically designed displays that use a number of confusing graphic symbols. Furthermore, the CRT-based digital control uses serial presentation of data whereas the conventional analogue control uses parallel presentation of data. Parallel data presentation is, however, claimed to be superior to serial in human cognitive function (Haber, 1981; Salvendy, 1987).

In this study, we attempted to survey and identify MMI problems using structured questionnaires developed in a preliminary study (Chung and Choi, 1988). The questionnaire surveys were conducted with control room operators working at a major iron and steel making company in Korea. Based on the survey results, two major parameters, namely colour consistency and grouping, were selected, and their effect upon operator performance was investigated using a laboratory experiment. Colour consistency refers to the use of the same colour for display labels and their corresponding graphic symbols. Many studies have been carried out to investigate the effect of colour codes on human visual search and identification tasks (Schontz et al., 1971; Christ, 1975; Danchak, 1976; Cakir et al., 1980; Macdonald and Cole, 1988), but little has been done with colour consistency. Grouping is normally used to improve the readability and to highlight the relationship among different groups of data. Grouping of related information or a tabular form of related data could increase the operator's understanding or facilitate comparisons among related parameters (Danchak, 1976; Cakir et al., 1980; Stewart, 1976).

## A structured questionnaire survey

### Questionnaire survey structure

A structured questionnaire was constructed to survey and identify potential MMI problems by gathering subjective and empirical views from operators of CRT-based process-control systems. The content of the questionnaire is based on published ergonomics guidelines and recommendations in designing CRT displays (Cakir et al., 1980; Gertman et al., 1982). The questionnaire is composed of six dimensions: general acceptance, content density, content integration, display format, cognitive fidelity, and environmental considerations (Figure 1). Most of the items in the questionnaire have a 5-point scale, with 5 being the most problematic in man-machine interface. The questionnaire also includes a few "yes-no" type questions.

### Selection of respondents

The representative types of CRT-based systems at the company under study are CENTUM by Yokogawa Co., MICREX by Fuji, VAX/77 by BBC, TOSDIC by Toshiba, etc.. All the operators currently working with the CENTUM type system, or having experienced using such a system, were selected as questionnaire respondents. The total number of operators who participated in this survey is 218 and 118 from Plant A and Plant B, respectively.

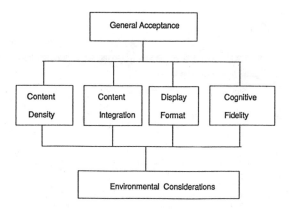

*Figure 1. Structure of the questionnaire.*

**Questionnaire survey results**

The results of the questionnaire survey with the CENTUM type CRT-based process control operations indicate several potential MMI problems in cognitive fidelity and display format. The responses for typical questionnaire items which indicate potential MMI problems are presented in Table 1.

The responses regarding general acceptance indicate that new operators are having difficulties in learning the system, and that most of the operators recognise a need for description of the control labels in the Korean language. As for content integration, the operators feel that the amount of information presented on the display is more excessive than the existing system, although it is all necessary. This suggests a need for simplification of the content of the display. Concerning display format, a need for the redesign of the symbols currently used is evident in the questionnaire responses. Frequently read data do not seem to stand out more clearly than infrequently read data. Responses to the questions regarding cognitive fidelity indicate that many operators experience difficulties in reading or interpreting the information presented on the CRT display.

## A laboratory experiment with a simulated display

Based on the survey results, several key design parameters of information presentation were identified to enhance the operator's visual search and identification capability in process control operations. A laboratory experiment was carried out to investigate the effect of two display design parameters (grouping and colour consistency) upon operator performance in alarm detection on a process control CRT display.

The experiment included two types of display: Heat Pattern Display (HPD) and Control Group Panel Display (CGPD). New display designs were proposed for these two types of display based on the aforementioned two design parameters: colour consistency and grouping. Therefore, four display designs were used in this experiment for HPD and CGPD: (i) Original Design; (ii) Colour Consistency Design; (iii) Grouping Design; and (iv) Mixed Design, in which both colour consistency and grouping are taken into account.

*Table 1. Responses for typical questionnaire items*

**Display Format**

1. If some of the symbols were shaded a different color, they would be easier to recognize or read.

| Strongly Disagree | Disagree | Neutral | Agree | Strongly Agree |
|---|---|---|---|---|
| 0.89% | 13.73% | 16.72% | 63.58% | 5.08% |

2. Frequently read data stand out more clearly than infrequently read data.

| Strongly Agree | Agree | Neutral | Disagree | Strongly Disagree |
|---|---|---|---|---|
| 0.29% | 17.86% | 20.83% | 55.93% | 5.09% |

3. It is clear that display could be better organized.

| Strongly Disagree | Disagree | Neutral | Agree | Strongly Agree |
|---|---|---|---|---|
| 0.30% | 4.22% | 17.77% | 67.77% | 9.94% |

## Subjects

Ten subjects voluntarily participated in this experiment. They were undergraduate students at the Pohang Institute of Science and Technology. The students' ages ranged from 21 to 22 years with a mean of 21.3, and they had all experienced working with personal computers and workstations for 2-3 years. They were also tested for visual acuity and colour vision, and were found to have normal vision.

## Apparatus

The equipment and software used in the experiment were as follows:
1) Vision tester (Lafayette Instrument Co. Model 14019).
2) Apple Macintosh II system and 14-inch high resolution RGB colour monitor (640x480 pixels).
3) Simulated display designs programmed in True Basic language.

## Procedure

The subjects were given a forty-minute tutorial session in which they familiarized themselves with display screens, operational procedures, and the use of controls. After the tutorial session, each subject went through a sixty-minute test session. A simulated screen display was generated using a Macintosh II system. The screen formats used in the experiment represented the ones currently used in the control operations of the cold rolling mill process in the aforementioned steel-making company.

*Table 2. Three way ANOVA*

| Source of Variation | Sum of Squares | Degrees of Freedom | Mean Square | Computed F value | Probability Pr > F |
|---|---|---|---|---|---|
| Color Con. (CC) | 110.7 | 1 | 110.7 | 15.66 | < 0.001 * * * |
| Grouping (G) | 263.0 | 1 | 263.0 | 37.22 | < 0.001 * * * |
| Complexity (C) | 16447.0 | 2 | 8223.5 | 1163.75 | < 0.001 * * * |
| CC * G | 11.8 | 1 | 11.8 | 1.67 | 0.197 |
| CC * C | 34.7 | 2 | 17.3 | 2.45 | 0.087 |
| G * C | 226.6 | 2 | 113.3 | 16.04 | < 0.001 * * * |
| CC * G * C | 15.4 | 2 | 7.7 | 1.09 | 0.337 |
| Error | 6260.9 | 886 | 7.1 | | |
| Total | 23370.1 | 897 | | | |

Note : * * * = Significant Effect with $\alpha = 0.05$

In the test session, the subject is seated approximately 1 foot away from the screen. The program starts a couple of seconds after the subject is prompted, and then process data are updated constantly. When an alarm sounds, the subject is required to detect the abnormal status and respond to it by pressing predefined keys. The performance was measured in terms of reaction time from the onset of the alarm sound until the completion of the predefined response procedures. A total of 36 trials were performed for each subject (2 colour consistency levels X 2 grouping levels X 3 task complexity levels X 3 replicates). The sequence of these trials was completely randomized to minimize learning effects. The computer program automatically recorded the trial sequences and corresponding reaction times for each subject.

**Laboratory experiment results**

The experimental design utilized three factors and three replicates. A three-way analysis of variance (ANOVA) was performed on two colour consistency, two grouping and three task complexity levels to determine their effects on each subject's mean reaction time (Table 2). All the main effects were found to be significant for the three factors, and interaction of grouping and colour consistency was present (p.<0.001). Figure 2 portrays the mean reaction times in a graphic form. In addition, a two-way ANOVA examined whether or not the presence of colour consistency and grouping improved subject performance for each task complexity level. Both design parameters were found to affect significantly the reaction time at the high task complexity level (p.<0.001). On the other hand, neither factor showed significant effects at the medium and low task complexity levels (p.>0.1).

Bonferroni's multiple comparison was performed to examine the differentiated effects of colour consistency and grouping at the high task complexity level (Table 3). The test revealed that mean reaction times (RTs) with the presence of colour consistency were reduced by 1.842 seconds (10.5%) relative to that with the original design. It is noted that RT with grouping was 1.287 seconds (8.2%) faster than that with colour consistency, and that when the combination of colour consistency and grouping was taken into account, mean RT was 1.926 seconds (12.2%) faster than

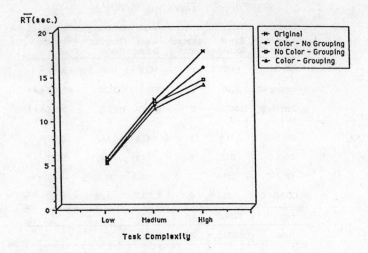

Figure 2. Mean reaction times for display type and task complexity

Table 3. Results from the Bonferroni's Multiple Comparisons

| Display Type | Lower Confidence Limit | Upper Confidence Limit | | Difference Between Means |
|---|---|---|---|---|
| O — C | 0.690 | 2.993 | *** | 1.842 |
| O — G | 1.986 | 4.272 | *** | 3.129 |
| O — M | 2.598 | 4.939 | *** | 3.768 |
| C — G | 0.102 | 2.472 | *** | 1.287 |
| C — M | 0.715 | 3.318 | *** | 1.926 |
| G — M | - 0.564 | 1.842 | | 0.639 |

Note: O = Original
C = Color Consistency and No Grouping
G = No Color Consistency and Grouping
M = Color Consistency and Grouping
*** = Significant Effect with $\alpha = 0.05$

when colour consistency alone was considered. However, it is interesting to see that Grouping Design and Mixed Design did not show a significant difference, so that the improvements are not additive.

## Discussion and conclusions

Based on the structured questionnaire survey results, two key display design parameters, namely colour consistency and grouping, were identified as major sources of MMI problems in a CRT-based process control system. A laboratory experiment was conducted to investigate the efficacy of the ergonomically designed displays taking the aforementioned two design parameters into account.

Statistical analysis showed that there were significant colour consistency and grouping effects when task complexity was high. This implies that when the information is presented in a consistent and well-organized manner, the promptness of the control room operator's response is greatly improved for complex tasks. In further analysis, the grouping effect was found to be significantly greater than the colour consistency effect. Thus, the grouping of the relevant items is an effective technique in a display design. This result is in agreement with previous research findings (Stewart, 1976; Cakir et al., 1980). Moreover, combination of these two design parameters showed a further improvement in operator performance. This result suggests that when several display design parameters are optimally integrated to generate better screen formats, a great improvement in operator performance can be achieved.

The importance of display design parameters is again emphasized in CRT-based process control operations. An ergonomically designed display will critically improve accuracy and reaction time, especially when immediate human decision making is necessary in urgent situations.

## References

Cakir, A., Hart, D. J. and Stewart, T. F. M., Visual Display Terminals, New York: John Wiley and Sons, Ltd., 1980.

Christ, R. E., Review and Analysis of Color Coding Research of Visual Displays, Human Factors, 17, 542-570, 1975.

Chung, M. K. and Choi, J. H., Ergonomic Considerations for the Design of Production Process-Control Systems, Technical Report, Research Institute of Industrial Science and Technology, Pohang, Korea, 1988.

Danchak, M. M., CRT Displays for Power Plants, Instrumentation Technology, 23(10), 29-36, 1976.

Gertman, D. I., Blackman, H. S., Banks, W. W. and Petersen, R. J., CRT Display Evaluation: The Multidimensional Rating of CRT-generated Displays, NUREG/CR-2942, 1982.

Haber, R. N., The Power of Visual Perceiving, Journal of Mental Imagery, 5, 1-40, 1981.

Macdonald, W. A. and Cole, B. L., Evaluating the Role of Colour in a Flight Information Cockpit Display, Ergonomics, 31, 13-37, 1988.

Rasmussen, J., Information Processing and Human-machine Interaction, New York: North-Holland, 1986.

Salvendy, G., Handbook of Human Factors, New York: John Wiley and Sons, 1987.

Schontz, W. D., Trumm, G. A. and Williams, L. G., Color Coding for Information Location, Human Factors, 13, 237-246, 1971.

Stewart, T. F. M. Displays and the Software Interface, Applied Ergonomics, 7, 137-146, 1976.

# 55. HUMAN ENGINEERING ANALYSIS OF THE CHERNOBYL ACCIDENT

Vladimir M. Munipov

*The All-Union Scientific and Research
Institute of Industrial Design
Moscow, USSR*

**Keywords:** *accidents; human error; nuclear power plants; system reliability; ergonomics intervention*

## Introduction

The accident at the fourth unit of the Chernobyl nuclear power plant (NPP) is still being studied by Soviet and foreign specialists in different fields. This accident is often compared with the one that happened at the American NPP "Three Mile Island" (TMI) on March 28, 1979. The Chernobyl accident was caused by a combination of design deficiencies, operation shortcomings and human errors. Therefore, it would be wrong to blame only operators or designers. Nevertheless, only six members of the plant management were convicted on the grounds of having violated safety regulations at potentially explosive facilities which caused deaths. The Chairman presiding over the court said some words to the effect of proceeding with the investigations as regards "those who failed to take measures to improve the plant design". He also mentioned responsibility of the Department officials, local authorities and medical services. But in fact, it was clear that the matter was closed. Nobody else was held responsible for the greatest disaster in the history of technology. It would be naive to think that a director, chief engineer or any other personnel could really be the main culprits in such a case. There exists a certain international safety standard which calls for adequate protection of nuclear plants against all possible errors or incidents including direct fall of a large aircraft.

## Report on the TMI accident

The special committee appointed by the US President to investigate the TMI accident named the operator actions as its direct cause. The operators committed five fatal errors. They failed to identify the state of the plant and aggravated the accident by their actions. The presidential committee recommended revising the programmes for operator training, setting up special training courses, defining the scope and intervals of emergency training and developing unbiased criteria for assessing operator skills.

After the TMI accident a human engineering programme was developed in the USA with a view to eliminating the existing shortcomings of the NPP control room design, construction and testing. The working conditions were checked at all operating plants.

## Resistance and secrecy in safety supervision bodies in nuclear operations

The author of this paper reported the above information to the USSR State Committee for Supervising Safety of Works in Nuclear Power but was rudely advised "to mind his own business and let the others indulge in important state affairs". This happened in 1985, a year before the Chernobyl accident.

The system of great secrecy which was common practice in the Soviet nuclear energy industry could be pointed out as one of the causes of the Chernobyl accident. Certain scientists and groups of scientists were given an exclusive right to determine what was the truth in nuclear power. This monopoly was reliably protected by the adopted policy of secrecy. As a result, the assertion of the Soviet scientists of the utter safety of the NPPs remained unchallenged for 35 years. Secrecy covered the incompetence of the civil nuclear leaders. However, it has recently become known that the secrecy extended to the TMI accident as well. The operating personnel of the Soviet NPPs did not have full information on this accident. Only selected information which did not contradict the official policy concerning NPP safety was made known. Thus, it becomes clear why the report presented by the author of this paper was not welcomed in the State Supervisory Committee.

Accidents at nuclear plants began virtually with the onset of nuclear energy. The first accident at an American plant occurred in 1951, and at a Soviet one in 1966. No Soviet nuclear accidents were ever made public except for the accidents at the Armyanskaya and Chernobyl (1982) nuclear power plants which were casually mentioned in "Pravda". Concealing the true state of affairs, failing to learn lessons based on accident analysis, the leaders of the nuclear power industry were heading straight for Chernobyl (1986). The road was made easier by the fact that, given unprecedented growth in nuclear capacity, there existed a distinct shortage of trained reactor operators and use was often made of unskilled personnel. A simplified idea of the operator activities was implanted and the risk in operating NPPs was underestimated. As a consequence, by the early 1980s the wages at thermal power stations exceeded those at nuclear power plants.

The policy of secrecy admirably suits many people today as well. It seems that given the possibility, the Chernobyl disaster would have been hushed up despite perestroika, the new way of thinking, etc. According to the first reports of the accident, the normal reactor state was being restored. This outrageous lie was encouraged in every possible way. The panic-stricken citizens of Kiev were forced to take part in the May-Day demonstration, while just at that time a well-known scientist, a member of the State Committee set up to investigate the causes of the accident, openly declared that by the beginning of May the total radioactive release was almost thrice that of Hiroshima. This statement would never have been made public in those days. A special censorial body was set up under the aegis of the State Committee for Nuclear Energy and not a single publication concerning the accident could be issued without its approval. The nuclear authorities were in a hurry to start up other Chernobyl units shut down after the accident and did their best to protect themselves against future criticism. An attempt to direct the investigation to the wrong track in order to conceal the true causes of the accident was disclosed. It is not quite clear who inspired this attempt and what they hoped to gain. Actually, much still remains unknown about the causes and consequences of the accident.

## The Chernobyl 1986 accident

### Faulty planning of test

On April 25, 1986 the fourth unit of the Chernobyl NPP was being prepared for routine maintenance. An experiment involving inoperative safety systems under total loss of power was to be performed during the outage. The test programme was approved by the Chief Engineer. The power during the test was supposed to be generated by the run-out energy of the turbine rotor (inertia-induced rotation). When still rotating the rotor generates electric power which could be used in an emergency. Total loss of power at a nuclear plant causes all mechanisms to stop including the pumps which provide the coolant circulation in the core, which in turn results in core meltdown, that is, an accident. The above experiment aimed to test the possibility of using some other available means of generating power. It is not forbidden to carry out such tests at operating plants provided there is an adequate procedure and additional safety precautions. The test programme must be prepared beforehand and submitted for approval to the Chief Designer and Chief Engineer of the reactor, and the State Supervisory Committee. The programmes must provide a back-up power supply for the whole test period. In other words, the loss of power is only implied but never made actual. The test may be performed only after the reactor is shut down, that is, when the scram button is pushed and the absorbing rods are inserted in the core. Prior to this, the reactor must be in a stable controlled condition with the reactivity margin specified in the operating procedure. At least 28-30 absorbing rods must be inserted in the core.

The programme approved by the Chief Engineer of the Chernobyl plant did not satisfy the above requirements. Moreover, it called for the shut-off of the Emergency Core Cooling System (ECCS), thus jeopardizing the plant safety for the whole test period (about 4 hr). When developing the programme, the authors took into account the possibility of the ECCS initiation, which would have prevented them from completing the runout test. The bleed off method was not specified in the programme since the turbine no longer needed steam. Clearly, it took people completely ignorant in reactor physics and highly presumptuous at that to develop and carry out such a programme. The Chernobyl administration fell into this description. Such people were clearly also among the nuclear power leaders, which would account for the fact that the above programme submitted for approval to the corresponding authorities in January 1986 was never commented upon by them in any way. The dulled feeling of danger also made its contribution. Due to the policy of secrecy the opinion was formed that nuclear power plants were safe and reliable, and their operation was accident-free. Lack of response to the programme did not, however, alert the director of the Chernobyl plant. He decided to proceed with the test using the uncertified programme though this was not allowed.

## Faulty operating of test

When performing the test, the personnel were violating the programme itself, thus creating further possibilities for an accident. The Chernobyl personnel committed six gross errors and violations. I would like to dwell on some of them. According to the programme the ECCS was made inoperative, it being one of the greatest and fatal errors. The feedwater valves had been cut off and locked beforehand so that it would be impossible to open them even manually. The emergency cooling was deliberately put out of action in order to prevent possible thermal shock resulting from the cold water entering the hot core. This decision was based on the firm belief that the reactor would hold out. In its turn, this brief was due to poor understanding of reactor physics and the failure to foresee the worst course of events. The "faith" in the reactor was strengthened by ten years comparatively trouble-free operation of the plant. Even a rather grave warning, the partial core meltdown at the first Chernobyl unit in September 1982, was ignored.

According to the test programme the rotor run-out was to be carried out at a power level of 700-1000MW. Such run-out should have been performed as the reactor was being shut down. But the other, disastrous, way was chosen: to proceed with the test with the reactor still operating. This was to ensure the purity of the experiment.

In certain operating conditions it becomes necessary to change or turn off a local control for clusters of absorbing rods. When turning off one such local system (it being specified in the procedure for low power operation) the senior reactor control engineer was slow to correct the imbalance in the control system. As a result, the power fell below 30 MW which led to fission-product reactor poisoning (xenon, iodine). In such an event, it is next to impossible to restore normal conditions. It was necessary to interrupt the test and wait a day till the poisoning was overcome. The Deputy Chief Engineer for Operations, not a specialist in the nuclear field, did not want to interrupt the test and by shouting forced the control room operators to begin raising the power level which had been stabilized at 200 MW. The reactor poisoning continued and a further power rise was not permissible due to the small operating reactivity margin of only 30 rods for a large power pressure-tube reactor (RBMK). The reactor became practically uncontrollable and explosive. While taking it out of the "iodinepit" the operators had withdrawn several rods of the reactivity margin thus making the scram system ineffective. Nevertheless, it was decided to proceed with the test. The operator behaviour was motivated mainly by the desire to complete the test as soon as possible.

## The problem of the absorbing rod design fault

To get a better understanding of the causes of the accident, it is necessary to point to the major design deficiencies of the absorbing rods of the control and scram system. The core height is 7 m, while the absorbing length of the rods amounts to 5 m with 1 m hollow parts above and below. The bottom ends of the absorbing rods which go under the core when fully inserted are filled with graphite. Given such a design, the control rods enter the core followed by the 1 m hollow parts and finally the absorbing parts. In total, there were at Chernobyl 211 absorbing rods, 205 of which were fully withdrawn. Simultaneous insertion of so many rods initially results in reactivity burst due to the fact that at first the graphite ends and hollow parts enter the core. In the case of a stable controlled reactor such a burst is nothing to worry about but in the event of a combination of adverse conditions such an addition may prove fatal since it leads to prompt neutron reactor runaway. This design deficiency can cause operator errors.

## The accidents sequence

The course of events was as follows. With the onset of steam flashing in the reactor coolant pumps which led to reduced flow rate in the core, the coolant boiled in the pressure tubes. Just then the shift supervisor pushed the button of the scram system. In response, all the control rods (which were withdrawn) and the safety rods dropped into the core. But the first to enter the core were the graphite and hollow ends of the rods which cause reactivity growth. They entered the core just as intensive steam began to generate. The rise in the core temperature produced the same effect. There combined three unfavourable conditions for the core. The prompt neutron reactor runaway began. This was due primarily to gross design deficiencies of the RBMK. It should be recalled that the ECCS had been made inoperative, locked and sealed.

Further events are well known. The reactor was damaged. The major part of the fuel, graphite and other in-core components was thrown outside. The radiation level in the vicinity of the damaged unit amounted to 1000-15000 r/hr. There were some distant or sheltered areas where the radiation level was considerably lower.

At first the personnel failed to realize what was happening and kept on saying, "It is impossible!...Everything was done properly". There still remained blind faith in the reactor. The explosion was presumed to have happened in the emergency tank of the control and scram system. This information was reported to Moscow. The plant management and operators could not begin to think that the explosion could have been caused by the ends of the absorbing rods, being the major means for reactor protection.

## Human engineering issues

### Need to recognize and design for human error

Analysing the Chernobyl accident I cannot but recall my first lectures on human engineering which dwelt on one of its principles, namely, "foolproof design". This principle calls for developing a system in such a way that it can protect itself against possible human errors. Each time this explanation was given at a lecture there was an exclamation from the audience, "Comrade lecturer, don't you know the main power of a socialist society?""Yes, I know," was the answer, "it is high consciousness of the masses.""Then what are you calling for?", this time the tone became menacing. This discouraged the lecturer and henceforth he took care to avoid this matter.

This came to my mind when I was reading the report "Chernobyl Accident and Its Consequences" prepared for the International Atomic Energy Agency (IAEA). It was stated in the report that the personnel behaviour was caused by the desire to complete the test as soon as possible. Judging from the fact that the personnel violated the procedure for preparing and carrying out tests, violated the test programme itself, was careless when performing the reactor control, operators were not fully aware of the processes taking place in the reactor and had lost all feeling of danger. According to the report, the reactor designers failed to provide safety systems designed to prevent an accident in the case of deliberate shut-off of the engineered safety means combined with violations of the operating procedure since they *regarded such a combination as unlikely* (stressed by the author). Hence, according to the report, the initial cause of the accident was a very unlikely violation of the operating procedure and conditions by the plant personnel.

To put it plainly, the designers considered the interference of "clever" fools in plant control unlikely and therefore failed to develop the corresponding engineered safety means. I have stressed the phrase in the report stating that the designers considered the above combination unlikely. Here a question arises: Have the designers considered all possible situations associated with human activity at the plant? If the answer is positive, then how was it taken into account in the plant design? Unfortunately, the answer to both questions is negative. As a result, the in-situ emergency training, theoretical and practical training were carried out mainly within the primitive control algorithm.

### Resulting remedies being implemented

At present additional measures are being developed and put into effect to ensure safe operation of current NPPs and improve the design and construction of future ones, including human engineering aspects. In particular, measures have been taken to make the scram system more fast-operating and to exclude any possibility of its deliberate shut-off by personnel. The absorbing rod design has been modified and their number increased. It was decided to abandon plans for constructing new reactors of the Chernobyl type. In accordance with the scientific and technical programme for the period of 1986-1990 "To develop and apply in industry the system of human engineering to support design, construction and operation of machines and equipment" approved by the USSR State Committee for Science and Technology, the All-Union Scientific and Research Institute of Industrial Design is working jointly with the Kiev Automatics Institute on the human engineering aspects of developing and testing NPPs. Human engineering analysis and assessment is being carried out in relation to the information support of reactor operators in normal and abnormal conditions and when field testing the engineered means of the NPP process control systems. Work is being carried out for vocational training of personnel in the All-Union Scientific and Research Institute of NPPs. There are plans to enlarge these studies considerably. However, much still remains to be done. It would require a lot of persistence and even civic courage to keep pace in this field.

**Resulting problems in trying to implement human engineering**

To solve many problems mentioned in this paper it is necessary to carry out combined research and development involving physicists, designers, industrial engineers, operating personnel, specialists in human engineering, psychology and other fields. Organizing such joint work presents great difficulties, one particular difficulty being the remaining monopoly of some scientists and groups of scientists on the truth in the field of nuclear energy and of the operating personnel on the information concerning NPP operation. It is impossible to carry out really scientific studies given such deliberately scarce information. Without comprehensive information, it is impossible to give a human engineering diagnosis of a NPP and, if necessary, propose ways of eliminating the disclosed shortcomings as well as a system of preventive measures.

If in Western countries a human engineer enjoys full rights in nuclear power, in this country such a situation is still a long way off. Thus, the USSR Ministry of Nuclear Power replied to the official inquiry that in 1990-2000 there was no need for specialists in human engineering with secondary and higher education as there were no corresponding requests from nuclear plants and enterprises.

Yet, nearly total absence of human engineering support for the NPP design and construction is not the worst aspect; the worst is the total absence of expressions of need for such support. An academic, V.A. Legasov, once declared with great conviction; "I advocate respect for human engineering and sound man-machine interaction. This is a lesson that Chernobyl taught us".

## *The human engineering lessons are not yet learned*

The writer S. Zalygin advocated the need for proceeding with investigations into the causes of the Chernobyl accident, warning that it was not impossible for some minor neglected detail of this accident to some day become the main cause of another one. We must learn from experience, including tragic experience. An American scientist, James Auberg, the author of the book "Digging Out Soviet Disasters" recently published in the USA, noted that if the Soviets had thoroughly investigated the TMI accident, it is not improbable that the Chernobyl accident could have been prevented.

The following fact causes anxiety in this connection. There appeared a touching coincidence in the interests of the Soviet bureaucratic departments and the forces in the West called the "nuclear lobby"; this was revealed in their desire to diminish the scale of the Chernobyl tragedy and its consequences. I would like to give here only one example reported by the Japan Civil Centre of Nuclear Information. Not so long ago a delegation of the Japan State Committee for Nuclear Safety visited the USSR. On their return home the members of the delegation informed the media of the so-called "scientific findings", according to which the Chernobyl consequences are limited to the sadly known statistics about the death of over 30 persons and a closed 30 km zone in the vicinity of the plant. The Committee failed to find any other negative effects of the accident. The Japan Civil Centre doubted these findings and found other ways to obtain reliable information.

## *The risks are not only Russian but global*

The main lesson from the Chernobyl accident is that the nuclear age calls for a new culture and a new way of thinking and is certain not to tolerate ignorance. The new way of thinking is also necessary to enable a really scientific analysis of the Chernobyl accident including its human engineering aspects. Nothing worthwhile can be obtained insofar as the Chernobyl matter rests with the very people and departments guilty of this tragedy and doing their best to hush up its consequences.

It is necessary to combine the efforts of all countries to prevent a new Chernobyl not only in the USSR, but elsewhere in the world where NPPs are operated.

Nuclear accidents are not only of national but of global concern. Hence the need for scientific and technological co-operation and interaction in improving the safety of operating NPPs. A team of experts from the USA, France, UK, Japan and Finland revealed the following shortcomigns of Soviet NPPs: complicated plant control, the absence of modern means for personnel training, poor support of the operation by designers, outdated forms of operation manuals. Add to this the fact that we fall considerably behind the international level of engineered means of diagnosis, control and automation of operator activities.

Pondering over all this, one is sure to get alarmed. Therefore, it is necessary not only to state the facts, but to enlist foreign help to correct the above shortcomings as soon as possible. There is no time to be lost! According to official data, in 1989 there were 118 reactor scrams at Soviet NPPs, including 55 caused by human errors and 63 due to equipment failure.

# 56. AN EVALUATION OF OPERATOR WORKLOAD IN NUCLEAR POWER PLANTS

Yoshiaki Hattori[*], Ju-Ichiro Itoh[*], Teruaki Tomizawa[**] and Katsuhiko Iwaki[***]

[*]Systems Analysis Group, Nuclear Engineering Laboratory
Toshiba Corporation, Japan

[**]Control & Electrical Engineering Dept., Nuclear Energy Group
Toshiba Corporation, Japan

[***]Nuclear Power Design Div., Nuclear Power Construction Dept.
Tokyo Electric Power Co., Japan

**Keywords:** *mental workload; task analysis; subjective evaluation; rating; nuclear power plants*

## Introduction

In nuclear power plant operation, safety and reliability are essential issues. The human factor, in particular, is one of the most important issues in assessing overall plant reliability (Honeywell Inc. and Lockheed & Space Company, Inc., 1982; Essex Corporation, 1984). The heavy workload of human operators is a factor that may decrease plant operational reliability. Handing of complicated tasks to an automated system minimizes the chance of human error in that task. While it is important to evaluate the allocation of tasks for enhancement of overall reliability, no easy way of carrying out such an evaluation was available. Most task analysis studies in nuclear power plants are based on a qualitative approach and only a few studies have proposed a quantitative approach (Burgy et al., 1983).

This study was carried out to provide a relative rating method for the evaluation of operators' task workloads in nuclear power plants.

## Methods of workload rating

In nuclear power plants, operators' tasks are classified mainly into two categories; operation in the main control room and operation in the field.

### Operator workload ratings in the main control room

For tasks in the main control room, a relative workload rating method was developed. In this method, the workload is represented by the number of task elements per minute. A single action was regarded as a task element; i.e., the observation of a meter or a lamp, or the operation of a switch or a pushbutton. We call this the task element density. The task element density was measured for each operation in the plant; start up, shut down, and post-scram. The time taken to carry out the tasks was identified by field surveys. Observers measured the duration of each task in an actual plant. The number of task elements in each task was determined by counting them in the operation procedure documents of the subject plant. The overall task element density in the control room for every minute of task execution was also determined. By comparing these ratings, tasks carrying relatively heavy workloads were identified.

A questionnaire on the need for automation was also used. The subjects were 136 operators of nuclear power plants. Operators were asked to give their requirements for automation using three ranks (high, medium, low) for 101 different operation tasks.

## Operator workload ratings in field operations

For field operations, the tasks were classified as moving only, monitoring equipment or panels, operation of local equipment, communication with the main control room, and extra movements to fetch tools or room keys. These task categories were rated by relative values determined according to engineers' judgements and the ratings were reviewed by expert operators. The assigned rating values are shown in Table 1. Tasks in difficult or radiation environments were rated as, respectively, twice or four times as large as the basic values.

Table 1. Ratings for field tasks

| task classification | | basic ratings |
|---|---|---|
| moving | | 1 |
| waiting | | 1 |
| monitoring | inspection | 2 |
| | measurement | 3 |
| operation | large valves | 10 |
| | general valves | 5 |
| | rack-out at MCC* | 5 |
| | large breaker | 7 |
| | change lamps | 4 |
| | change chart papers | 4 |
| communication | | 6 |
| extra move (to fetch tools or keys) | | 3 |

| environment | rating factors |
|---|---|
| in difficult environments | X 2 |
| in radiation environments | X 4 |

\* MCC : Motor Control Center

The workload involved in field operations during routine procedures was identified by field surveys. Field data were collected in such a way that an experimenter followed an operator and verbalized both his task and its time using an audio tape recorder. Based on the collected data, the location of the recorded data and the operators' walking route were identified according to the geometrical arrangement of the plant. Before rating workload, the task classifications and the locations of observed operators' actions were determined in the form of a field task diagram along with elapsed time from the start. For each task, the workload rating shown in Table 1 was assigned and cumulative rating was also made for comparison with the basic load of simple walking.

## Results

As a result of evaluating the workload in nuclear power plants, a relative workload rating for each operational task was determined.

Using field surveys of main control room operations and by checking operational procedure documents, the task element density was determined for each task involved in plant start up, shut down, and post-scram. Figure 1 shows the example of task element density for the condenser vacuum up task. Sub-task time lines are shown simultaneously. This example shows that a concurrence of several sub-tasks produces a maximum workload for the condenser vacuum up task.

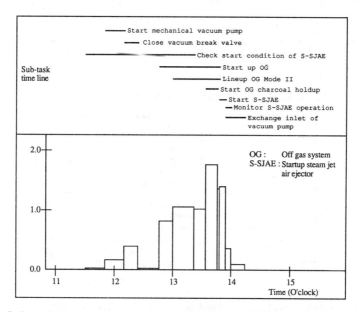

*Figure 1. Sub-task time line and task element density for 'condenser vacuum up' task*

Tasks with a heavy workload (high task element density) were identified. One of these was control rod position management. The rod position task was often carried out simultaneously with other tasks and overall operation usually depends on rod control. Task elements involved in withdrawing rods are shown in Table 2. As a result of the field survey, the task of withdrawing rods from start up to 10% power was found to consist of 1448 withdrawal operations, 88 one-notch withdrawals and 76 coupling checks over 607 minutes.

The workload in post-scram procedures for gently taking a plant into a stable state was also identified as high. While most of the procedures are automated by plant interlocks to ensure safety, operators were expected to perform them manually to keep the equipment sound.

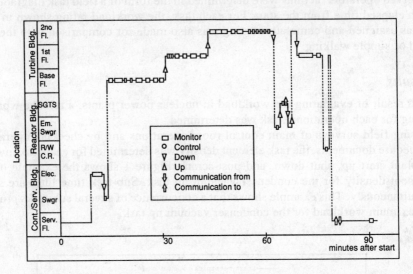

Figure 2. An example of a field task diagram (routine procedures in turbine building)

Table 2. Task elements in withdrawing control rods

| operation | task elements |
|---|---|
| withdrawing rods | rod selection<br>withdrawal operation<br>observation of rod position<br>assurance of withdrawn rod |
| one-notch withdrawal | one-notch withdrawal operation<br>check lamp off |
| coupling check | withdrawal operation<br>check annunciator on |

Several other tasks were identified as relatively heavy workload tasks. The improvement of these relatively heavy workload tasks by task reallocation (i.e., allocation to automatic control) will enhance overall operation reliability with a minimum of effort.

Relatively low workload tasks were also identified, because these tasks are candidates for automation to relieve operators of tedious tasks.

The results were reviewed by operators using the questionnaire. Their subjective judgements were consistent with the results.

Figure 2 shows the example result of a field operation task. The example illustrates the routine procedures with regard to a turbine building. The workload rating is shown in Figure 3. The cumulative rating curve lies in the region bounded by reference curves for one per minute and three per minute. Communication with a main control room imposes high workload on field operators. From the results, high workload tasks were determined and many design changes were proposed to enhance the field work.

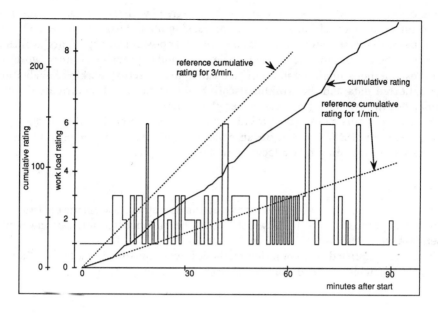

Figure 3. Workload rating for a field task (routine procedures in turbine building)

## Discussion

To determine task allocations, both the human workload and available technology for automation are important factors. Where a heavy workload is imposed on operators during a certain task, and if the task does not include human decision making, the task should be automated. On the other hand, where the workload is too low in a task, the operator cannot concentrate his attention on it for long, so such a task should also be automated (Pulliam et al., 1983).

In the nuclear power field, safety is an important factor in designing task allocation. Tasks that include critical decision making are always assigned to operators (IEC 45A/WG 48, 1986). The operator workload should not be increased by monotonous and mechanical operations, since operators have to concentrate on decision making.

Since workload is always a major factor in designing operation task allocation, a portable workload rating method is desirable to enable a nuclear plant engineer to evaluate it.

This simplified relative rating approach described here is efficient if the purpose of the evaluation is to determine priority for task improvements. Without quantitative data, plant designers have failed to recognize the high workload imposed by plant operations. Workload estimations are critical because other factors, such as criticality for safety and requirements for human decision making, can be estimated by plant designers.

In rating workload in the main control room, all task elements are regarded as equivalent. Because the ratio of element categories comprising the tasks do not vary significantly, that assumption is valid whenever the discussion is within relative workload evaluations.

The workload rating of field operations is highly dependent on the subjective judgement of engineers themselves. While our results also reflected this subjective rating of task categories, expert plant operators agreed with the ratings in interviews. The results provide preliminary data for discussing plant design enhancement.

This paper describes our trial estimation of nuclear power plant operators' workload. There may be more elaborate methods to describe and to predict operators' workload. But those methods might be more difficult to apply to actual workload evaluations. The collected data and the workload scaling in this study will contribute to future approaches for evaluating nuclear power plant operations.

The results of this study have provided the basis for the design of controls for the next generation of BWR plants. Because of its simplicity, this rating method is easily applicable to evaluating task allocation in any power plant.

## Conclusions

A relative workload rating method for evaluating operators' tasks in the nuclear power plants was proposed. The tasks which should be enhanced in nuclear power plant operation were determined based on this method.

Using the simplified relative rating method, it was confirmed that this method was useful and efficient to determine actual necessity for enhancement in nuclear power plants.

## Acknowledgments

This study was carried out in cooperation with the operators at the Fukushima power stations of the Tokyo electric power company. The authors acknowledge the help of those operators, especially those who gave their expert advice.

## References

Burgy, B. et al. 1983, Task Analysis of Nuclear Power Plant Control Room Crews, NUREG/CR-3371.

Essex Corporation 1984, Human Factor Guide for Nuclear Power Plant Control Room Development, EPRI NP-3659.

Honeywell Inc. and Lockheed & Space Company, Inc. 1982, Human Engineering Guide for Enhancing Nuclear Control Rooms, EPRI NP-24121.

IEC 45A/WG 48 1986, International Electronic Commission Technical Committee 45: Nuclear Instrumentation, Sub-Committee 45A: Reactor Instrumentation, Working Group 48: Control Room Design Standard for Control Rooms of Nuclear Power Plants.

Pulliam, R. et al. 1983, A Methodology for Allocating Nuclear-Power-Plant Control Functions to Human or Automatic Control, NUREG/CR-3331.

# 57. DEVELOPMENT OF THE TEAM ACTIVITY DESCRIPTION METHOD (TADEM)

Kunihide Sasou[*], Akihiko Nagasaka[*] and Takeo Yukimachi[**]

[*]Human Factor Research Center
Central Research Institute of Electric Power Industry, Japan

[**]Dept. of Science and Technology, Keio University, Japan

**Abstract**

Three aspects, i.e. operators' communications, actions and observation points, are very important when considering team activities for coping with abnormal operating conditions at a nuclear power plant. This paper develops a method for describing these aspects in detail and with relatively low dependence on subjective judgements and discusses its effectiveness. As a result, the effectiveness of this method is confirmed. It seems to provide a new approach for analysing team activities when coping with abnormal operating conditions at a nuclear power plant.

**Keywords:** *task analysis; nuclear power plants; team work; communication; training; simulators; methodology; work procedures*

## Introduction

A nuclear power plant is operated by a team. The behaviour of the team, namely the team activities, should be considered when discussing the reliability of nuclear power plant operations. However there is only one study (Becker, 1987) which positively determined the actual team activities. In that study, several flow diagrams were developed to describe the plant parameters, the team strategies and so on. The communications among the operators were also very important, but the contexts of the communications were not described in the flow diagrams developed in that study.

This paper then focuses on three aspects - operators' communications, actions and observation points - which are important in the consideration of team activities and develops a method to describe the team activities from the above three aspects in detail and with a relatively low dependence on subjective judgements.

## Experiments using a training simulator

To obtain some examples of the team activities for coping with abnormal operating conditions, experiments using a full-scope training simulator were conducted with four instructor-operators assuming that the abnormal operating conditions happened during full-power operations or during the plant start-up without informing them of the conditions. Each instructor-operator was assigned his role (the shift foreman, the reactor operator, the turbine operator and the auxiliary operator). The team activities and the plant parameters during the simulator experiments were recorded by four video tape recorders. The operators' conversations were also recorded by portable tape recorders to record conversations which were not videotaped.

The instructor-operators are very skilful in coping with the simulated abnormal conditions in the training curricula and they should cope with them easily. The following complex simulated abnormal operating conditions which were not used in the training curricula were devised:

Condition 1 - Circulation water pump trip and condensate pump trip without warning
Condition 2 - Grand steam pressure lowering and control rod drift out
Condition 3 - Condensate rubber seal water level lowering and primary hot well water level lowering.

## Team activity description method (TADEM)

Based on the results obtained from the simulator experiments, this paper improved the method previously proposed (Yukimachi, 1989) and developed a method called "Team Activity Description Method (TADEM)" (Sasou et al., 1990). This method divides the team activities into communication flow, action flow and position flow in order to describe them. Figure 1 shows an example of the results of the team activities when coping with simulated condition 3. The features of this method are:

1   it is not limited by the number of operators,
2   it does not lose the sequence of speech and actions,
3   it makes it possible to describe the context of speech in detail.

**The communication flow**

The communication flow expresses the operators' speech, i.e., the shift foreman, the reactor operator, the turbine operator, the auxiliary operator in the control room and the equipment operators in the reactor building, the turbine building and so on.

The segment of the speech and the time elapsed should be described in the space above the broken lines with nodes (O). Speech by telephone should be marked with " ☎ " and the announcements to the equipment operators on a bullhorn should be enclosed with " 「」 ". The arrows should be drawn from the speaker to the listeners.

Writing the speech segments may lose their meanings since they depend on conversational accent. The context for each speech should be classified in the communication categories and the code should be written to express the meaning of each speech. But speech by telephone or an announcement on a bullhorn should not be classified because the aims of such speech (for example, informing of the plant's operating status or giving instructions) are obvious.

"ANNOUNCE" is the category for speech which provides information about the plant's operating status (for example, reading the electric power or giving a warning).

"INQUIRY" is the category for speech which requests information about the plant's operating status.

"YES" is the category for speech which expresses a repeat of or an agreement to an instruction.

"NO" is the category for speech which expresses a disagreement to an instruction.

In addition, the speech included in the above four categories should be classified in the category for the present condition "CURRENT" or the future condition "ANTICIPATION".

Speech not classified in the above categories should be classified in "OFFER" for speech requiring agreement of the actions, "SELF" for soliloquies, "ENLIVEN" for speech reducing the tension, "MANDATE" for speech giving instructions, or "PROMPT" for speech requiring instructions.

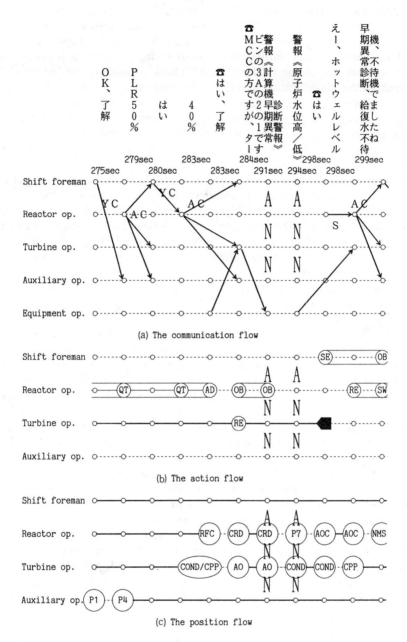

Figure 1. Example of the team activities when coping with simulated condition 3

## The action flow

The action flow expresses the elementary actions of the eyes and hands of operators in the control room.

The elementary actions during each speech should be classified into the 10 action categories (observe, detect, quantify, check, read, select, switch, adjust, write and telephone) and the action symbols should be entered on the action flow. It is easy to identify the kind of elementary action of the hands (for example, adjust or switch) from the videotaping. But it is difficult to identify the kind of elementary action of the eyes (for example, observe or quantify). If the operators continue to observe several indicators on the control panel without saying anything, it is not possible to identify whether the action is classified as "check", "quantify" or "observe". This study defines the classification criteria of the elementary actions of the eyes. First, the action symbols for "detect", "quantify", "check" and "read" should be entered only when speech related to the elementary actions is identified (for example the category for "The electric power is increasing" is "check"). Secondly, the action symbols for "observe" should be entered when the operators continue to observe without saying anything.

The notations for the continuous action are needed because many human actions are continuous. In general, human actions are classified into actions by the hands or by the eyes. There are some cases which apply to either of the two and some to both at the same time. As mentioned above, this paper defines the classification criteria for the elementary actions of the eyes. The actions classified in "observe" are only continuous ones. The notations of the continuous actions are defined so that continuous observations can be indicated by drawing parallel lines around the action symbols and continuous elementary actions of the hands can be indicated by drawing a line between the action symbols.

**The position flow**

What operators are saying and doing are described on the communication flow and the action flow. However, where operators are and what the operators are looking at are not described. The points the operators are looking at are very important for considering behaviour when collecting information. The position flow expresses the observation points.

The control panel is divided at every system to give a position code. The codes for the observation points the operators are looking at with interest should be entered on the position flow from the videotaping.

The operators sometimes change the observation points quickly or look at the same point for a long time without moving. If the operators change the observation points quickly without saying anything, all the position codes cannot be entered on the position flow. The following notations are then used in these cases. If the operators change the observation points quickly, the position code should be \*\*\*/+++. For example, if the operators glance quickly back and forth between the Grand Steam Seal System (GS) and the Air Off Take System (AO), enter "GS/AO". If the operators do not move, the position code they are looking at with interest should be entered at the start point and the end point on the position flow and a line should be drawn between them.

*Discussion on the effectiveness of this method*

The aim of this method is to describe team activities in detail in order to evaluate them. But important elements might be lost by describing them. Therefore, the operators' workloads are characterized from two aspects, the information processing workloads and the physical workloads based on results of team activities obtained by this method in order to confirm the effectiveness of the method.

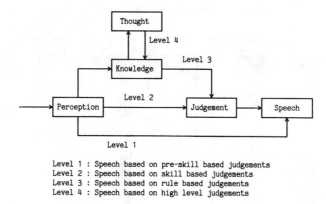

Figure 2. *A model of the differences in speech levels*

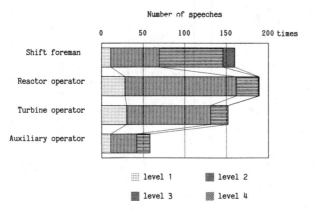

Figure 3. *The relationship between the operators and the speech levels for all conditions combined*

## The information processing workloads

From the contexts of the conversations, there seem to be some differences in the thinking processes. The simple model shown in Figure 2 is developed in order to determine the differences in the thinking processes from the operators' speech levels. This model consists of five elements (perception, knowledge, thought, judgement and speech) and divides the process from perception to speech into four levels according to the degree of difficulty.

Figure 3 shows the relationship between the operators and the speech levels for all simulated conditions combined. This shows that the shift foreman's speech based on levels 3 or 4 is more frequent than on levels 1 or 2, a large proportion of the speech by the other operators is on level 2 and the speech by the operators who take charge of the important events in the simulated conditions is of the higher level and the amount of speeches increases (in these simulated conditions, the reactor operator was the busiest of all and the auxiliary operator was the least occupied).

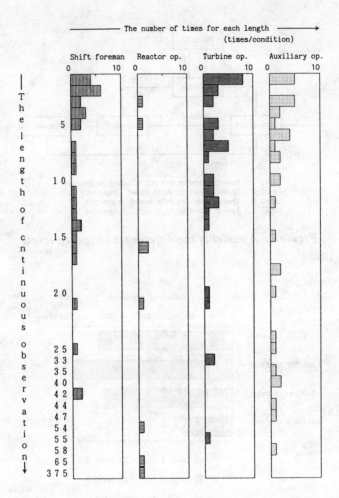

*Figure 4. The physical workloads by the length of continuous observations for simulated condition 2*

Based on the results of the team activities obtained by this method, the relationship between the operators and their information processing workloads could be determined.

### The physical workloads

If an abnormal event occurs and the plant's operating status continually changes, the operators will constantly watch several indicators on the control panel and act quickly. This paper determines the differences in the physical workloads from the action flow and the position flow.

First, based on the action flow, the differences in the physical workloads from the length of continuous observations are discussed. This "length" means the numbers of nodes on the action flow during which an action is executed, not the time elapsed. Figure 4 shows an example of the relationship between the length and number of continuous observations for the simulated condition 2.

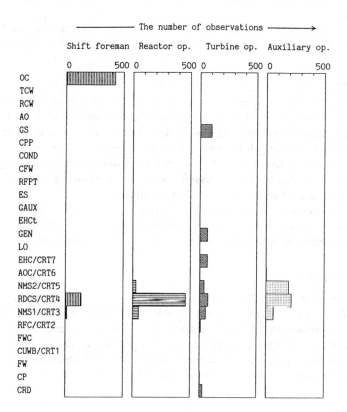

*Figure 5. The physical workloads by the number of observations at each observation point for simulated condition 2*

Figure 4 shows that the reactor operator continues to observe for a very long time. In this simulated condition, one of the control rods drifted out. The reactor operator then hastily tried to re-insert it but failed. That is to say, this event in the simulated condition made the reactor operator continue to observe for a long time.

In contrast, the shift foreman does not continue to observe for a long time. As described above, the speech level of the shift foreman is higher than the level of the others. The shift foreman thus gathers information about the plant's operating status from the others and he speaks based on the collected information.

Secondly, there are differences in the physical workloads due to the number of observations at each observation point. Figure 5 shows an example of the relationship between the operators and the number of observations at each observation point for simulated condition 2 based on the position flow.

From Figure 5, the shift foreman is in front of the operating console (OC) which is the centre of the control room for a very long time. That is to say, the shift foreman gathers the information to identify the plant's operational status, in front of the operating console (OC), which is the most suitable place to gather information and he speaks based on the collected information.

This shows that the other operators are observing the points around the main panel, especially RDCS/CRT4. As mentioned above, in this simulated condition one of the control rods drifted out and the reactor operator could not re-insert it. Therefore, all the operators gathered around RDCS/CRT4 where there was a display of the control rods patterns.

Based on the results of the team activities obtained by this method, the relationship between the operators and their physical workloads could be determined.

## Conclusion

This study focuses on the operators' communications, actions and observation points which were important in clarifying the team activities and develops the "Team Activity Description Method (TADEM)" which makes it possible to describe them in detail and with a relatively low dependence on subjective judgements.

To confirm the effectiveness of this method, the information processing workloads and the physical workloads are discussed, based on the results of team activities when coping with simulated abnormal operating conditions.

As a result, the effectiveness of the "Team Activity Description Method" is confirmed and it seems to provide a new approach for analyzing team activities when coping with abnormal operating conditions at a nuclear power plant.

## Acknowledgments

Syuzo Kawashima, Kouji Miyakita and Genpachi Saitou, BWR Operator Training Center Corporation, provided valuable technical background and insight into the simulated experiments. Hiroshi Kamiyama, Tokyo Electric Power Company, provided valuable technical background and insight into the development of this method. Special thanks are due to the instructors, BWR Operator Training Center Corporation, who participated in the experiments on which this study is based.

## References

Becker, G. 1987, Simulator Experiment as a Method of Giving Feedback to Improve Man-Machine Interaction in Nuclear Power Plants, IAEA Specialists' Meeting on Human Factors, RISO

Page, S. J. 1983, A Study of Control Room Team Activities during Simulated Plant Disturbances, Ergonomics Conference, Cambridge, UK

Sasou, K., Nagasaka, A. and Yukimachi, T. 1990, An Improvement of the Method to Describe Team Activities - On the Team Activity Description Method (TADEM), The Japanese Journal of Ergonomics, Vol.26, No.5, 251-260

Yukimachi, T. 1989, A Tentative Study on an Analytical Method for Team Activities with a Training Simulator, the Present Conditions and Views of the Human Factors Study on Nuclear Industries, Atomic Energy Society of Japan, pp.73-77

Yukimachi, T. 1989, A Study on the Analytical Method for Team Activities during Plant Disturbance, The Japanese Journal of Ergonomics, Vol.25, No.6, pp.357-366

# Part X
# Job Design

# 58. MACROERGONOMICS: A SOCIOTECHNICAL SYSTEMS APPROACH FOR IMPROVING WORK PERFORMANCE AND JOB SATISFACTION

## Hal W. Hendrick

*College of Systems Science*
*University of Denver*
*U.S.A.*

**Abstract**

Microergonomics is viewed historically as consisting of the development and application of two distinct technologies: Human-machine interface technology, concerned with human physical and perceptual characteristics; and user-machine interface technology, concerned with human conceptual and information processing characteristics. The emergence of a new technology that focuses on the organization-machine interfaces or macroergonomic level is reviewed. Empirically developed models of how key characteristics of a system's technology, personnel subsystem, and external environment impact on organizational structure to affect work performance and job satisfaction, and the use of these models in macroergonomic design are described.

**Keywords:** *job design; sociotechnical systems; human-machine interface; organizational design; cognitive complexity*

## *Introduction*

Ergonomics, or human factors as it is better known in North America, is now entering its sixth decade as an identifiable profession. During the first several decades, ergonomics was concerned with studying human physical and perceptual characteristics and translating these data into recommendations for the design of human-machine interfaces. This development and application of human-machine interface technology has had a tremendous impact on the safety, efficiency and comfort of many systems throughout the world. Of particular note has been the application of this interface technology to the design of transportation systems and to individual workstations and workspace arrangements in factories and offices.

Late in the third decade of our profession's history, as we began to work on computers, the way people think and process information became increasingly critical to system design. Software structure, as well as knowledge about how people process and use information became important ergonomics considerations for increasing the functional utility of computer-based systems. This second focus of ergonomics influences systems effectiveness through the development and application of user-machine interface technology.

The application of both human-machine and user-machine interface technologies will continue to be the major part our profession and to contribute to the betterment of system design for generations to come. It should be noted, however, that these two ergonomics technologies tend to focus on individual workstations and subsystems, or the microergonomic aspects of systems. With the progressively increasing automation of factory and office systems, we have begun to realize that it is entirely possible to do an outstanding job of microergonomically designing a system's components and subsystems, yet fail to reach relevant system effectiveness goals because of inattention to the macroergonomic design of the system. When this happens, individual and team work performance, system safety, motivation, and job satisfaction are all likely to be degraded.

During the past decade we have witnessed an emerging third focus of ergonomics and the related development of a new ergonomics technology. This new focus has been at the macroergonomic or overall sociotechnical systems level. It is concerned with the research, development and application of an organization-machine interface technology. I have labelled this new focus descriptively as macroergonomics.

Macroergonomics begins with an assessment of the system from the top-down, using a sociotechnical systems approach to organizational and system design. The notion here is that one cannot effectively design specific components of a sociotechnical system without first making scientific decisions about the overall structure of the organization, including how it is to be managed.

## *Organizational dimensions of macroergonomics*

The structure of an organization may be thought of as having three major components: Complexity, formalization, and centralization (Robbins, 1983).

### Complexity

Complexity refers to the degree of differentiation and integration that exists within an organization. Organizational structures involve three major types of differentiation. These are horizontal differentiation, or the extent of departmentalization and specialization of units; vertical differentiation, or the number of hierarchical levels; and spatial dispersion, or the degree to which an organization's facilities and personnel are geographically separated from the headquarters. Increasing any one of these three dimensions will increase the organization's complexity.

As the degree of complexity increases, the need for integrating devices also increases. This is true because with greater differentiation of an organization's activities, the difficulty of communication, coordination, and control increases. Some of the more common integrating mechanisms that can be designed into an organization's structure are formal rules and procedures, liaison positions, committees, system integration offices, and information and decision support systems. Vertical differentiation, in itself, also serves as an integrating mechanism for horizontally differentiated units.

### Formalization

From an ergonomics perspective, formalization can be defined as the degree to which jobs within organizations are standardized. In highly formalized organizations, jobs are designed so as to allow for little employee discretion over what is to be done, when or in what sequence tasks are to be performed, and how they will be accomplished. In organizations having formalization, the design of the human-machine and user-machine interfaces allows for considerably greater use of one's mental capacities.

### Centralization

Centralization refers to the degree that formal decision-making is centralized in an individual, unit, or level (usually high in the organization) thus permitting employees (usually low in the organization) only minimal input into decisions affecting their jobs. In general, centralization is desirable (a) when a comprehensive perspective is required, such as in strategic decision-making, (b) when operating in a highly stable and predictive environment, (c) for financial, legal and other decisions where they clearly can be done more efficiently when centralized, and (d) when significant economies clearly can be realized. Decentralization is to be preferred (a) when operating in a highly unstable or unpredictable environment, (b) when the design of a particular manager's job will result in taxing or exceeding human information processing and decision-making capability, (c) for enhancing employee motivation and job satisfaction, and gaining greater employee commitment, (d) when "grass roots" inputs to decisions are wanted, and (e) for providing greater training opportunities for low-level managers.

## Sociotechnical system considerations in macroergonomics

To more descriptively convey the nature of complex, human-machine systems, Emory and Trist (1960) coined the term sociotechnical systems. The sociotechnical systems concept views organizations as open systems engaged in transforming inputs into desired outcomes. Organizations are viewed as open because they have permeable boundaries exposed to the environments in which they exist.

Organizations bring two major sociotechnical system components to bear on the transformation process: technology in the form of a technological subsystem, and people in the form of a personnel subsystem. These two subsystems interact with one another at every human-machine and user-machine interface. The two subsystems thus are interdependent and operate under joint causation, meaning that both subsystems are affected by causal events in the environment. Joint causation gives rise to the related key sociotechnical system concept of joint optimization. Since both the technological and personnel subsystems respond jointly to causal events, optimizing one subsystem and then fitting the second one to it will result in suboptimization of the joint system. Joint optimization thus requires joint design.

As inferred above, the design of a sociotechnical system's structure involves consideration of the key characteristics of three major sociotechnical system components: the technological subsystems, the personnel subsystems, and the external environment. Each of these sociotechnical system components has been studied in relation to its effect on the elements of the fourth major component, organizational structure, and empirical models have emerged.

### Technological subsystem

To date, the most thoroughly validated and generalizable model of the technology-structure relationship is that of Perrow (1967). Perrow utilizes a knowledge-based concept of technology. In his classification scheme, Perrow begins by defining technology as the action that one performs upon an object in order to transform it. This action requires some form of technological knowledge; thus, technology can be categorized by the required knowledge base. Using this approach, he identified two underlying dimensions of knowledge-based technology: Task variability, or the number of exceptions encountered in one's work; and task analyzability or the type of search procedures one has available for responding to task exceptions (rational/logical vs. experience and intuition). These two dimensions, when dichotomized, yield the following matrix. Each cell represents a different knowledge-based technology as shown in Figure 1.

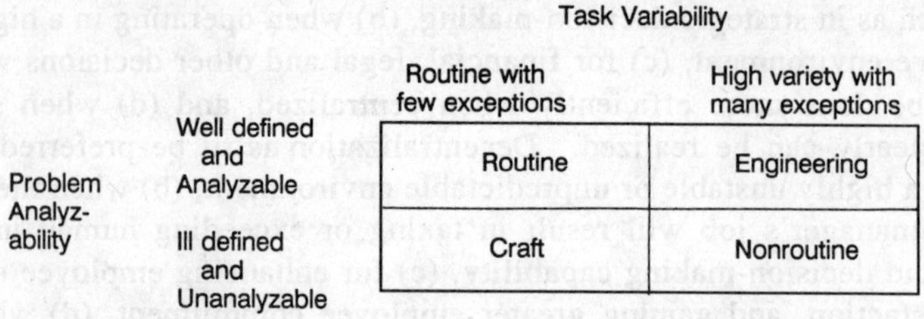

|  |  | Task Variability | |
|---|---|---|---|
|  |  | Routine with few exceptions | High variety with many exceptions |
| Problem Analyzability | Well defined and Analyzable | Routine | Engineering |
|  | Ill defined and Unanalyzable | Craft | Nonroutine |

*Routine* technologies have few exceptions and well defined problems. Mass production units most frequently fall into this category. Routine technologies are best accomplished through standardized coordination and control procedures, and are associated with high formalization and centralization.

*Nonroutine* technologies have many exceptions and difficult to analyze problems. Most critical to these technologies is flexibility. They thus lend themselves to decentralization and low formalization.

*Engineering* technologies have many exceptions, but they can be handled using well-defined rational-logical processes. Accordingly, they lend themselves to centralization, but require the flexibility that is achievable through low formalization.

*Craft* technologies involve relatively routine tasks, but problems rely heavily on experience, judgement, and intuition for decision. Problem-solving thus needs to be done by those with the particular expertise. Therefore, decentralization and low formalization are required for effective functioning.

*Figure 1. Perrow's knowledge-based technology classes*

## Personnel subsystem

At least two major characteristics of the personnel subsystem are important to an organization's structure. These are the degree of professionalism and psychological characteristics of the workforce.

*Professionalism* refers to the amount of education and training possessed by the personnel in the organization or its constituent units (and, presumably, required by their respective jobs). It should be noted that formalization can occur either on the job, or off (Robbins, 1983). When done on the job, formalization is external to the employee. In contrast, professionalism creates internalized formalization of behaviour through a socialization process that is an integral part of education and training. Ergonomically, the greater the degree of professionalism the lower the degree of formalization that should be designed into the system, and vice versa.

*Psychological factors.* I have found that the most useful integrating model of psychological influences on organizational design to be that of cognitive complexity. Harvey, Hunt and Schroder (1961) have identified the higher-order structural personality dimension of concreteness-abstractness of thinking or cognitive complexity as underlying different conceptual systems of perceiving reality. Regardless of our culture, we all start out in life conceptualizing reality concretely. As we gain experience we become more abstract or cognitively complex in our conceptual functioning. For a variety of reasons, adults plateau at different levels of complexity in their development. Concrete adult functioning consistently has been found associated with a high need for structure and order for stability and consistency, closedness of beliefs, authoritarianism, absolutism, paternalism and ethnocentrism. Relatively concrete persons tend to see their views, values, and institutional structures as static and unchanging. In contrast, cognitively complex adults tend to have a low need for structure and order, openness of beliefs, and relativistic thinking; they tend to be less authoritarian and to have a dynamic conception of their world - they expect their views, values, norms and institutional structures to change. My evidence suggests that relatively concrete work groups and managers function best under comparatively high vertical differentiation, centralization, and formalization, and vice versa (Hendrick, 1981; Hendrick, 1979).

**Environment**

The success of an organization depends on its ability to adapt to its external environment. Of the various characteristics of sociotechnical system environments, by far the most critical to macroergonomics are the degree of change and complexity (Duncan, 1972). The degree of change refers to the extent the environment remains dynamic or stable over time. The degree of complexity refers to whether the components of the organization's environment are few, or many in number. These two dimensions in combination determine the environmental uncertainty of an organization. With a high degree of uncertainty, a premium is placed on an organization's ability to be flexible and rapidly responsive to change; with low uncertainty, maintaining stability and control becomes most important. Thus, the greater the environmental uncertainty, the greater is the need for low vertical differentiation, decentralization of tactical decision making, and for low formalization coupled with a relatively high degree of professionalism; conversely, organizations with environmentally certain environments are most efficient with relatively high vertical differentiation, centralization and formalization (Burns and Stalker, 1961; Emory and Trist, 1960; Lawrence and Lorsh, 1969).

## Integrating micro with macro ergonomic design

Through a macroergonomic approach to determining the optimal design of a system's organizational structure, many of the characteristics of the jobs to be designed into the system already are determined. For example, horizontal differentiation decisions prescribe how narrowly or broadly individual jobs must be designed; decisions concerning the degree of centralization will determine the degree of decision-discretion to be included in a given job's design; the degree of formalization will dictate the extent to which functions are to be routinized in the job. Each of these job design dimensions impacts on the attendant hardware and software design applications of the system's technology - particularly the attendant design of the human-machine and user-machine interfaces. For example, the degree of formalization and centralization ergonomically designed into a given position determines the information requirements for that position; this, in turn, drives the design of the information and decision support systems for that workstation, and attendant controls, displays and workspace arrangements.

In summary, effective macroergonomic design drives much of the microergonomic design of the system, and thus ensures optimal ergonomic compatibility of system components with the system's overall structure. The result is greater assurance of optimal system functioning and effectiveness, including productivity, safety, comfort and intrinsic employee motivation.

## References

Burns, T., and Stalker, G. M. (1961). The management of innovation. London: Tavistock.

Duncan, R. B. (1972). Characteristics of organizational environments and perceived environmental uncertainty. Administrative Science Quarterly, September, 315.

Emory, F. E., and Trist, E. L. (1960). Sociotechnical systems. In C.W. Churchman and M. Verhulst (Eds.), Management science: models and techniques (vol. 2). Oxford: Pergamon.

Harvey, O. J., Hunt, D. E., and Schroder, H. M. (1961). Conceptual systems and personality organization. New York: Wiley.

Hendrick, H. W. (1981). Abstractness, conceptual systems, and the functioning of complex organizations. In G. W. England, A. R. Nagandhi, and B. Wilpert (Eds.), The functioning of complex organizations (pp.25-50). Cambridge, MS: Oelgeschiager, Gunn & Hain.

Hendrick, H. W. (1979). Differences in group problem-solving behavior and effectiveness as a function of abstractness. Journal of Applied Psychology, 64, 518-525.

Lawrence, P. R., and Lorsh, J. W. (1969). Organization and environment. Homewood, IL: Irwin.

Perrow, C. (1967). A framework for the comparative analysis of organizations. American Sociological Review, April, 194-208.

Robbins, S. R. (1983). Organizational theory: the structure and design of organizations. Englewood Cliffs, NJ: Prentice-Hall.

# 59. OPERATOR CONTROL IN MODERN MANUFACTURING

### John R. Wilson

*Institute for Occupational Ergonomics*
*Department of Production Engineering and Production Management*
*University of Nottingham, Nottingham NG7 2RD, England*

**Abstract**

Over a period of some years a number of investigations have been made into technical and organisational changes in manufacturing and service industries. The focus of the studies has embraced both improvements in work effectiveness and also the development of jobs and roles which give greater autonomy to operatives. In this work, a broad view has been taken of an appropriate framework for the studies, in terms of legitimate areas of concern for ergonomists. This paper examines some of the relevant issues, especially in the context of cell-based flexible manufacturing with modern approaches to organisation and management.

**Keywords:** *job autonomy; advanced manufacturing technology; just-in-time manufacturing; job design; supervision*

## Introduction

If they were to assess the single most important attribute of work that gives satisfaction and of work that does not stress, then many researchers would talk about being "in control". Workers themselves would talk about "knowing what I'm doing", about "being given some freedom", about "feeling I'm in charge of something". Work redesign scientists would mention autonomy as a vital job characteristic, about experienced responsibility for work outcomes, about self-regulation and so on. This key to good work design can be seen in many approaches, from autonomous or semi-autonomous work groups (AWG/SAWG), to matrix-style organisation structures, to vertically enriched jobs.

It is also apparent that there will be a considerable interaction between the concept of autonomy or control in individual or group work, and modern approaches to manufacturing. The latter include the philosophy of just in time (JIT), systems of cell-based or group technology manufacturing, and techniques such as Optimised Production Technology (OPT). Furthermore, when we look at local control, that is control over manufacturing processes at operator or cell level, then there is a further interaction with the human-machine interface. An operator's ability to control a process, and to make on-line decisions, will depend upon the form, content, quality and timeliness of the information available.

There is then, in manufacturing industry, a strong relationship, and optimistically a symbiotic one, between the content of jobs, the quality of human-machine interfaces, and philosophies and approaches to production control. To make a substantial contribution to effective and quality manufacturing practices, ergonomics will need to integrate its theories, models and practices from its psychosocial, cognitive and management components.

In this paper some of the issues relevant to operator local control of manufacturing processes are outlined. Most of them represent problems or ideas still to be investigated and developed; it is a wide open field.

People will be required to organise and control factories in the foreseeable future; this is really not in doubt. The concept of total automation, of human-less factories, belongs to science fiction of the nightmarish variety. Many reasons can be put forward for this. Amongst them have been cited the following (Wilson, 1989):

1. Despite vast leaps in microprocessor technology development, the technical capabilities of advanced manufacturing technology (AMT) and computer integrated manufacturing (CIM) are not all that is sometimes claimed for them.
2. Automated systems would have difficulty in dealing with high level information, with information from the environment that is incomplete, abstract or disordered. The ability of people to deduce, induce, adapt or purely to make inspired guesses, and then to respond appropriately, will be very difficult if not impossible to replace.
3. Whatever might be possible as regards human resource policies inside an organisation, there will be a need to communicate and interact with people in the supply, production and consumer chains outside.
4. The innate abilities of people will ensure that they will be required to enable effective and reliable production systems. Indeed they are often required to overcome the limitations of inadequate systems, whose designers and planners would probably have preferred to have eliminated all operators in the first place.
5. The knowledge and skills of a company's staff comprise a valuable, and perhaps the most important, company resource. Sinclair (1989) has valued this resource at about 30% of a company's market valuation. It thus makes sense to utilise such a resource to the maximum extent possible.
6. Use of such knowledge and skills of the staff, and increasing work content in terms of skills, autonomy and variety, helps meet the requirements of good job design. To do so will in turn produce benefits for both organisational and individual well-being.

Given the above, then work design practitioners will argue not just **against** the concept of the person-less factory but also **for** a human-centred approach to manufacturing systems design. This implies that manufacturing be developed with a view to complementarity, coordination and communication between people and between people and machines. Systems must make best use of the qualities of modern technology **and** of an enterprise's people, but within complete, integrated roles and functions. What should be aimed for is a system of decision support: people determine when and where to intervene, optimise, innovate, diagnose and rectify, guided by appropriate information from the system. Together, the person and the technology determine what inputs should be made, carry them out, monitor their effects and adapt inputs accordingly.

The control implicit in this role for operators at a local level will be primarily supervisory control (Barfield et al., 1986). Planning is done in the scheduling of production, maintenance, and machine utilisation; in inventory control; and in capacity planning. Teaching is largely in the form of programming - NC part programming or robot programming or in the setting of job priority and dispatching rules. Monitoring will be of single but also of simultaneous processes, such as several material flow lines, and of the variety of information sources needed for successful system functioning. Intervening may take place in emergencies, to optimise systems, change batches and so on. A final element of supervisory control, the learning of operators (Sheridan, 1988), is also a well-recognised building brick of "world class" manufacturing enterprises.

## Local control - issues for ergonomics

### Autonomy and supervision

Local control - at the level of the operator in the AMT process or flexible manufacturing cell (FMC) - is a major component of the general work redesign approach of introducing autonomy into work roles. The more that jobs require people to exercise control over their own work, the greater are said to be the psychological benefits to them (Wall, 1986). Autonomy itself is at the core of frameworks for individual and group work redesign as well as of the job design models and theories underlying such redesign. There is a need for ergonomists to embed local control of processes within work restructuring which supports autonomy at work.

Control or autonomy is intimately related to supervision and supervisory practices, especially in a (semi-) autonomous work group change. Cordery and Wall (1985) believe that supervisory behaviour is the key route to implement autonomy at work. It is possible however that a broader brush approach is needed, with several mechanisms and practices set up to promote and support autonomous working. Autonomy is not a simple outcome; it does not just happen. Like a participative process it has to be directed - even if subtly. It also depends upon good in-company communication, and not the typical approach of filtering and censoring information as it goes up the company hierarchy, and restricting and inflating it as it comes downwards. Autonomy also is not unitary; it has several different possible foci and each demands a different emphasis and procedures. Workers can have control over methods, targets and schedules, timing, criteria for acceptance of output and so on. Attempts to provide supervisory control at cell operation level should be seen to promote, not counteract, other efforts at autonomy in planning and organisation.

### Task and interface design

Work redesign which meets the needs of company and workforce embraces such initiatives as operator local control, skilled and reliable performance, feedback of the results of actions taken, and support for successful local decision making. For all these there must be provision of the right sort of information in the right way at the right time. Off-line this will involve considering training systems, formal and informal operating and emergency procedures, communication structures and participation networks. On-line, information will come through process and system displays. Considerable work is needed in the development of these displays; human factors work on human-computer interfaces for AMT is in its infancy compared to what has been carried out for office automation or in continuous process control industries.

The interface, both the physical interface and the cognitive interface, will affect the everyday work content and role of an operator. The degree of control and breadth of influence exerted will be determined by: the task behaviours (skill-, rule- or knowledge-based, see Rasmussen, 1986) that the controls and displays will support; the extent to which the displays match, enhance and strengthen the operators' mental models (Wilson and Rutherford, 1989); and the degree of display understanding and interpretation we can promote. There is a direct relationship between the quality of human-machine interfaces and the work system outcome variables of worker performance and attitudes. There is also an indirect relationship in the way that the interface interacts with job content and characteristics, thereby affecting performance and attitudes (Wilson and Grey, 1990). Specification of the interface can limit or enhance job content; the needs of good job design should therefore influence display and control development.

## Operator knowledge and decision making

People are required to control manufacturing systems, but they are not infallible. Far from it. Task, job and equipment design must account for any human decision making and problem solving deficiencies and for potential errors in skilled task performance.

Industrial staff will interact with controls and displays of advanced manufacturing processes or cells, with scheduling system information displays, with CAD output, with cell balancing data output and so on. Design and organisation of these systems must be in the light of knowledge of human abilities to make decisions, to reason and to make accurate and reliable control actions. Personnel (whether in roles of operating, maintenance, set-up, etc.) will interact with the technical, information and social systems implied within AMT, at all levels. Ergonomists need to investigate and provide empirical and experimental information on people's abilities to optimise or 'tweak' systems; on their fault detection, diagnosis and rectification performance; on their strategies for intervention; and on human reliability in all these functions and more.

Related to this is the issue of how to encapsulate design and production knowledge within a manufacturing data base; this is particularly pertinent for CIM on a large-scale. One research requirement is to establish or adapt methods for acquiring and formalising (without corrupting) the knowledge of design engineers, production engineers, etc. This must then be incorporated into the CIM information systems. The second research requirement involves the related question of how best to aid the designers of advanced systems and subsequently also how to aid designers **using** the systems, specifically CAD/CAM. Many inputs must be made by human factors professionals here. These include: studies of the cognitive psychology and the ergonomics of CAD itself; examination of how designers actually design; provision of human factors design guidelines to be used by designers; and study of the best ways in which to convey design information through the information systems (including CAD) in a modern factory.

## Influence of production systems

It is natural that many developments in job and work design occur in the context of changing technologies. Such technology change can provide the impetus, environment, and vehicle to change jobs also; work designers can ride in on the back of technological change. We need to know what features of AMT and production control might be harnessed to assist in job design, and which features might hinder. Of course, much depends upon the approach and desires of management and their convictions as to the benefits to be gained from autonomy and local control over work.

Job design changes in manufacturing industry will be context specific, often production change driven and always production change oriented (Susman and Chase, 1986). Newer approaches to production, Total Quality Management, group technology, or organisation for JIT, will provide both opportunities and also their own restrictions for changes in work design.

Group technology for instance can provide a very convenient vehicle on which job redesign, in terms of autonomous groups, can hitch a ride; increased autonomy can logically parallel group technology. Other generally desirable job characteristics can be enabled also - for instance task identity or skill variety. The increase in application of JIT principles in manufacturing industry also has a number of implications for job design, in particular posing a challenge in terms of its compatibility with the objectives of job enrichment and job enlargement. At the least JIT may serve to highlight the restrictive nature of those practices of scientific management which encourage high task specialisation, minimal training, individual financial incentives and directive supervisory styles.

Two positions may be taken over the effects on workers of changes in production control, and especially the effects of JIT. It may be that increased workforce autonomy, skills and involvement will be needed to make JIT work. Policies such as worker flexibility and participation, together with a higher level of shopfloor training, "facilitative" supervision and a move away from high individual incentive payment systems may be appropriate to the application of JIT on the shopfloor. In this sense, there would be at least a superficial similarity between the objectives of job redesign and the personnel-related requirements of successful JIT implementation. On the other hand it may be that the successful use of JIT will mean reductions in all production variances, and hence will interfere with some of the "building bricks" of true worker involvement and control. As yet there is a paucity of analyses of these alternative scenarios but literature is beginning to emerge (see Wilson, 1990, for a more complete review).

## Conclusions

The ways in which new manufacturing systems - organisational and technical - are implemented, and in which the work of relevant staff is redesigned, will go a long way to determining the success of manufacturing companies in the future. Within a human-centred approach to systems design and implementation we must consider such factors as: the content of people's jobs and their roles, in terms of aspects like autonomy, feedback, and significance; the technology itself; the physical environment and workplace; support procedures and facilities; and work context factors (Wilson and Grey, 1990). To these must be added the interface through which operators receive information about the systems and enact control, and the philosophy and system of production control employed.

The notion of control lies at the heart of, first, most current developments to job redesign (which embrace the primary role of autonomy), second the requirements of a modern, flexible, quality-oriented firm, third the concepts of decision support systems and dynamic allocation of function at the interface, and fourth (in the view of some if not all scientists) manufacturing management based on just-in-time practices. Good work redesign will enable shopfloor control, for workers to self-organise individually or in groups and to develop their own targets and means of meeting them. For a company to react quickly and innovatively to demands from the environment could necessitate that management place greater trust and power in the hands of operators. To make Total Quality Control and JIT work, to produce the claimed benefits, may further emphasise this need for local control. Whether these changes in manufacturing would naturally have a restrictive or a expansionist effect on the autonomy of workers is the subject of some debate, and of on-going research. What they do offer is the chance to institute a system of local control, at operator level, in manufacturing enterprises. This is the challenge for ergonomists.

## References

Barfield, W., Hwang, S-L. and Chang, T-C., 1986, Human-computer supervisory performance in the operation and control of flexible manufacturing systems. In: Flexible Manufacturing Systems: Methods and Studies (ed: A. Kusiak). Amsterdam: Elsevier, 377-408.

Cordery, J.L. and Wall, T.D., 1985, Work design and supervisory practice: a model. Human Relations, 38, 425-441.

Rasmussen, J., 1986, Information Processing and Human-Machine Interaction. Amsterdam: North Holland.

Sheridan, T.B., 1988, Task allocation and supervisory control. In: Handbook of Human-Computer Interaction (ed: M. Helander). Amsterdam: North-Holland, 159-173.

Sinclair, M.A., 1989, Human Factors and Organisational Issues in AMT. Paper for ACOST Emerging Technologies Committee. Loughborough: HUSAT Research Centre.

Susman, G.I. and Chase, R.B., 1986, A sociotechnical analysis of the integrated factory. Journal of Applied Behavioural Science, 22, 257-270.

Wall, T.D., 1986, The Human Side of New Manufacturing Technology. Memo No. 808 from MRC/ESRC Social and Applied Psychology Unit, Dept. of Psychology, University of Sheffield.

Wilson, J.R. and Grey, S.M., 1990, But what are the issues in work redesign? In: Work Design in Practice (eds: C. Haslegrave, J. Wilson, E. Corlett). London: Taylor and Francis, 7-15.

Wilson, J.R. and Rutherford, A., 1989, Mental models: theory and application in human factors. Human Factors, 31, 617-634.

Wilson, J.R., 1989, People and modern manufacturing. Paper presented at the 4th Annual Conference of the Brazilian Ergonomics Society, Rio de Janeiro.

Wilson, J.R., 1990, Control and jobs in manufacturing systems: the effect of JIT/TQM initiatives. Proceedings of the 2nd International Conference on Human Aspects of Advanced Manufacturing and Hybrid Automation, Honolulu, pp.501-508.

# Closing Remarks

# 60. SOME FUTURE DIRECTIONS FOR ERGONOMICS

## E. Nigel Corlett

*Institute for Occupational Ergonomics*
*University of Nottingham*
*University Park, Nottingham NG7 2RD England*

**Abstract**

The importance of the contribution of good working conditions to human health and performance is outlined. From this position, the importance of an ethical standpoint for ergonomists in their work is discussed. Given that the performance of people depends on more than the early ergonomic measures of anthropometry, work physiology and information manipulation, the need for a cost benefit analysis to incorporate the important work influences is presented.

**Keywords:** *quality of working life; job design; cost benefit analysis; ethics; future of ergonomics*

Throughout the conference it has been shown that there is a widespread recognition of the integrative nature of people. Our experiences are modified internally so that our responses are more varied than a one-to-one correlation with the stimulus. In any design or other ergonomics decision, therefore, it is necessary to bear in mind that, whether we are dealing with handle size or an organization, it must be considered in terms of whether or not it matches human requirements on many dimensions. We are not technologically or organisationally or economically centred in our emphasis as ergonomists, but human centred. Our studies show, and our experience confirms, that where we create conditions which match people's requirements, the results in terms of performance, health and attitudes are better than where such matching has not been done.

Cynics will oppose this view, taking the "practical" viewpoint that it is not economic. What is economic, of course, depends on who is paying. Where the costs, eg of ill-health, pollution or inefficient work can be taken over by society in general, then it may be cheaper but history has a way of hitting back at those who do not learn its lessons. In the West the conditions of the nineteenth century cast a long shadow on the future, the social, political and industrial consequences of which are still much in evidence in the attitudes and behaviour of management, unions and politicians. The almost total lack of understanding of the consequences of neglecting human needs in relation to industry is now evident in Eastern Europe and in several third world countries. The scale of pollution and ill-health has loaded such a burden onto their societies that economic advance is threatened, whilst people's resentment at their treatment is not conducive to efficient work performance.

These are broad generalisations, but have relevance to the work of the individual ergonomist. The problems of technology transfer are not confined to the third world, for the ergonomist they are always present, together with the associated problems of cultural change. The recognition of our technical contribution is difficult to win in commerce and industry. It is equally hard to achieve a recognition that a working culture, which sees people as expendable units of production whom it is appropriate to assess on a simple direct costs basis, is not suitable for a modern society.

Where we stand now, as a profession, is not yet satisfactory, in terms of our economic, social or philosophical beliefs. We are not a contribution of methods engineering for modern management, although one of our goals is very similar - the creation of efficient working conditions. Our definition of efficiency, however, is difficult and must be different for the reasons outlined above - ergonomics deals with the whole person in the context of the whole environment.

So we must have faith in our principles and exercise our knowledge and opportunities to provide pressure for change. Again history tells us that it will not be easy, there will be pioneering companies but the majority will be reluctant to modify their approaches. It will be a long haul. But if we ourselves do not change, then it will be even longer, and in several Western nations it is already evident that many work situations are out of step with what society leads their workpeople to expect for their lives. This differentiation between expectations and experience hinders efficient, long-term operation of industry.

To change these situations needs more than determination and a persistence at sticking to our human-centred perspective on work and working conditions. We are involved in a "technology transfer" as well as system change. We must work to transfer our knowledge to practice, to develop the "art" of our subject rather than just the knowledge base. If we are to see the benefits of ergonomics spreading widely across employment and society in general we need to pursue two objectives.

The first of these is to "give ergonomics away", to transfer our knowledge and methods to others who are closer to the places where changes have to be made, so that they do much of the ergonomics for themselves. Ergonomists must see that they have a teaching role so that ergonomics is seen, and implemented for what it really is, rather than what people, as a result of a lack of understanding, think it is. Of course there will be occasions when ergonomics is mis-used, or badly used, but with experience it will improve and the knowledge itself is a good protection from abuse. Until ergonomics is widely practised by other than professional ergonomists it is likely to remain something to be added on at the end. What is more, until it is more widely realised what ergonomics really is, we will not get it introduced into design specifications, where human performance and requirements become a normal part of the design criteria. Until this becomes the accepted practice, the implementation of ergonomics will always be an uphill fight.

The second target for our endeavours is the creation of ways to demonstrate that our interventions are beneficial to employers, employees and others. All too often ergonomics is seen in welfare terms, and as a cost with little return. "We can't afford to make them comfortable" not only demonstrates a lack of knowledge of ergonomics, it demonstrates a lack of understanding of, and identity with, the very people on whom the organisation is totally dependent for its continuity and success.

To change such a view, exhortation is of little use. We need solid evidence in a form which management can understand, and a strategy for putting it across. An example of such a strategy has been outlined in an earlier paper in this volume. For the structure and coverage of the evidence, however, we have again to take a broader view than is common. We need to recognise the costs to the organisation of poor quality, absenteeism, protracted labour negotiations, obsolescence and resistance to change and all those other factors which prevent the organisation responding to competition from technology, other companies, the market or customer demands. The need is for a procedure for analysing the costs and benefits of change or operation where, for example, the stresses on people are taken into account as a cost with the other more conventional costs, and the improvements in attitudes, in health and in motivation as benefits (Corlett, 1988). To look only at the increase in output or quality is to ignore the basic ergonomics concept (and practical reality) of the persons being the central point of interest and of assessing other factors in relation to human responses.

One model for such a cost benefit analysis used the concept of matching between people's requirements on the one hand and technical and organisational ones on the other. The accordance existing between these was measured by a range of instruments and cost models developed by Schiro (1985). It was recognised that ergonomic changes, whilst having direct effects on errors, output, health or safety for example, would in turn reflect into the personnel, production, management or quality sub-systems of the organisation. These have long-term results on the profitability and continuity of the business. This whole chain of events is mediated by the interaction between the ergonomic quality of the changes and the human responses to these, illustrated in simplified form in Figure 1.

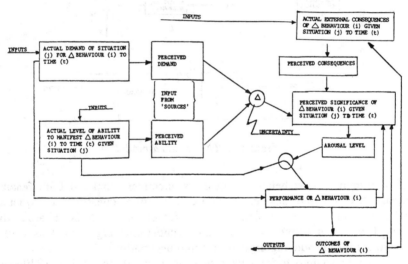

Figure 1. The process model

We can see here a typical interactive model of behaviour, where the output is affected not just by the input, but by the beliefs held by the individuals concerning their own abilities, goals, their motivation etc, where we match task demands to the human factors - by modifying tasks etc as well as changing situations so that effort is seen to be rewarding, the improved accordance can allow a better level of performance. The paper by Hendrick in this volume, covering what he describes as macroergonomics, reinforces and enlarges on this theme.

This qualitative explanation can yield to measurement. Work by Macy and Mirvis (1976) at the Social Research Institute of the University of Michigan conceived the idea of a profile of attitudes linked with measures of performance. The costs of changing the profile could be calculated, as could the changes in performance, and both of these could be linked with the particular attitude dimensions which had been modified. Appropriate social science models provided possible causal links for the observed results, as well as a rationale for deciding which attitude dimensions it was appropriate to change to achieve the performance alterations needed.

The simplified diagram of Figure 2 shows the concepts underlying this process of matching. The level of accordance between performance factors, as well as with motivational/social psychological factors are assessed. The impact of creating both a more possible job (by removing physical constraints for example) and one where the interest, responsibility and authority are within the envelope of adult needs (motivational accordance) lead to performance improvements which themselves reflect back into aspects of well-being which reinforce the main components, leading to further improvements and stability of the system.

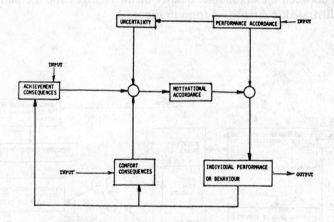

Figure 2. The reduced model

Such a system of cost benefit analysis is unconventional but both feasible and practical. It could give answer to such disconcerting questions as "what are the benefits from improving job satisfaction?" for example. It is an area which, if developed, would improve the lot of the occupational ergonomist as well as the opportunities for introducing ergonomics into occupations.

What it could also do is to force us to face more directly the ethical problems which underlie all our activities. We would spell out those as yet intangible factors which ill-informed people believe are matters only of goodwill, or for the soft-hearted. It could be shown that treating people as people just like ourselves, and distributing authority to the points where it was most effectively implemented for the benefit of the organisation, were realistic, common-sense directions for organisations wishing to utilise their resources efficiently.

## References

Corlett, E N (1988) Cost-benefit analysis of ergonomic and work design changes. In: International Reviews of Ergonomics (ed. D J Oborne). Taylor and Francis, London.

Macy, B A and Murvis, P H (1976) A methodology for assessment of quality of work life and organisational effectiveness in behavioural-economic terms. Administrative Science Quarterly, 21, 212-216.

Schiro, S G (1985) Predicting the potential benefits of work design changes. Unpublished PhD thesis, University of Birmingham, UK.

# Author Index

Aaras, A. 115
Aberg, E. 358
Abeysekera, J.D.A. 297
Ahonen, E. 280
Akiya, S. 316, 352
Aoyama, T. 243

Balliett, J. 156
Bao, S. 99
Buckle, P. 108
Bullock, M.I. 93

Cai, R. 327
Chavalitsakulchai, P. 99
Choi, J.H. 373
Chou, F.S. 80
Chung, M.K. 373
Cohen, B.G.F. 203
Coray, K.E. 203
Corlett, E.N. 179, 417

Dainoff, M.J. 156
Darby, F.W. 364
Dupuis, C. 297

Erneling, L. 308

Feng, A. 222
Feng, G. 31
Fu, Z. 222

Gao, C. 327
Gardner, A.W. 65
Gormley, J. 145

Haslegrave, C.M. 151
Hattori, Y. 387
Helander, M.G. 193
Hendrick, H.W. 403
Hiraoka, Y. 226, 234
Hirose, N. 316
Hiruta, S. 263
Hisanaga, N. 263
Hisashige, A. 133
Holmer, I. 297

Horie, Y. 250
Hsu, S.H. 85
Huang, H. 304
Huang, J. 263

Ishida, Y. 170
Ishihara, S. 346
Itoh, J.I. 387
Iwaki, K. 387
Iwasaki, T. 352
Izumi, S. 255

Jeyaratnam, J. 333
Jung, E.S. 373

Karwowski, W. 68
Kee, W.C. 333
Kishida, K. 269
Klen, T. 280
Koda, S. 133
Kononen, U. 280
Koshi, K. 316
Kumashiro, M. 216, 255
Kuorinka, I. 55
Kwon, Y.G. 289

Lee, D.H. 210
Lee, S.D. 35
Li, T. 222, 304
Liu, X. 75
Liu, Z. 304
Lu, D. 31, 327
Luopajarvi, T. 126

Ma, A.W.S. 80
Megaw, T. 42
Mikami, K. 255
Miyao, M. 346
Mukund, S. 193
Munipov, V.M. 380

Nagasaka, A. 393
Nygard, C.H. 126

Oda, M. 226, 234

Ohkubo, T. 25
Okuda, H. 226, 234
Ong, C.N. 333
Ono, Y. 263
Orring, R. 186, 308
Oshima, M. 18

Park, K.S. 210
Patkin, M. 145
Piirainen, J. 280
Piotrkowski, C.S. 203

Saito, S. 316, 340
Sakamoto, H. 60
Sasou, K. 393
Seo, Y.J. 216
Shackel, B. 3
Shahnavaz, H. 99
She, Q. 327
Shibata, E. 263
Shimaoka, M. 263
Soued, A. 142
Stubbs, D. 108
Sugiyama, S. 49

Takeuchi, Y. 263
Tanaka, J. 226, 234
Taptagaporn, S. 316, 340
Tomizawa, T. 387

Umemura, M. 243

Venalainen, J. 280

Wang, M.J.J. 80, 164
Westlander, G. 358
Wilson, J.R. 409
Wu, Z. 222

Yang, L. 327
Yoshizawa, H. 226, 234
Yu, Y. 304
Yukimachi, T. 393

Zhang, C. 222
Zhang, G. 327
Zhang, X. 222
Zheng, L. 222

# Keyword Index

Absenteeism   99, 115
Accident prevention   3, 75
Accidents   68, 75, 80, 85, 133, 142, 380
Advanced manufacturing technology   409
Ageing   18, 255, 263, 333
Aircraft   308
Arousal   216, 327

Back pain   99, 126, 133, 145
Biomechanics   151, 170
Brick making   75

Cabin attendants   308
Catecholamines   327
Chair design   156
Chemical hazards   42
Cleanrooms   289
Clothing   289
Clothing materials   164
Cognitive complexity   403
Colour coding   373
Communication   393
Commuting   243
Cosmic radiation   308
Cost benefit analysis   108, 417
Critical fusion frequency   250, 255, 327, 352
CRT displays   340, 346

Data entry   327, 333
Design methods   55
Designers   243
Developing countries   31, 35, 99
Discomfort   55, 156, 164, 179, 269, 289, 297, 304, 308, 316, 333, 340
Display design   373
Display format   373
Display polarity   340, 346

Education in ergonomics   31, 93, 179
Electromyography   115, 280

Environment   42, 289, 297, 304, 308, 340, 364
Environmental temperature   304

Ergonomics intervention   49, 80, 108, 179, 380
Ethics   65, 417
Ethnic groups   333

Factor analysis   216
Fatigue   55, 226, 234, 255, 269, 308
Fault diagnosis   85
Fault tree analysis   75
Forestry   280
Furniture   364
Furniture industry   80
Future of ergonomics   3, 18, 25, 35, 49, 417

Glare   316
Gloves   164
Grip strength   164

Health and safety   55, 60, 65, 222, 308
Health care   60, 65, 93, 126
Health education   60
Heart rate   133, 210, 243, 280
Heat stress   297
Heavy manufacturing industries   216
History of ergonomics   3, 18, 25, 31, 35, 49
Home life   226, 234, 269
Hospitals   222
Human error   75, 85, 380
Human-computer interface   186, 373
Human-machine interface   3, 403

Illumination levels   340
Industrial relations   216
Information technology   3, 18
Inspection   316
Interviews   186

Job analysis   93, 133, 269
Job attitudes   234, 255
Job autonomy   409
Job characteristics   358
Job design   93, 255, 403, 409, 417
Job satisfaction   308

Job variety 358
Just-in-time manufacturing 409

Kitchens 263

Leisure 3
Liquid plasma displays 333

Machine guarding 75, 80
Maintenance 85, 186
Manikin 297
Manual materials handling 42, 133, 142, 145, 151
Mechanical materials handling 142
Medical staff 222
Mental fatigue 226, 234
Mental health 60, 226
Mental workload 210, 243, 250, 352, 387
Methodology 18, 179, 186, 193, 203, 297, 393
Motor skills 145
Muscular strength 151
Musculoskeletal disorders 93, 99, 108, 115, 126, 145, 263, 308, 333

Nuclear power plants 380, 387, 393
Nurseries 263
Nursing 126

Offices 203, 243, 304, 358
Organizational design 403
Oxygen consumption 280

Perception of movement 68
Physical workload 126, 133, 179, 210, 255, 269, 280
Posture 42, 93, 115, 126, 133, 145, 151, 156, 170, 179, 203, 269, 280, 289
Press work 255
Process control 373
Productivity 280, 327
Psychosocial factors 203, 333
Pupil size 316, 340, 346

Quality of working life 417
Questionnaires 186, 203, 216, 226, 234, 373

Rating 193, 387
Regression analysis 210
Research in ergonomics 25, 31, 35
Rheolimbgram 304
Risk 55, 68, 280
Risk perception 68
Robotics 68

Safety 68, 75
Safety helmets 297
Scaling 193
Scenario analysis 80
Seating 156
Simulators 393
Sinus arrhythmia 210
Skin temperature 250, 304
Sociotechnical systems 403
Standards 31
Static workload 115, 156, 179
Status information 85
Stress 42, 126, 203, 216, 222, 226, 234, 308, 327, 333
Stress management 60
Subjective evaluation 193, 203, 387
Supermarkets 269
Supervision 409
Surveys 179, 364
System characteristics 18
System design 3
System reliability 380

Task analysis 387, 393
Teachers 333
Team work 393
Telecommunications 186
Training 393
Trade unions 216
Truck drivers 133

Usability 186
User acceptance 186

Vehicle design 3
Visual accommodation 316
Visual display terminals 327, 333, 340, 346, 352, 358, 364
Visual fatigue 316, 340, 352

Welding 255
Word processing 333
Work experience 222, 255
Work procedures 85, 393
Working hours 263, 269, 308
Workplace 3, 68, 263, 269, 289, 308, 364
Writing 170